cSUR-UT Series: Library for Sustainable Urban Regeneration
Volume 2

Series Editor: Shinichiro Ohgaki, Tokyo, Japan

cSUR-UT Series: Library for Sustainable Urban Regeneration

By the process of urban development in the 20th century, characterized by suburban expansion and urban redevelopment, many huge and sophisticated complexes of urban structures have been erected in developed countries. However, with conventional technologies focused on the construction of structures, it has become difficult to keep urban spaces adaptable to environmental constraints and economic, social and cultural changes. In other words, it has become difficult for conventional technologies to meet social demands for the upgrading of social capital in a sustainable manner and for the regeneration of attractive urban space that is not only safe and highly efficient but also conscious of historical, cultural and local identities to guarantee a high quality of life for all. Therefore, what is needed now is the creation of a new discipline that is able to reorganize the existing social capital and the technologies to implement it.

For this purpose, there is a need to go beyond the boundaries of conventional technologies of construction and structural design and to integrate the following technologies:

(1) Technology concerned with environmental and risk management
(2) Technology of conservation and regeneration with due consideration to the local characteristics of existing structures including historical and cultural resources
(3) Technologies of communication, consensus building, plan making and space management to coordinate and integrate the individual activities initiated by various actors of society

Up to now, architecture, civil engineering, and urban engineering in their respective fields have, while dealing with different time-space scales and structures, accumulated cutting-edge knowledge and contributed to the formation of favorable urban spaces. In the past, when emphasis was put on developing new residential areas and constructing new structures, development and advancement of such specialized disciplines were found to be the most effective.

However, current problems confronting urban development can be highlighted by the fact that a set of optimum solutions drawn from the best practices of each discipline is not necessarily the best solution. This is especially true where there are relationships of trade-offs among such issues as human risk and environmental load. In this way, the integration of the above three disciplines is strongly called for.

In order to create new integrated knowledge for sustainable urban regeneration, the Center for Sustainable Urban Regeneration (cSUR), The University of Tokyo, was established in 2003 as a core organization of one of the 21st Century Centers of Excellence Programs funded by the Ministry of Education and Science, Japan, and cSUR has coordinated international research alliances and collaboratively engages with common issues of sustainable urban regeneration.

The cSUR series are edited and published to present the achievements of our collaborative research and new integrated approaches toward sustainable urban regeneration.

S. Takizawa (Ed.)

Groundwater Management in Asian Cities

Technology and Policy for Sustainability

Springer

Satoshi Takizawa, Ph.D.
Professor
Department of Urban Engineering
The University of Tokyo
7-3-1 Hongo, Bunkyo-ku,
Tokyo 113-8656, Japan

Cover photo: a hand pump in Tokyo, Japan; © Tomoko Takeda

ISSN 1865-8504
ISBN 978-4-431-78398-5 e-ISBN 978-4-431-78399-2

Library of Congress Control Number: 2008923163

Springer is a part of Springer Science+Business Media
springer.com
Printed in Japan

Typesetting: Camera-ready by the editors and authors
Printing and binding: Shinano Inc., Japan

Printed on acid-free paper

Preface

According to United Nations statistics, the world's urban population currently outnumbers the world's rural population. For the growing number of urban dwellers, water supply systems are the most important type of infrastructure to support their daily lives. Because it is difficult to build reliable water supply systems in developing countries, many people and industries must utilise groundwater underneath their own properties. Consequently, the adverse impacts of groundwater over-abstraction have become obvious in many cities in economically developing countries in Asia. Those cities are now facing the difficult problem of how to limit groundwater abstraction without negative impacts on the economy or the natural environment. Groundwater quality in these urban areas is also deteriorating, mainly due to various human activities. Deterioration of water quality is especially grave in the case of unconfined aquifers, which are the main source of drinking water for people without connections to piped water-supply systems. Meanwhile, in some developed countries in Asia, the urban groundwater table is rising, thanks to decades-long bans on groundwater abstraction. Because rising groundwater negatively impacts underground infrastructures, a new scheme for urban groundwater management in these cases is required.

This book is the product of a 4-year study on sustainable urban groundwater management in Asian cities carried out as a strategic study of the 21st Century COE (Centers of Excellence) Program on Sustainable Urban Regeneration. It summarises the current problems of urban groundwater use, reviews groundwater management practices in various cities, and presents technical tools that may help to deal with urban groundwater problems. Urban groundwater is a renewable and efficient resource if it is wisely managed. Hence, I hope this book will provide an opportunity to share knowledge on urban groundwater management and will inspire better governance of urban groundwater resources in the present and future.

Shinichiro Ohgaki
Project Leader, COE Program on Sustainable Urban Regeneration
The University of Tokyo

Contents

List of Authors

Chapters 1, 2 and 15:

Dr. Satoshi Takizawa, Professor, Department of Urban Engineering, Graduate School of Engineering, the University of Tokyo. His research fields are urban water systems, management of groundwater in cities and water supply in Asian developing countries. He has been the primary investigator of the Research Group on Sustainable Groundwater Management in Urban Areas, which is the birthplace of this publication.
Address: 7-3-1 Hongo, Bunkyo-ku, Tokyo 113-8656, Japan.
E-mail: takizawa@env.t.u-tokyo.ac.jp

Chapter 3:

Dr. Yanjun Shen, Research Professor, Center for Agricultural Resources Research, Chinese Academy of Sciences. He is a geographic hydrologist, studying hydrologic processes on different scales through observation, remote sensing, and modeling. Most of the results presented in this book were obtained during his research fellowship at Prof. Oki's laboratory in the Institute of Industrial Sciences, the University of Tokyo.
Address: 286 Huaizhong Rd., Shijiazhuang 050021, China.
E-mail: yjshen@sjziam.ac.cn

Dr. Taikan Oki, Professor, Institute for Industrial Science, the University of Tokyo. Prof. Oki has carried out research on global-scale hydrology and water balance models. The model presented in Chapter 3 is based on this research.
Address: 4-6-1 Komaba, Meguro-ku, Tokyo 153-8505, Japan.
E-mail: taikan@iis.u-tokyo.ac.jp

Chapter 4:

Dr. Tomochika Tokunaga, Associate Professor, Department of Environment Systems, Graduate School of Frontier Sciences, the University of Tokyo. Dr. Tokunaga's research interests include poromechanics and long-term subsurface environmental processes. He has been involved with nu-

merous geology and groundwater-related research projects both in Japan and abroad.
Address:5-1-5 Kashiwanoha, Kashiwa-shi, Chiba 277-8563, Japan.
E-mail: tokunaga@k.u-tokyo.ac.jp

Chapter 5:

Dr. Takeshi Hayashi, Associate Professor, Faculty of Education and Human Studies, Akita University. He is a hydrologist who has studied the groundwater flow system by using physical and chemical properties of groundwater. The methodologies for analysis of shallow groundwater is one of his research interests.
Address: 1-1 Tegata Gakuen-machi, Akita City, Akita 010-8502, Japan.
E-mail: thayashi@ed.akita-u.ac.jp

Chapter 6:

Mr. Masaatsu Aichi, Graduate Student, Department of Geosystem Engineering, Graduate School of Engineering, the University of Tokyo. He has developed a numerical model to predict the effect of groundwater abstraction and conservation of Tokyo's land subsidence. The simulation results presented in Chapter 6 are the outcome of his graduate research.
Address:5-1-5 Kashiwanoha, Kashiwa-shi, Chiba 277-8561, Japan.
E-mail: aichi@geoenv.k.u-tokyo.ac.jp

Chapter 7:

Mr. Keisuke Kuroda and Mr. Tetsuo Fukushi, Graduate Students, Department of Urban Engineering, Graduate School of Engineering, the University of Tokyo. They are carrying out a study on groundwater quality and management in Tokyo. The data presented in this book were obtained during their graduate research.
Address: 7-3-1 Hongo, Bunkyo-ku, Tokyo 113-8656, Japan.
E-mail: k_kuroda@env.t.u-tokyo.ac.jp ; t_fukushi@env.t.u-tokyo.ac.jp

Chapter 8:

Dr. Hiroyuki Katayama, Associate Professor, Department of Urban Engineering, Graduate School of Engineering, the University of Tokyo. His research field is health-related microbes in a water, with special emphasis on detection methods of viruses in water environment.
Address: 7-3-1 Hongo, Bunkyo-ku, Tokyo 113-8656, Japan.
E-mail: katayama@env.t.u-tokyo.ac.jp

Chapter 9:

Dr. Satoshi Ishii, Research Fellow, the Center for Sustainable Urban Regeneration established under the 21st Century Center of Excellence (COE) Project "Sustainable Urban Regeneration." Chapter 9 contains the results of his research while he was involved with the COE Project.
Address: 7-3-1 Hongo, Bunkyo-ku, Tokyo 113-8656, Japan.
E-mail: ishii@csur.t.u-tokyo.ac.jp

Chapter 10:

Dr. Ryozo Ooka, Associate Professor, Institute of Industrial Science, the University of Tokyo. He is specialized both in experimental and computer analysis of the heat environment in and around buildings. Recently, his research has focused on the use of groundwater's energy potential for air conditioning of buildings.
Address: 4-6-1 Komaba, Meguro-ku, Tokyo 153-8505, Japan.
E-mail: ooka@iis.u-tokyo.ac.jp

Mr. Yujin Nam, Graduate Student, Institute of Industrial Science. He is a member of Dr. Ooka's research laboratory, and has been carrying out a study on the potential of Tokyo's groundwater as an energy source.
Address: 4-6-1 Komaba, Meguro-ku, Tokyo 153-8505, Japan.
E-mail: namyujin@iis.u-tokyo.ac.jp

Chapter 11:

Dr. Futoshi Kurisu, Associate Professor, Research Center for Water Environment Technology, Graduate School of Engineering, the University of Tokyo. His research field is bioremediation of contaminated soil and groundwater and microbiology of biological water and wastewater treatment processes.
Address: 7-3-1 Hongo, Bunkyo-ku, Tokyo 113-8656, Japan.
E-mail: kurisu@env.t.u-tokyo.ac.jp

Chapter 12:

Ms. Yatsuka Kataoka, Senior Policy Researcher, the Institute for Global Environmental Studies (IGES).
Mr. Tetsuo Kuyama, Researcher, IGES.
They are members of an IGES research project on sustainable ground management policy in Asian cities, which has been carried out as joint research with universities and environmental institutions from six Asian countries. The information and data presented in Chapter 12 are based on the results of the research project.

Address: 2108-11 Kamiyamaguchi, Hayama-machi, Kanagawa 240-0115, Japan.
E-mail: kataoka@iges.or.jp ; kuyama@iges.or.jp

Chapter 13:

Mr. Micha Sigrist, Researcher, Department of Urban Engineering, Graduate School of Engineering, the University of Tokyo. He received a diploma in natural sciences (earth sciences) from ETH Zurich. During the writing of this book, he has carried out a study on groundwater contamination in Ho Chi Minh City, Vietnam.
Address: 7-3-1 Hongo, Bunkyo-ku, Tokyo 113-8656, Japan.
E-mail: sigrist.micha72@yahoo.de

Chapter 14:

Dr. Tran Thi Viet Nga, Researcher, Department of Urban Engineering, Graduate School of Engineering, the University of Tokyo.
The data presented in Chapter 14 were obtained in her Ph.D. study on groundwater contamination and its effects on the water supply in Hanoi City, the University of Tokyo.
Address: 7-3-1 Hongo, Bunkyo-ku, Tokyo 113-8656, Japan.
E-mail: tranvietngajp@yahoo.com

Chapter 15:

Ms. Tomoko Takeda, Graduate Student, Department of Urban Engineering, Graduate School of Engineering, the University of Tokyo.
The results presented in Chapter 15 were mostly obtained while she carried out research on mitigation of health impacts of fluoride intake among the residents in the Chiang Mai Basin.
Address: 7-3-1 Hongo, Bunkyo-ku, Tokyo 113-8656, Japan.
E-mail: takeda@env.t.u-tokyo.ac.jp

Part I Groundwater Problems in Urban Areas

1. Introduction

Satoshi Takizawa

Department of Urban Engineering, the University of Tokyo

1.1 Groundwater Problems in Urban Areas

Ancient human settlements were often established where there were artesian wells and springs. Ever since, people's lives in towns and cities around the world have been supported by groundwater. It was in the 20th century, however, that groundwater over-abstraction brought about problems such as depletion of groundwater resources and serious land subsidence. In dry regions like some parts of China, India and United States, agriculture has used the largest amount of groundwater. However, in countries where agriculture is more dependent on rainfall and irrigation systems, groundwater is exploited extensively by industries and urban development. In Japan, groundwater over-abstraction had become a serious environmental concern by the 1960s because it had brought about extensive land subsidence in the major cities. For example, in the Koto Delta region in Tokyo, the land has subsided over four metres since 1900.

There are some cities solely dependent on groundwater for their water supplies such as Hanoi City, Vietnam and Kumamoto City, Japan. Other cities, such as Jakarta, Bandung, Kathmandu, Ho Chi Minh City, and Tianjin, have piped water supply systems that use surface water, but these cities are heavily dependent on groundwater to supply drinking water and water for industrial use. In these cities, the protection of groundwater is essential for the well-being of residents and for industrial development, but over-abstraction and contamination of groundwater have become major environmental problems. In most of these cities, groundwater is contaminated with nitrate or ammonia nitrogen due to the excessive application of fertilizers and/or the infiltration of wastewater. Recently some industries, especially high-tech industries, have used chlorinated solvents without any precaution to protect groundwater. As a result, there are numerous cases of groundwater contamination by industrial facilities. In addition to anthropogenic contaminants, naturally occurring contaminants have emerged as serious threats to the safe water supply in some cities. Arsenic and fluoride are naturally occurring contaminants that affect tens of millions of people who drink groundwater containing these contaminants. Arsenic and fluoride in groundwater are reported in many countries including Bangladesh, India and China, but so far there has been no ultimate solution to these problems.

Groundwater over-abstraction is another common problem in many Asian cities. During the rapid economic growth phase of cities, both domestic and industrial water demands increase dramatically, outpacing the increasing rate of investment for water infrastructure. Hence, groundwater has been extensively overexploited in such cities. Drawdown of groundwater tables, land subsidence and saltwater intrusion are commonly found problems in cities where groundwater abstraction far exceeds the rate of natural replenishment. Nonetheless, stopping or even reducing groundwater usage is politically and practically difficult because groundwater sustains the growth of the local economy. Although sustainability is the most important concept for any environmental issue, there have been only a few cities that have succeeded in managing groundwater resources in sustainable ways. Japanese cities, such as Tokyo and Osaka, are examples of those which succeeded in controlling groundwater over-abstraction, but even these cities have not yet found a way to get rid of various contaminants that infiltrate into the aquifers.

1.2 Analysing and Addressing the Groundwater Problem

With the recognition of the importance of groundwater in urban development and sustainability, a research group on urban groundwater management was established in 2003 as a strategic research group of the 21st Century Center of Excellence Project entitled "Sustainable Urban Regeneration" (Project Leader: Professor Shinichiro Ohgaki, University of Tokyo). The Research Group on Sustainable Groundwater Management in Urban Areas comprised some twenty professors, research fellows and graduate students with different academic backgrounds such as hydrology, hydrogeology and environmental engineering. The research group took interdisciplinary approaches by looking at groundwater's different features, discussing the problems from various viewpoints, and holding an open symposium to disseminate the findings and identify issues to be elaborated further. For example, more than 200 government officers, groundwater engineers, researchers and those who are concerned about groundwater issues joined the symposium held in Tokyo in February 2007. The backgrounds and affiliations of the participants are diversified, which indicated that there are a wide variety of people who are concerned about the future management of groundwater in urban areas.

This book is one in a series of publications emanating from the COE Project on Sustainable Urban Regeneration, and contains the outcome of the research activities of the Research Group on Sustainable Groundwater Management in Urban Areas. Most of the authors of this publication are members of this research group, but others were invited to write chapters that are complementary to the rest of the book's contents. The aims of this publication are to share the knowledge and experiences of groundwater problems, to elucidate the policies that have been taken to abate those problems, and to disseminate the latest scientific knowledge on groundwater monitoring and management.

This book is intended for a diverse audience including: graduate students and professors who are interested in groundwater problems in urban areas, engineers and consultants dealing with the groundwater problems, policy makers and government officers in charge of groundwater management, and any stakeholders involved in urban groundwater issues.

1.3 Outline of Chapters

This publication comprises fifteen chapters, which are grouped into four parts. Part I, Groundwater Problems in Urban Areas, is a four-chapter introduction that provides background information regarding groundwater problems in urban areas, groundwater contamination, groundwater dependence of world cities, and the hydrogeological processes of urban areas. Part II, Novel Methods for the Analysis of Urban Groundwater Problems, comprises three chapters, and presents the methodologies for the analyses of land liquefaction, land subsidence, and groundwater contamination processes. Part III, Efficient Use of Groundwater Resources, comprises the next three chapters. In this part, the potential of groundwater to reduce energy consumption and CO_2 emission in urban areas is discussed by comparing the present piped water supply systems with decentralised water supply systems using groundwater resources. Various examples of using groundwater's heat potential are also elucidated. Remediation technologies for contaminated groundwater can improve the efficiency of groundwater use, and thus a discussion of them is also included in this part. Part IV, Groundwater Problems in Asian Countries, is made up of four chapters presenting case studies on groundwater management and contamination in Bangkok, Ho Chi Minh City, Hanoi and the Chiang Mai Basin.

Following this introductory chapter, Chapter 2 discusses the relationship between urban development and groundwater and the current problems of groundwater management in Asian cities. Although economic development and urbanisation of many Asian cities have been supported by clean and plentiful groundwater resources, people who used groundwater paid little attention to the abundance and quality of groundwater resources until they finally faced serious problems of land subsidence and groundwater contamination. This is because many people commonly think that there is no limit to groundwater resources and they take it for granted that they have the right to abstract groundwater eternally. For better management of urban groundwater, it is indispensable to change the minds of those who presently enjoy unrestrained access to groundwater resources. In another section of Chapter 2, the characteristics of the contaminants commonly found in urban groundwater and their removal technologies are presented. Nitrate is the most commonly detected contaminant in urban groundwater, but in some cities, naturally occurring contaminants such as arsenic and fluoride are affecting the health of the residents. Since the protection of groundwater quality is closely related with land use regulations, it is quite important to integrate groundwater conservation policy into urban planning processes.

Chapter 3 has three sections: in the first, a global perspective on urban water demand is projected using United Nations population statistics, and the importance, as well as the difficulty, of meeting the increasing demands on the urban water supply is elucidated. In the second section, a new index of cities' dependence on groundwater is introduced as the Urban Groundwater Dependence Index (UGDI). A high-resolution digital map of the world with a 0.5-degree x 0.5-degree grid was produced and the population, water demand and availability of surface water were estimated in each part of the grid. In this method, urban areas are defined as the areas with high population density. The areas with urban water demands in excess of 40% of the surface water resources were defined as highly dependent on groundwater, and those cities with water demands more than 100% of the surface water resources were defined as very highly dependent on groundwater. The study revealed that there are about one billion people living in cities that are either highly or very highly dependent on groundwater; 40% of whom are living in Asian cities. In the third section of Chapter 3, the past and present problems of groundwater use in some Chinese cities are presented. The North China Plain holds a large population living in large cities, but due to its semi-arid climate, the water supply in this region has been a major constraint on development. Groundwater has been extensively exploited in such cities as Tianjin and Shijiazhuang, which in turn brought about serious land subsidence and groundwater depletion. Groundwater quality has also deteriorated due to poor protection of aquifers. Due to a lack of irrigation water, farmers have used wastewater for irrigation since the mid-1960s. High levels of nitrogen in the wastewater resulted in extremely high concentrations of nitrate in the groundwater. These examples indicate the importance of precautionary approaches for the protection of groundwater resources.

In Chapter 4, the hydrogeological setting of the Tokyo Metropolitan Area, which includes the administrative boundary of Tokyo and the surrounding municipalities, is elucidated as an example of a region that has experienced both groundwater over-abstraction and later recovery of the groundwater table as a result of strict regulation. Although Tokyo has experienced serious land subsidence due to drawdown of the groundwater table of confined aquifers, the groundwater table of unconfined aquifers has not changed significantly over the last few decades. Over-abstraction of confined groundwater, however, has not only brought about drawdown of the groundwater table but also has changed the oxidation-reduction (redox) conditions of the aquifers by extracting anoxic pore waters from the overlaying semi-permeable clay layer. As a result, the number of accidents due to breathing oxygen-depleted air has increased. At the end of this chapter, technologies for the monitoring of groundwater and land subsidence are presented. The prospects and difficulties of controlled groundwater pumping for stabilising the groundwater table are also discussed.

Chapter 5 describes the mechanisms of soil liquefaction and the effect of changing groundwater tables on soil liquefaction. Earthquakes are a serious problem in Japan and some other Asian countries. Soil liquefaction caused by earthquakes has destroyed urban infrastructure in the past. Since soil liquefaction takes place as a result of interaction between soil and unconfined groundwater, bringing down the shallow groundwater table can reduce the risk of soil liquefaction. In this regard, the importance of water mains and sewer pipes as the major source and sink of shallow groundwater is illustrated, using Tokyo as an example.

Chapter 6 demonstrates the effectiveness, as well as the limits, of groundwater flow modelling to predict land subsidence due to over-abstraction of groundwater. The modelling of land subsidence is explained in detail, and the method for computer modelling of the process is elucidated using mathematical equations. A nested modelling approach, which combines a coarse regional model with a fine local model, is shown to be quite effective when we simulate the hydrogeological process of a small area within a larger recharge area. The model was applied to predict the changes of groundwater in Tokyo. The simulation results showed good agreement with the observed data, though further elaboration and fine-tuning of the model would make it possible to predict land subsidence more accurately.

In Chapter 7, both naturally occurring and anthropogenic groundwater contamination processes are elucidated. Both arsenic and fluoride are affecting the health of hundreds of millions of people, whereas nitrate and chlorinated organic solvents are the most commonly detected anthropogenic contaminants in urban groundwater. In the latter half of this chapter, the results of groundwater monitoring in Tokyo are presented. The contribution of different sources to groundwater recharge is estimated using data obtained from various sources. The effectiveness of taking the mass balance of both groundwater recharge processes and infiltration of contaminants for the estimation of contaminant sources is demonstrated in a case study of Tokyo.

Chapter 8 explains microbial contamination of groundwater. The outbreaks of waterborne disease due to ingestion of groundwater contaminated with infectious microbes, as well as the potential sources of microbial contamination of groundwater, are presented in this chapter. There are several factors that affect the fate and transport of infectious microbes when they are released into aquifers. The principles of microbial transportation and the die-off process are elucidated together with the methods for the detection of human health-related microbes.

The environmental impact of urban water systems is estimated in Chapter 9. Supplying water in urban areas requires a large amount of energy; hence, it releases a lot of carbon dioxide. Reducing the environmental impact by cutting off the CO_2 emission is a highly committed mandate of responsible countries, and mega-cities should make efforts to cut down CO_2 emissions. In this chapter, the CO_2 emission of the existing water supply system using surface water in Tokyo was compared with the CO_2 emission of a potential water supply system using groundwater. Of the four different scenarios of water supply using groundwater, the one to supply both public and private buildings with drinking and toilet flushing water would cut the CO_2 emission by the largest amount, whereas limiting the use only to drinking water or limiting the customers only to public buildings decreased the efficiency of groundwater use for cutting CO_2 emissions. It was also demonstrated that the effectiveness of using groundwater varies significantly among the 23 words in Tokyo because of the differences in numbers of large buildings in each ward. Although several factors, such as the cost of laying dual-pipe water supply systems, are excluded from the evaluation in this chapter, the results illustrate where and how groundwater could be used for the reduction of environmental impacts of urban water systems.

Chapter 10 presents the potential and practical application of groundwater's heat energy. Groundwater is a more efficient medium for heat exchange than ambient air because the groundwater temperature is almost constant throughout the year. Whether a groundwater heat pump (GWHP) is more energy efficient than any conventional air conditioning systems is dependent on the energy consumption of the pumps used for groundwater abstraction. It is demonstrated here that a GWHP system can be best applied where the groundwater level is fairly high and requires less energy for abstraction.

In Chapter 11, clean-up technologies of contaminated soil and groundwater due to organic solvents and fuel are elucidated. Among the various technologies for environmental remediation, this chapter focuses on bioremediation. In Japan, bioremediation is applied to 23% of clean-up projects of contaminated sites by volatile organic carbons. The pros and cons of bioremediation are discussed, and different methods of bioremediation in terms of the methods to enhance biological activities, e.g. biostimulation and bioaugmentation, are elucidated. At the end of this chapter, the recent research advancements on degradation of chlorinated ethylenes and aromatic hydrocarbons are summarised.

Chapters 12 to 15 are case studies on groundwater management and contamination in selected Asian countries. Chapter 12 focuses on groundwater management policies in Asian mega-cities. Three cities are selected for the comparative study: namely, Bandung, Bangkok, and Osaka. In these cities the largest amounts of groundwater is abstracted by the industrial sector. In Bandung, 80% of groundwater is pumped by industries and the rest is for domestic uses. Groundwater abstraction in Bandung increased nearly six-fold between 1970 and 1996, the year just before the Indonesian currency crisis. Following the crisis, Bandung's industrial output plummeted; thus, groundwater abstraction has dropped significantly since 1997. This coincidence indicates the connection between the industrial sector and the over-abstraction of groundwater resources.

In Bangkok, land subsidence due to over-abstraction of groundwater became obvious in the 1980s. The Thai Government has revised the Groundwater Act several times since the 1970s, but until the 1990s the government was mostly unsuccessful in controlling the total amount of groundwater abstraction. In the revised Groundwater Act of 1997, it was stipulated that both a groundwater abstraction charge and a groundwater conservation fee would be levied on groundwater users. The rates of these are set at a slightly higher price in regions covered by a public water supply, but at lower prices in areas without a public water supply. Although the groundwater charge plus conservation fee is slightly higher than the public water supply, which uses surface water, many industries still prefer groundwater because they think the public water supply is not very reliable in times of drought. There are also industries moving out to nearby areas without a public water supply so that they can pump groundwater at a cheaper cost.

The final case study in chapter 12 concerns Osaka City, Its historical problems of serious land subsidence and a set of policies to stop groundwater use which has successfully reduced the amount of groundwater pumping since the 1970s. Osaka, like many other Japanese cities, built an industrial water supply system in the 1960s which made it possible to supply industrial water to factories at a cheaper price than the municipal water supply. Subsidies from both municipal and central governments were important incentives to make the industries change their old habits of abstracting groundwater. The promotion of water recycling within factories, increased wastewater treatment fees, and also public concern regarding the protection of the environment were also important factors that changed the pattern of groundwater over-use by industries. Hence, based on the Japanese experience, it is fair to say that a good combination of policies can make the control of groundwater over-abstraction more effective.

Chapter 13 deals with groundwater management problems in Ho Chi Minh City (HCMC), Vietnam. HCMC is the largest city in Vietnam, and its industries and population are growing rapidly. There was no regulation of groundwater abstraction in HCMC until 2007. Hence, over-abstraction of groundwater has brought about severe drawdown of groundwater tables. Concurrently, groundwater pollution has also become a serious problem. Groundwater systems in cities such as HCMC that lie on alluvial deposits are quite complex, and require extensive scientific studies before we can understand the groundwater availability and mechanisms of groundwater contamination. In this chapter, the possible mechanisms of groundwater contamination in HCMC are elucidated.

Chapter 14 concerns arsenic contamination in Hanoi City, Vietnam. Although Hanoi City is almost totally dependent on groundwater resources, the availability of clean groundwater is limited. In and around Hanoi City there are organic-rich layers, out of which pore water might have been squeezed into aquifers due to the over-abstraction of groundwater. As a result, some groundwater sources have become rich in organic matter, ammonia nitrogen and arsenic. This situation is quite similar to the problem of arsenic contamination in such places as Bangladesh. Arsenic can be removed by co-precipitation with iron relatively easily, but if the arsenic concentration is rather high, it is very difficult to remove arsenic from groundwater. Therefore, good management practices of water treatment facilities are very important for the efficient removal of arsenic from contaminated groundwater.

In Chapter 15, the health effects caused by groundwater contamination by fluoride are illustrated. In Lamphun Province, Thailand, the maximum fluoride concentration reached 16mg/L, which is more than 20 times higher than the Thai Driking Water Standard of 0.7 mg/L. The health effects of ingesting excess fluoride are widespread among the population living in this region. In order to reduce the health risks, most of the local people drink either bottled water or rain water. Since the people have switched to these alternative sources, the drinking waters turned in by the local residents for analysis contained quite small amounts of fluoride. However, it was revealed that their food, especially rice, is a major source of fluoride intake, and thus reduction of fluoride intake solely from drinking water is not enough to protect the health of the local people. It was found that residents should be given scientifically sound and comprehensive information on the health risks of fluoride, and information about its various intake sources.

1.4 Acknowledgements

We appreciate the cooperation of numerous people who kindly helped us to take groundwater samples in Tokyo and other Asian cities, others who provided us information and data on groundwater resources, and also many people who gave us suggestions and comments on our research findings. Chapter 12 contains the findings of the research project on freshwater resources carried out by the Institute for Global Environmental Studies (IGES). A part of Chapter 13 concerns research conducted by the CREST project headed by Dr. Kengo Sunaga of Yamanashi University.

Last but not least, we appreciate Dr. Reiko Hayashi, Research Fellow with the Centre for Sustainable Urban Regeneration, who helped us edit this book. Without her kind encouragement and suggestions, this book would not have been published.

2. Groundwater Use and Management in Urban Areas

Satoshi Takizawa

Department of Urban Engineering, the University of Tokyo

2.1 Introduction

Rapid urbanisation and industrial development in the 20th century could not have been realised without an affluent water supply. Groundwater plays an especially important role during the early stage of urban development when large-scale surface water development has not yet been accomplished. The heavy dependence on groundwater resources, however, has brought about problems such as land subsidence and saltwater intrusion in coastal cities. Even though many decision makers are aware of these problems, regulating groundwater abstraction is difficult because there are many people and industries that can be negatively affected by such regulations. The economic growth rates of these cities would also be limited by reducing the quantity of groundwater abstraction. Although in some countries groundwater resources are traded like commodities, it would be not easy to deal with groundwater in such a way in Asian countries because of the people's perception of natural resources and the conventional right to exploit groundwater resources beneath their properties. The problems pertaining to urban groundwater use arise not only from a small number of big industries, but also from a large number of residents, whose collective actions bring about serious consequences. Hence, a good combination of regulation, economic incentives and a supply of alternative water is necessary to control groundwater over-abstraction and contamination in urban areas. In this chapter, the current problems of groundwater use and management in Asian cities are presented. Some typical contaminants found in groundwater and their treatment processes are also elucidated.

2.2 Urban Development and Groundwater Resources

2.2.1 Groundwater Problems in Developing Asian Countries

The consequences of urban development on groundwater resources differ depending on the stage of development, the local climate, the aquifer structure and the land use patterns of the cities. In arid regions, Foster et al. (1999) explained the changes in the groundwater table as follows; in the early stage of urban development, the groundwater table goes down because groundwater is abstracted for both domestic and industrial uses. In the next stage of urban development, however, the groundwater table of unconfined aquifers goes up because water is brought into the city from external sources such as river water diversion or seawater desalination plants when the city grows even larger. In this stage, both leaking water from water mains and infiltration of wastewater bring up the groundwater table in the city area. Rushton and Al-Othman (1994) reported the problems of rising groundwater in Riyadh, Saudi Arabia. The same problems are taking place in some other cities in Saudi Arabia, including Jedda City, where streets are inundated by overflowing groundwater, especially after rainfall.

In monsoon Asia, where annual rainfall is comparatively abundant, natural groundwater recharge rates are by far greater than in those countries in the arid regions. Hence, in monsoon countries, groundwater resources are renewable if they are abstracted at rates equal to or smaller than the natural recharge rates. However, as the cities grow larger, groundwater over-abstraction often brings about land subsidence because many cities are located in river delta regions with quaternary deposits (Yusuf et al. 2007).

For example, Hanoi City is totally dependent on groundwater resources for both its municipal and industrial water supply. Households also have dug wells and tube wells to augment the water supply from the water company. In addition to the water quality problems in the southern region of Hanoi City (see Chapter 14), the extensive groundwater pumping has brought about severe land subsidence. Fig. 2.1 shows one of the land subsidence monitoring stations in Hanoi City. Although land subsidence has been monitored, so far no regulations to control groundwater abstraction have been implemented because there is no alternative water source in Hanoi City.

Fig. 2.1. Land Subsidence Monitoring Station, Hanoi, Vietnam.

Ho Chi Minh City (HCMC) also has experienced rapid drawdown of groundwater tables, though there is no monitoring network of land subsidence. The major users of confined groundwater are the industries that are rapidly growing in and around HCMC. In order to alleviate the impact of over-abstraction, HCMC has introduced a groundwater tax system in conjunction with a ban on the installation of new wells since 2007. In HCMC, the Department of Natural Resources and Environment (DONRE) is primarily responsible for the protection of the environment and natural resources, including groundwater resources. Under DONRE, the Division of Water Resources is responsible for the allocation and regulation of water resources mainly in terms of water quantity, whereas the Environmental Protection Agency is responsible for water quality monitoring and protection. Before the implementation of groundwater regulations, these two official bodies worked together to hear the opinion of industries and decision makers. They have invited a panel of international experts and policy makers to the stakeholders meetings that are jointly organised by Institute of Global Environmental Strategies (IGES), Japan. Because the groundwater table in HCMC is rapidly drawing down and the quality is deteriorating, the People's Committee finally agreed to set regulations on groundwater use. The economic incentive of the new tax is quite limited, however, because the tax is only 40 to 160 VND/m^3 (approx. 0.25 to 1 cent/m^3) compared with the piped water tariff of 4500 VND/m^3 (approx. 28 cent/m^3). It may be of more importance that any kind of regulation was imposed on the hitherto free access to the groundwater resources rather than the actual amount of the tax. It is also very important for HCMC to make alternative water sources available to the industries.

In such landlocked cities as Kathmandu, Nepal and Bandung, Indonesia, groundwater is an important source for municipal water supply and domestic wells because clean surface water is available only on the hillsides of the valleys. In these cities, the public water supply is limited due to a lack of water resources. Surface waters flowing through the middle of the valleys accumulate all the wastewater and, therefore cannot be easily used for water supply. Hence, the water company must go outside of the city's central district to the vicinity of the mountains, to get clean water (Fig. 2.2). Since the available volume of groundwater is limited and surface water is extremely contaminated, the water supply in Kathmandu is unreliable. Hence, most of the residents are on their own to get drinking and domestic water; that is, they are dependent on shallow, contaminated groundwater from spouts, dug wells or tube wells (Fig. 2.3).

Groundwater in unconfined aquifers is commonly used by households, but it is often contaminated with both domestic and industrial wastewater because unconfined aquifers are vulnerable to these types of contamination. In the city centre, houses are built close together, so it is not easy to protect the unconfined aquifers below from contamination by domestic wastewater. In such places, it is almost impossible to implement any wellhead protection program. There is no water treatment after pumping the groundwater except iron and manganese removal, which are also often not sufficient anyway due to too high iron and manganese concentrations. There are spouts in some places of Kathmandu City, where people come to clean their clothes and dishes, wash their bodies, and get water for drinking and cooking. These spouts take water from very shallow groundwater, and as such, are often contaminated with coliform bacteria. Drinking such water is a potential public health threat (Fig. 2.4).

There is a fossil aquifer beneath Kathmandu City which is utilised by hotels, breweries and other industries. Most of these businesses have their own water treatment plants, but the fossil groundwater contains extremely high concentrations of iron, ammonia and organic matter (Khatiwada et al. 2002). Arsenic concentration is also elevated in some areas. Therefore, the water treatment processes must be designed to be able to treat such highly contaminated water, and they should be well maintained to achieve a high removal efficiency of those contaminants (see Sec. 2.4.1). A lack of skilled and experienced operators can easily put those water treatment plants into a faulty state, and make the treated water quality unacceptable for domestic and industrial uses.

Fig. 2.2. Deep well built on the hillside of the Kathmandu valley

Fig. 2.3. Dug well commonly found in households in Kathmandu City

Fig. 2.4. A boy drinking spout water in Kathmandu City

2.2.2 Groundwater Management in Japanese Cities

Groundwater and Japanese Urban Development

Before the middle of the 19th century, Tokyo, then called Edo, was one of the largest cities in the world with an estimated population of about one million. At this time, water supply systems took water from upstream of the Tama and Kanda Rivers and brought the water all the way down to the city centre by gravity flow. Although they were marvellous civil engineering works, the water supply systems were not connected to each household. Hence, there were many shallow wells in and around the houses. For centuries, those shallow wells were also important sources of water for the residents of Edo. Because those shallow wells were not protected against microbial contamination, contaminated wells infected hundreds of thousands of people since the mid-19th century, when infectious microbes such as cholera and shigella were brought to Japan. After experiencing several outbreaks of waterborne diseases, the importance of a piped water supply system was recognised and slow sand filters were constructed using surface water. Ever since, surface water has become more important as the source of the municipal water supply to Tokyo.

During the 1950s, the volume of groundwater abstracted by the industries and buildings in Tokyo increased rapidly as the city recovered from the aftermath of World War II. Post-war industrial development had just started and water demand by the booming industries soared. Using new well drilling rigs and powerful pumps, industries exploited deep aquifers to get abundant groundwater. As a result, the amount of groundwater abstracted everyday increased every year in the 1950s and 1960s. In addition, natural gas deeply buried underneath Tokyo was exploited to supply energy to the burgeoning industries. As a consequence, the ground level in Tokyo, Osaka and Nagoya, as well as some other cities, subsided more than a few meters, to below the mean sea level. The fear of tsunamis, as well as the huge cost of protecting those sub-zero meter regions triggered an important policy change from the free, unlimited rights to exploit groundwater resources, to strict regulation of groundwater abstraction.

Regulation of Groundwater Abstraction

Land subsidence was first observed in the 1910s and in the 1920s in Tokyo and Osaka, respectively (Fig. 2.5). It was only since the 1950s, however, that land subsidence had brought about damage to buildings and that the fear of flooding urged the construction of levees to protect the Tokyo lowland from tsunamis. These problems drew the public's attention and the call for countermeasures against over-abstraction of groundwater emerged. Consequently, several regulations were implemented to control groundwater abstraction. The laws promulgated on the national level were:

- The Industrial Water Law of 1956
- Law Concerning the Regulation of Pumping-up of Underground Water for Use in Buildings of 1965

The former law regulates groundwater abstraction by industries in the designated subsiding areas, whereas the latter law regulates the groundwater use by the buildings in cities (see also Chapter 12). These laws prohibit drilling boreholes larger than 21cm^2, and for boreholes between 6 and 21cm^2, these laws limit the screen depths of boreholes below 400–550m. In addition to these laws, municipal governments formulated ordinances to control the pumping of groundwater. In Tokyo, the average pumpage rate of groundwater from boreholes smaller than 6 cm^2, which is not regulated by the aforementioned laws, is limited to less than 10 m^3 per day.

After three decades of regulation, the groundwater tables in the major Japanese cities have come back, and are still going up in some regions. There are, however, still some other cities and regions where groundwater over-abstraction is causing land subsidence. The purposes of groundwater use in those regions are mainly agricultural and for snow-melting, which are different from the purposes of groundwater use in the major cities.

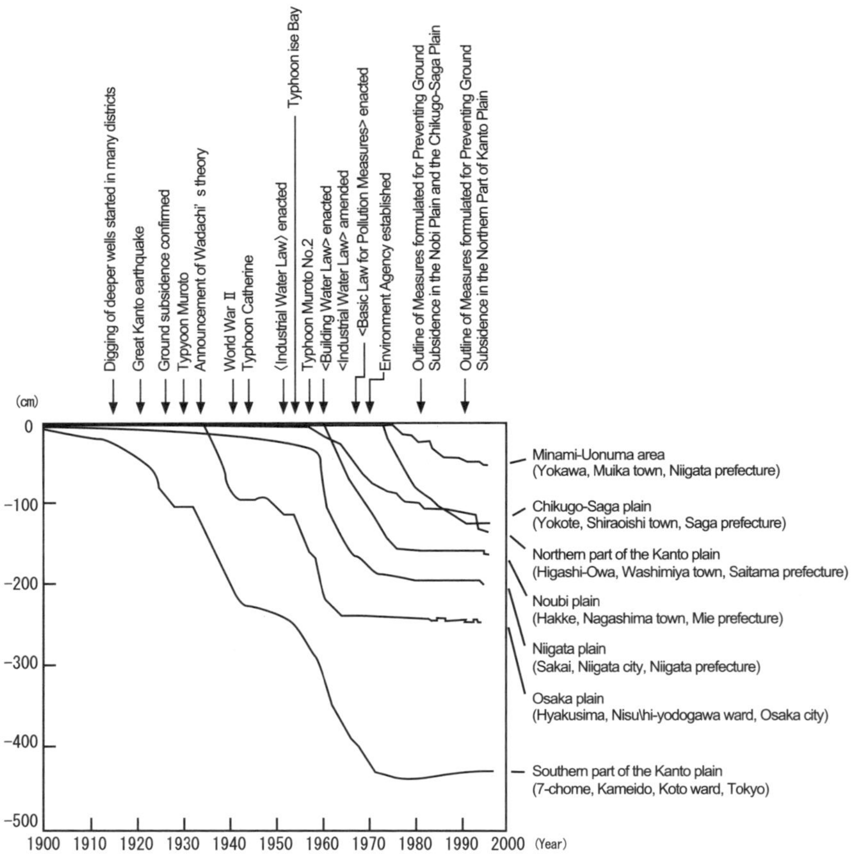

Fig. 2.5. Land subsidence in Japanese cities since 1900 (Ministry of the Environment, Japan, 2007)

2.3 Groundwater Contamination in Urban Areas

2.3.1 Abundance and Vulnerability of Groundwater Resources

The abundance of groundwater in urban areas is dependent on the local hydrogeological conditions. Confined aquifers can have a higher yield of groundwater than unconfined, shallow aquifers, which are significantly affected by urban and suburban development. Hence, confined groundwater is mostly used by industries and businesses, whereas unconfined groundwater is used by households. Unconfined aquifers are vulnerable to anthropogenic contamination, and the water quality is affected by human activities (see Chapter 13). Thus, it can be said that unconfined aquifers in urban areas are now at risk of various kinds of contamination because there are no wellhead protection programs in Asian countries. In congested neighbourhoods in Asian cities, dug wells and shallow tube wells are constructed in the vicinity of pit latrines and septic tanks, which contaminate shallow groundwater. Hence, the people who are dependent on unconfined groundwater are at high risk of waterborne diseases. So far, boiling water before drinking seems to be the only way to assure the microbial safety of unconfined groundwater. However, there are other kinds of contaminants such as nitrate that can hardly be removed by boiling.

Confined aquifers are not always free from contamination. Confined aquifers are often not completely separated throughout the whole groundwater basin, so contaminants can move down from the overlaying unconfined aquifer through the connected channels. Abandoned boreholes can bring contaminated surface waters directly down into the confined aquifers. There are numerous wells that are not completely sealed between the aquifers. Those wells can become connecting channels between the aquifers. Naturally occurring contaminants can be detected both in unconfined and confined aquifers depending on the geological distribution of the contaminants. Saltwater intrusion takes place both in confined and unconfined aquifers; but the extent of saltwater intrusion is much more serious when it takes place in unconfined aquifers because saltwater intrusion into unconfined aquifers is limited only to the locations very near to the sea. In addition, in places near the sea such as the Mekong Delta, ancient tidal fluctuations have left salty groundwater in some parts of confined aquifers. Therefore, people can find unconfined groundwater to be salty even if their wells are away from the sea. In alluvial deposits, organic-rich carbonaceous layers, which are also called peat layers, are present. Those organic-rich layers can be the source of organic matter and nitrogen that are sucked out into the aquifer when the hydraulic pressure of the confined aquifer becomes too low due to over-abstraction of groundwater. In addition, organic matter and ammonia nitrogen consume dissolved oxygen in groundwater and make that groundwater anaerobic. Under anaerobic conditions, oxidised metals such as iron and manganese oxides can be easily reduced and immobilised into groundwater. This is further problematic if iron oxide contains high amounts of toxic substances such as arsenic.

2.3.2 Impact of Land Use Change on Groundwater Quality

Geological and Geographic Conditions and Groundwater Quality

The basic constituents of groundwater quality are determined by the natural weathering process, which is an interaction between groundwater and rock minerals. In addition to the major ions such as sodium, potassium, calcium, magnesium, chloride, sulphate and bicarbonate, groundwater in tropical countries often contains high amounts of silica, iron and manganese. Hence, groundwater is often filtered to remove iron and manganese before use.

Inland cities built in valleys have well-defined, separated groundwater basins, whereas coastal cities have complex aquifers with multiple layers of highly permeable and impermeable layers. Alluvial deposits contain various elements, some of which are either toxic to human beings or make the groundwater difficult to use. In addition, coastal aquifers are vulnerable to saltwater intrusion if groundwater tables are drawn down.

Groundwater Contamination Due to Industrial Development

In the early stage of urban development, most industries are small-scale and there is no clear distinction between residential and industrial areas. Regulations on the effluents from industries are lax so that surface waters are often contaminated with effluents from industries. Unconfined aquifer can be easily contaminated due to infiltration of those contaminated surface waters. Some of these small businesses could be direct sources of groundwater contamination. For example, gas stations and commercial laundries are known to be the potential point sources of groundwater contamination due to poor management of chlorinated detergents and gasoline tanks constructed below the ground level.

Not only small businesses, but also large industries could be sources of groundwater contamination due partly to negligence and limited knowledge of the characteristics of organic solvents. In order to reduce the potential contamination hazards by industries, city planners should pay attention to the location of industries to avoid areas with vulnerable aquifers.

Groundwater Contamination Due to Urban Development

In many Asian cities, the urban population has exploded before the basic infrastructure, such as water supply and wastewater systems, has been built. In such cities, pit latrines and leaking septic tanks are the major sources of groundwater contamination. Wastewaters also infiltrate into unconfined aquifers from unlined open channels. Due to many sources of groundwater contamination within a city, it is very difficult to come up with a working plan to protect unconfined aquifers. Urban development projects disturb the aquifer structures, pump much groundwater during construction, and sometimes inject a variety of chemical materials into the subsoil. All construction projects can have significant impacts on groundwater flow and quality.

Water supply systems in urban areas can either increase or decrease groundwater resources. In many large cities, where water is brought in from external water resources, leaking water from water mains is one of the major sources of groundwater recharge. On the other hand, leaking sewer pipes are also known to be one of the major sources of groundwater contamination in urban areas. In cities where groundwater is pumped to be the source of a city's water supply, groundwater resources are decreased by the expansion of water supply systems.

Groundwater Contamination Due to Gardening and Agriculture

Fertilisers and agricultural chemicals used in gardens and farming plots within urban areas may contaminate groundwater. Although the areas of each garden or farming land in a city may be small, the use of fertilisers and agricultural chemicals may have significant effects on groundwater quality due to the close distance to the points of water use. However, their significance has not yet been well elucidated. In peri-urban farmlands, more chemical fertilisers are used when a city's population and economy grow beyond a certain point because farmers shift to cultivate high-price, cash products such as flowers, tea and fruits that require more fertilisers. They tend to spray more fertiliser than needed believing that the productivity may go up if they spray more fertilizer.

2.4 Groundwater Treatment Technologies

2.4.1 Removal of Naturally Occurring Contaminants

Iron and Manganese

Iron and manganese are two common contaminants found in groundwater in many places. The concentration of iron sometimes goes up to tens of milligrams per litre in tropical regions, where laterisation is the dominant mechanism of chemical weathering. Khatiwada et al. (2002) detected up to 12.1 mg/L of iron and 1.2 mg/L of manganese in groundwater in Kathmandu City. Nga et al. (2002) found more than 22 mg/L of iron in groundwater in Hanoi City.

Iron and manganese can be removed by aeration and sand filtration processes (Fig. 2.6). These processes are more efficient in tropical countries where the groundwater temperature is always higher than 25°C. Groundwater containing iron and manganese is usually anoxic; hence groundwater is first pumped to the aeration tray to increase dissolved oxygen. Then, iron and manganese are oxidised in the contact chamber and also by autocatalytic reactions in the sand filter; consequently, oxidised iron and manganese are filtered by the sand filters that contain coarse sand grains coated with iron oxyhydroxide or manganese oxyhydroxide. Because iron and manganese concentrations in some tropical cities are very high, as mentioned before, it is necessary to use the whole filter depth to remove iron and manganese. Hence, the grain size of sand and the filter depth are usually 1.0–1.2 mm and 1.5–2.0 m, respectively, in these iron and manganese removal processes, which is significantly greater than sand grains and filter depths used in the sand filtration processes for surface water treatment. The removal efficiency of these processes is fairly high, more than 95% if the process is designed suitable to the raw water quality and is maintained properly.

In some countries, such as Japan, where the water quality standards for iron and manganese are very stringent and water temperature is not as high as in tropical countries, chemical oxidizing agents such as chlorine and manganese peroxide are used to enhance oxidation of iron and manganese before filtration.

Fig. 2.7 shows a photograph of Phap Van Water Treatment Plant, which is one of the eight water treatment plants in Hanoi City. Table 2.1 describes groundwater and treated water quality of the selected six water treatment plants in Hanoi City. The treatment process of these plants follows the same process as shown in Fig. 2.6. Iron was removed to less than 0.1 mg/L in five out of six plants, showing the high removal efficiency of this process.

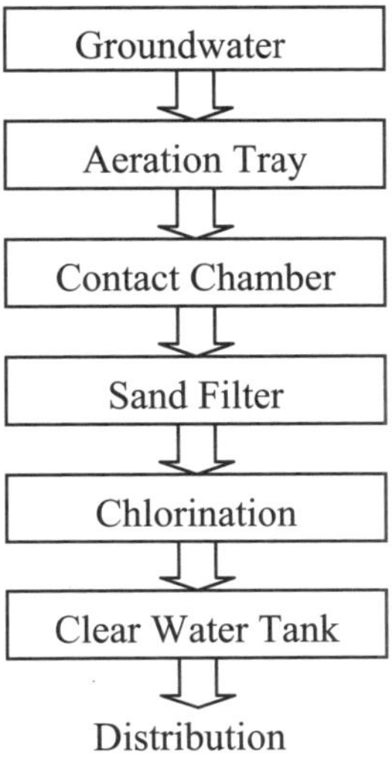

Fig. 2.6. Iron and manganese removal process by aeration tray and sand filter

Fig. 2.7. Phap Van Water Treatment Plant, Hanoi, Vietnam.

Table 2.1. Removal efficiency rates of water treatment plants in Hanoi City.

Treatment Plant	NH4-N		DOC		Fe		As	
	GW	treated	GW	treated	GW	treated	GW	treated
I	0.1	< 0/1	1.5	0.4	0.7	< 0.1	0.001	0.001
IV	1.3	< 0.1	1.8	1.1	1.6	< 0.1	0.043	0.016
V	0.5	< 0.1	1.3	0.3	1.6	< 0.1	0.022	0.015
VI	5.3	5.1	4.0	3.6	9.7	0.5	0.038	0.008
VII	16.0	14.0	7.3	6.0	10.8	< 0.1	0.070	0.019
VIII	6.6	7.0	3.4	3.3	17.0	< 0.1	0.097	0.008
Standard		3.0		---		1.0		0.010

Concentration in mg/L. (Source: Nga et al. 2003)

Arsenic

Arsenic is commonly found in groundwater at trace concentrations, but in some regions, the levels of arsenic in groundwater are prohibitively high for potable uses. The conventional method of arsenic removal from groundwater is co-precipitation with iron followed by sand filtration, which is the same as the iron and manganese removal processes mentioned in the preceding section. The arsenic removal efficiency of these processes is reported to be 50 to 90%, but the removal efficiency is affected by many factors such as the concentration of iron and phosphorus in groundwater and the structure and maintenance of the filters. In addition, adsorbed arsenic could be leached out of the filters if the filter becomes anaerobic or the groundwater quality suddenly changes. Therefore, small household units are often abandoned due to operational failure within several months after installation. Disposal of the sludge containing arsenic is also a problem in this process. Arsenic adsorbed on iron oxide can be easily released into water under anaerobic conditions, which may contaminate local environments.

Table 2.1 shows the arsenic concentration in groundwater and treated water in the six water treatment plants in Hanoi City. The removal rates of arsenic for groundwater containing more than 0.02 mg/L of arsenic varied between 32% and 92%. Plant VIII shows the highest arsenic removal rate, to below the Guideline Value, despite having the highest arsenic concentration in the groundwater among these six water treatment plants. Although the arsenic concentrations of the treated waters were higher than the WHO Guideline Value of 0.010 mg/L in some water treatment plants, the operation of these plants may be adjusted to achieve a higher arsenic removal efficiency because currently they are not operated to achieve the maximum removal of arsenic.

Numerous other processes have been proposed for arsenic removal from groundwater. Those processes include membrane filtration, synthetic resins adsorption and synthesised iron oxide grains. The arsenic removal rate by reverse osmosis membranes depends on the dominant arsenic species in the groundwater, whereas synthetic resins require regeneration. In developing counties, therefore, these processes cannot be easily applied.

Fluoride

Fluoride is another naturally occurring contaminant that is known to undermine the health of tens of millions of people all over the world. The conventional methods of fluoride removal employ either adsorption or precipitation mechanisms, whereas membrane filtration technologies are new processes that are gaining popularity because of easy operation and very high fluoride removal efficiency. Since a case study on membrane filtration of fluoride-laden groundwater is presented in Chapter 15, only conventional methods of defluoridation are explained in this chapter.

Fluoride is an ion dissolved in water; hence, it is not easy to remove fluoride from water. It is often the case that attempts to remove fluoride from groundwater end up in failure. This is partly due to the difficulty of maintaining the defluoridation units. Because of the difficulties of removal, the World Health Organization (WHO) recommends looking for alternative water sources before thinking about fluoride removal processes (Fawell et al. 2006). There are several defluoridation processes that have been applied in the field so far. Since the principle of fluoride removal and the ease of handling these processes differ significantly, one should choose the best-fitting process to the local conditions in terms of groundwater quality, operation and maintenance. In developing countries, centralised large-scale treatment processes are not always the best choice. Decentralised, community-level treatment or household treatment processes may perform better in some places. Therefore, the scale and treatment site should also be taken into consideration when choosing the treatment process.

Defluoridation techniques can be divided into the following categories:

- chemical adsorption processes, including activated alumina, bone charcoal, and clay minerals
- chemical precipitation processes, including aluminium sulphate, lime, polyaluminium chloride, and phosphate compounds
- other advanced treatment processes, including reverse osmosis, electrodialysis, and distillation.

Chemical adsorption process employs the ion-exchange process between hydroxide (-OH) and fluoride (F^-). Some clay minerals, e.g. clinoptilolite, gibbsite, goethite, kaolinite, and zeolite, are known to have metal lattice hydroxyl-groups, which are exchangeable with fluoride. Bone charcoal, on the other hand, contains hydroxylapatite, $Ca_{10}(PO_4)_6(OH)_2$, of which hydroxyl-groups can be exchanged with fluoride. The major drawback of bone charcoal is the difficulty in producing good-quality bone charcoal near the points of use. Poorly produced bone charcoal impairs the taste and odour of produced water; hence, bone charcoal processes are often suspended due to difficulty in replacing used bone charcoal. A good quality activated charcoal can adsorb more than 4 mg/L of fluoride (Fawell et al. 2006). Aluminium oxide grains, or more commonly, activated alumina, are made from aluminium hydroxide by removing hydroxide groups to produce a high porosity and surface area. When activated alumina is filled in a column bed, it can remove fluoride and other impurities in water. This process is popular in India, but the removal efficiencies of activated alumina in the field application vary significantly. In some cases, this method is no better than the bone charcoal process. Hence, it is recommended to perform a field test before the full-scale implementation of an activated alumina process.

Chemical precipitation processes are subdivided into contact precipitation processes and co-precipitation processes. These processes require addiction of chemicals to produce fluoride-containing precipitates by making the saturation index of sparingly soluble fluoride salts higher than zero; that is, above the saturation concentration. In contact coagulation processes, calcium chloride, $CaCl_2$, and mono sodium phosphate, NaH_2PO_4, are added to produce precipitates of calcium fluoride, CaF_2, and fluoroapatite, $Ca_{10}(PO_4)_6F_2$. These chemicals are mixed with groundwater containing fluoride, and then filtered through a bone-charcoal contact-bed to accelerate the precipitation of calcium fluoride and fluoroapatite. The contact time and addition of the chemicals are the important operational parameters of a contact precipitation process. The design criteria for the treatment of groundwater containing 10 mg/L of fluoride are given by Fawell et al. (2006).

Co-precipitation processes include the Nalgonda process, which was developed in India. In this process, aluminium sulphate, $Al_2(SO_4)_3$, is added, together with lime, $Ca(OH)_2$, into groundwater and then mixed to produce aluminium flocs that incorporate fluoride and other negatively charged impurities in water. The flocs are then left for about an hour to be settled, and the supernatant is either decanted or siphoned out to get defluoridated water. Since aluminium sulphate is slightly acidic, lime works as alkaline to neutralise the water to maintain an optimum pH. Thus, the required dosages of aluminium sulphate and lime depend on fluoride concentration and alkalinity of groundwater. Compared with conventional water flocculation for the removal of suspended particles, a much higher dosage of aluminium sulphate, up to 1200 mg/L, is required in the Nalgonda method, which causes the problem of toxic sludge disposal. In addition, the removal efficiency is limited to about 70% (Fawell et al. 2006) so that it can be applied only if the fluoride concentration in groundwater is a few milligrams per litre. It is also important to ensure that the residual aluminium will not exceed the WHO guideline value of 0.2 mg/L.

Organic Matter and Ammonia Nitrogen

In some cases, groundwater contains very high amounts of organic matter and ammonia nitrogen. Nawa Raj et al. (2002) reported the maximum concentrations of ammonia nitrogen as 62 mg/L and dissolved organic carbon (DOC) as 63.6 mg/L in fossil groundwater beneath Kathmandu City. Kuroda et al. (2006) detected ammonia nitrogen and DOC up to 20 mg/L and 13 mg/L, respectively, in the groundwater in the eastern lowlands of Tokyo. Common characteristics of this groundwater are low dissolved oxygen and anaerobic environments, as indicated by low oxidation-reduction potential (ORP). Anaerobic environments are often brought about by the degradation of organic matter, which consumes dissolved oxygen and releases both dissolved organic matter and ammonia nitrogen. Ammonia nitrogen can be removed from groundwater by the following processes:

- Break-point chlorination
- Biological nitrification
- Ammonia stripping

Since break-point chlorination requires chlorine addition at about ten times the value of the ammonia nitrogen concentration, it can be applied only if the ammonia concentration in groundwater is low, e.g. a few mg/L. Biological nitrification converts ammonia to nitrate by means of microbial activities. Hence, it shows a stable performance in tropical regions where the groundwater temperature is high all year round. It should be noted that the nitrate causes adverse health impacts, and thus the WHO has set the Drinking Water Guideline Value at 50mg/L for nitrate (approximately 10 mg/L as NO_3-N). If the ammonia nitrogen concentration is close to 10 mg/L, biological nitrification is not recommended. The ammonia stripping method leaves no residual or reaction product; hence it could be applied to rather high ammonia concentrations. However, it requires pH adjustment both before and after air-stripping. Thus it can be applied only if there is no other method.

Dissolved organic matter can be removed by the following processes:

- Activated carbon adsorption
- Ozone and granular activate carbon (GAC)
- Membrane filtration (nanofiltration: NF/ reverse osmosis: RO)

While activated carbon adsorption is the simplest process among these three processes, the life of activated carbon is rather short if DOC in groundwater is high. Pre-ozonation may extend the life of activated carbon, but it requires quite a large investment for the construction and operation of the ozone and GAC process. Membrane processes such as nanofiltration (NF) and reverse osmosis (RO) are promising technologies that can achieve very high rates of organic matter removal. These processes can be applied to the treatment of organic matter-containing groundwater where they are economically feasible.

2.4.2 Removal of Anthropogenic Contaminants

Nitrate

Nitrate is one of the most commonly detected contaminants in groundwater. Nitrate originates from chemical fertilisers and animal manure in agricultural fields. Massive application of chemical fertiliser is often brought about by the conversion of rice paddy fields to cash crops such as tea and flowers. As a result, several hundred milligrams per litre of nitrate can be detected in groundwater flowing beneath the agricultural fields. Furthermore, groundwater in urban areas often contains a very high concentration of nitrate. Nawa Raj et al. (2002) reported that a maximum concentration of 23 mg/L of NO_3-N (102 mg/L as nitrate) in Kathmandu City. Nitrate concentration was found to be higher in shallow, unconfined aquifers, but lower in confined fossil aquifers, which is in a reducing environment. High concentration of nitrate is also reported in Tokyo. Kuroda et al. (2006) found nitrate nitrogen up to 19 mg/L of NO_3-N (84 mg/L as nitrate) in the western terrace region of Tokyo, whereas nitrate nitrogen was not detected in the eastern lowlands of Tokyo. These results indicate that the dominant nitrogen species depends on the redox state of groundwater; namely, in reducing groundwater, ammonia nitrogen is dominant, whereas in oxidizing groundwater, nitrate nitrogen is dominant.

Nitrate in groundwater cannot be removed by conventional water treatment processes such as coagulation and sand filtration because nitrate is a monovalent anion that is completely dissolved in water. Nitrate removal from groundwater can be performed by either physicochemical or biological methods as follows:

- Biological denitrification
- Electro-dialysis
- Membrane filtration: NF/ RO
- Zero-valent iron (ZVI) reduction

Biological denitrification takes place both in natural and anthropogenic environments such as activated sludge processes in wastewater treatment plants. Nitrate reducing bacteria reduces nitrate to di-nitrogen gas and, in some part, other gaseous forms such as nitrous oxide while oxidising electron donors. Either organic matter, such as ethanol, or inorganic matter, such as sulphur, is used as an electron donor. Although biological denitrification requires the addition of an electron donor, electro-dialysis and membrane filtration processes do not require any chemical addition. The drawbacks of these membrane processes are their high energy requirements and the problem of disposal of the concentrated nitrate solution.

Nitrate reduction by zero-valent iron is a novel process that makes use of a totally chemical oxidation-reduction process. Nitrate reduction by zero-valent iron depends on the particle size of ZVI and also the reaction conditions, such as pH. Recently, various methods of producing nano-scale ZVI have been proposed. Praseatti (2007) examined the effects of pH and iron/nitrate ratios on the reaction kinetics and the end products. It was found that pH is the most important factor that determines the reaction kinetics because protons are involved in the redox reaction.

Microbial Contamination

Unconfined shallow groundwater can be easily contaminated by pathogenic bacteria if the groundwater is not well protected. In many cities in developing Asian countries, due to a large number of pit latrines, domestic wastewater that is frequently contaminated by infectious microbes seeps directly into unconfined aquifers. Protecting wellheads is quite important to prevent microbial contamination, but in densely populated urban areas, wells are often located near the latrines. Dug wells are more susceptible to microbial contamination than tube wells. In Jakarta, it was found that after incidents of flooding, many shallow wells were contaminated with bacteria (Phanuwan et al. 2006). Boiling water before drinking seems to be the most certain method of disinfection of groundwater in each household. In small to medium water treatment plants, membrane filtration followed by chlorine disinfection may be the most assuring method for the removal of pathogenic microbes. Conventional sand filters are designed for the removal of iron and manganese from groundwater; hence the removal rates of microbes are not high enough. UV irradiation is very popular for small water systems in Asian countries; however, it should be noted that water can be contaminated in distribution pipes before it reaches households.

Organic Solvents

The most commonly used methods for the removal of volatile contaminants, such as chlorinated organic solvents, are air-stripping and activated carbon adsorption. Air stripping is economically more suitable where the concentration of organic solvents is rather high, whereas activated carbon adsorption can bring down the concentration of treated water to a lower level. The use of activated carbon adsorption is, however, not very popular because of the labour required to replace the used activated carbon. The cost of replacing activated carbon becomes fairly high if the groundwater contains high concentration of organic solvents and/or any other organic matter which also adsorbs on activated carbon. Other treatment methods, such as bioremediation, are elucidated in Chapter 11.

2.5 Conclusion

In many Asian cities, groundwater plays an important role to supply drinking water and water for industrial use. In cities such as Hanoi, Ho Chi Minh, Jakarta and Kathmandu, groundwater sources are shrinking due to over-abstraction, and the quality has deteriorated because of a lack of effective countermeasures against anthropogenic contamination. Naturally occurring contaminants such as arsenic and fluoride are also detected at high concentrations in urban groundwater. There are various treatment technologies for the removal of different kinds of contaminants from groundwater, as illustrated in this chapter. However, groundwater treatment requires a substantial investment and a high level of operational skill. It is often difficult to make such an investment or to maintain a small-scale water treatment unit. Hence, the most important thing is to protect the urban groundwater both in quantity and quality. In this regard, a good combination of management tools including groundwater monitoring, information dissemination and regulation must be explored in each city. Treatment and disposal of domestic wastewater in burgeoning cities in developing countries is a priority issue to protect urban groundwater. Urban planning to designate the location of potentially polluting industries is also very important for the protection of vulnerable aquifers. An efficient and cost-effective monitoring network for groundwater quality and quantity should be developed to understand the present status of groundwater resources and also for the regulation of groundwater abstraction.

References

Foster S, Lawrence A, Moris B (1998) Groundwater in urban development assessing management needs and formulating policy strategies. World Bank Technical Paper No.390

Khatiwada NR, Takizawa S, Nga TTV, Inoue M (2002) Groundwater contamination assessment for sustainable water supply in Kathmandu Valley, Nepal. Water Science and Technology 46(9):147-154

Kuroda K, Fukushi T, Takizawa S, Aichi T, Hayashi T, Tokunaga T (2006) Groundwater contamination in Tokyo and estimation of the contamination source (in Japanese). Journal of Water and Waste 48(9):37-45

Matsui Y, Takeda T, Takizawa S, Wongrueng A, Wattanachira S (2006a) Application of nanofiltration to fluoride removal from groundwater in Chiang Mai Basin, Thailand. In: Proceedings of the 32nd WEDC International Conference, Colombo, Sri Lanka, pp. 585–592

Matsui Y, Takeda T, Takizawa S, Wongrueng A, Wattanachira S (2006b) Evaluation of nanofiltration process for fluoride removal from groundwater in the Chiangmai Basin (in Japanese). Jour JSCE 62(4):403-414

Ministry of the Environment, Japan (2007) Conservation of groundwater environment

Nga TTV, Inoue M, Khatiwada NR, Takizawa S (2003) Heavy metal tracers for the Analysis of Groundwater Contamination: Case Study in Hanoi City. Water Science and Technology 3(1-2):343-350

Nga TTV, Takizawa S, Natural occurrence of arsenic in the organic-rich aquifer in Hanoi City: sources and mobilization processes. In: Furumai H (ed) Southeast Asian Water Environment, vol. 2. IWA Publishing, pp 189–196

Phanuwan C, Takizawa S, Oguma K, Katayama K, Yunika A, Ohgaki S (2006) Monitoring of human enteric viruses and coliform bacteria in waters after urban flood in Jakarta, Indonesia. Water Science and Technology 54(3):203-210

Prasetiati MAN, Takizawa S (2007) Chemical reduction of nitrate by nanoscale zero-valent iron. In: Proceedings of the International Symposium on Southeast Asian Water Environment, pp 268-275

Rushton, Al-Othman (1996) Control of rising groundwater levels in Riyadh, Saudi Arabia. In: Wilkinson WB (ed) Groundwater problems in urban areas. Thomas Telford

Thuong NT, Takizawa S, Dan NP, Truong BD (2007) Water pollution in inner-city canals in Ho Chi Minh City, Viet Nam, In: Furumai H (ed), Southeast Asian Water Environment, vol. 2. IWA Publishing, pp 107-114

Yusuf MA, Takizawa S, Katayama H (2007) Ground water storage in aquifer of Dhaka City and land subsidence: A major concern in potable water supply in Dhaka Metropolis in 2030. In: Furumai H (ed) Southeast Asian Water Environment, vol. 2. IWA Publishing, pp 205–212

3. Groundwater Resources in Urban Water Management

Yanjun Shen

Centre for Agricultural Resources Research, Chinese Academy of Sciences

Taikan Oki

Institute of Industrial Science, the University of Tokyo

3.1 Introduction

During the twentieth century, the world's urban population increased more than tenfold, while rural population increased only twofold (United Nations, 2004). Today, nearly half the world's population lives in urban areas compared to less than 15% in 1900, and the absolute number of people living in urban areas amounted to 3.04 billion in 2003 (United Nations, 2004). Water use in urban areas is directly related to economic development, social prosperity, and environmental conservation. Accompanying urbanisation or urban population growth, the water supply in urban areas has also increased rapidly. According to United Nations Food and Agriculture Organisation (UN-FAO) statistics, worldwide water usage for domestic and industrial purposes were 380 and 770 km^3, respectively, in 2000. Different from UN-FAO statistics, Bengtsson et al. (2006) estimated the global urban domestic water use in the world as 230 km^3/year, about 60.5% of the total domestic water use in 2000. If industrial demand is also accounted for, total urban water use would be much higher, amounting to 1000 km^3/y globally in 2000. Thus, providing sufficient water has become a big problem for urban water management.

For developed countries, the technology and infrastructure are relatively well developed to handle the intensive water requirements of urban areas. However, in the developing world, the urban population growth rate always exceeds the growth rate of investments for water infrastructures. Because surface water development requires a big investment and a long planning and construction period, groundwater becomes the major water source for urban water use. Therefore, problems pertaining to groundwater over-abstraction have taken place in those cities.

This chapter provides some information on groundwater use in worldwide urban water management. After an overview of urban development and urban water use in the world, a global perspective of groundwater in urban water use is presented. The geographic distribution of cities highly dependent on groundwater in their water supply systems, problems arising from intensive groundwater exploitation in developing countries, especially the cases of Chinese cities, and some suggestions for groundwater use in future urban water management are presented.

3.2 Urban Water Use in the World

3.2.1 Urban Development and Urban Population

Throughout history, all cities evolved from small settlements. As agricultural production became more efficient, small settlements grew to become cities. Haggett (1972) estimated that Babylon had a population of 80,000 between 3000 and 2500 B.C. It was estimated that in the ancient age, there were as many as 50–90 agriculturalists to every non-agriculturalist in a given population, indicating that by far the majority of city dwellers were bound to the land (Lazaro, 1979).

Rugg (1972) pointed out the urban population occupied only a small proportion of the total population before 1800; after that, cities grew greater by accumulating more population. Johnson (1967) showed that from 1800 to 1950, the world population grew 16.5 times from less than 40 million to greater than 700 million.

Fig. 3.1 integrates different sources of estimates on world urban population, showing the world urban population increased logarithmically from 1800 to 2000. An obvious trend is that the population living in large cities is growing faster than the total population. This resulted in increasing demand for water supply in these large cities.

As shown in Fig. 3.1, the world urban population was less than 50 million in 1800. After a century, the world urban population exceeded 200 million in 1900. In 1950, it was estimated the world urban population exceeded 700 million. According to the UN's estimation, the world urban population has reached 3.04 billion in 2003.

Figure 3.2 shows the global distribution of urban populations and urban ratio at the resolution of 0.5-degree grid cells in 2005. High urban populations are distributed in Europe, the eastern U.S., eastern China, Japan, India, southeast Brazil, and Guinea Bay. Most grid cells (0.5-degree by 0.5-degree) with urban population more than 1 million are seen in those areas. It should be noted that the urban population density in developing countries is very high and the highest value even can reach nearly 20 million per grid. However, some developing countries, such as east China, India, and those near Guinea Bay show relatively low urban ratios due to their large rural populations.

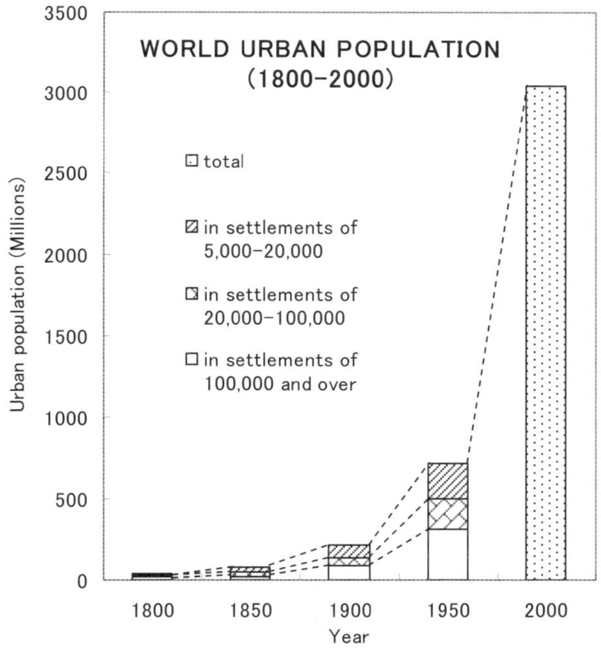

Fig. 3.1. World urban population growth (from Johnson 1967, updated with FAO's population estimation)

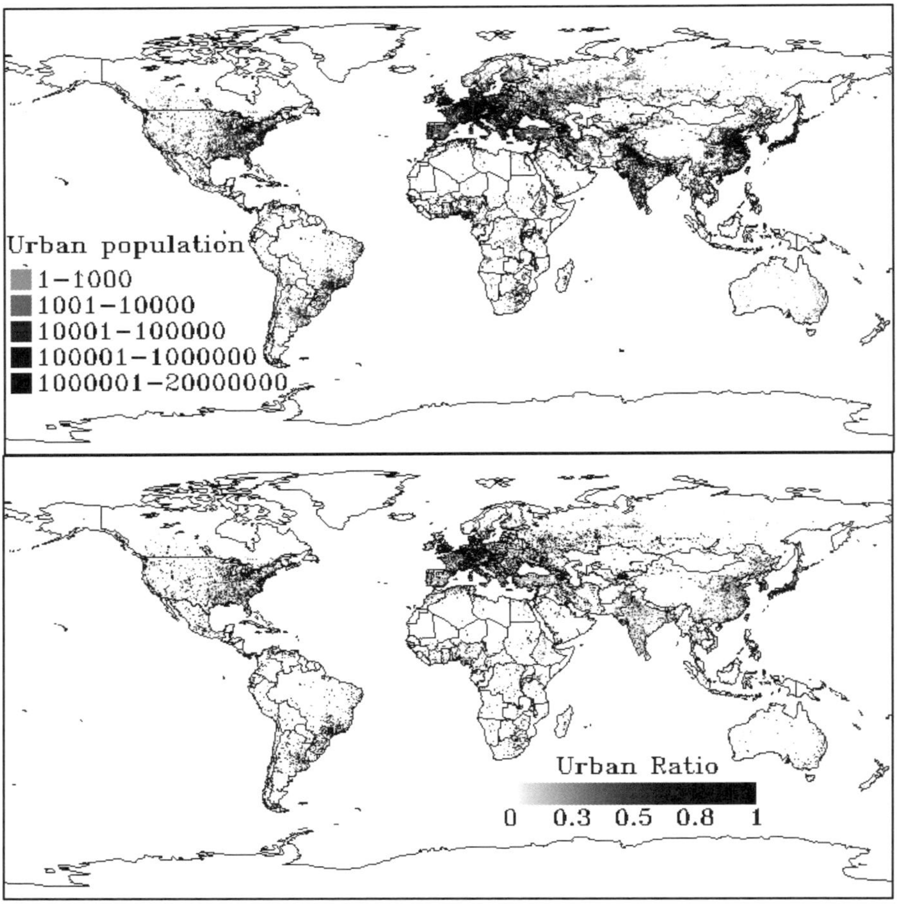

Fig. 3.2. Geographic distribution of world urban population and urban ratio in 2005 (Spatial resolution is 0.5-degree. Data source: Bengtsson et al. 2006)

3.2.2 Water Use in Urban Areas Worldwide

Water use in urban areas can be roughly divided into domestic use, industrial use, scenery use, and other uses. The per capita daily domestic water use increases along with economic development. In general, per capita water use reaches the maximum at some point of development, and then starts to decline. Bengtsson et al. (2006) estimated worldwide urban and rural water use, assuming that the urban domestic water use would reach its maximum when the national GDP per capita exceeds 3,200 USD/year. They estimated the maximum urban domestic water consumption as 330 litres/capita/day (lpcd). Table 3.1 shows the per-capita GDP and urban/rural domestic water use in selected countries on each continent in 1990. The domestic water use intensity is highly dependent on the economic development levels, even in the world's thirstiest region; for example, in the Middle East, the per capita urban domestic water use can reach 533 lpcd in UAE and 126 lpcd in Syria, which is still higher than many other countries in the region thanks to the relatively higher GDP/capita.

Table 3.1. Domestic water use of selected countries in 1990

Country	GDP/cap (USD)	Urban (lpcd)	Rural (lpcd)	Country	GDP/cap (USD)	Urban (lpcd)	Rural (lpcd)
USA	22609	629	349	Afghanistan	1568	177	64
Canada	20708	875	438	Kazakhstan	2407	256	93
Mexico	3171	326	124	Kyrgyzstan	675	100	39
Brazil	3157	325	124	Tajikistan	916	120	45
Paraguay	1254	149	54	Israel	11598	284	158
				Syria	991	126	47
Sweden	27767	364	202	Jordan	1232	147	54
UK	17139	315	175	UAE	16900	533	277
Netherlands	19772	237	132	Saudi Arabia	6778	211	110
France	21405	322	179	Iraq	2810	294	109
Germany	21634	204	113	Iran	2055	223	80
Italy	19414	451	250				
Spain	13054	358	199	Pakistan	359	73	31
Russia	3905	329	159	India	369	74	32
Poland	1604	180	65	Nepal	197	57	28
				Bangladesh	270	65	29
Libyan	7136	306	318	Myanmar	192	56	28
Egypt	765	107	41	Thailand	1476	169	61
Ethiopia	141	50	27	Malaysia	2336	249	90
Sudan	529	88	35	Singapore	11500	330	250
Somalia	125	47	26	Indonesia	594	93	37
Namibia	1664	186	67				
Guinea	449	81	33	Mongolia	1438	165	60
Kenya	354	73	31	China	312	69	30
Chad	292	67	30	China, HK	13326	330	250
Cameroon	938	122	45	Korea,	5579	421	234
Uganda	244	62	29	Japan	24041	443	246
Tanzania	160	52	27				
Zimbabwe	838	113	43	Australia	18359	512	284
South Africa	3009	312	117	New Zealand	12826	775	430

Fig. 3.3 shows the spatial distribution of urban domestic water use at 0.5-degree resolution. Even though the average incomes of India and China are relatively low, the urban domestic water uses of these two countries are dramatically high due to high urban population densities. This strong water demand leads to greater pressures on urban water supply.

Table 3.2 lists the world top 30 countries of urban water use in 2000. China, suffering a severe water shortage, is positioned in the second place with an annual urban water use of 191 billion m^3 per year, next to the world's highest urban water user, USA, at 271 billion m^3. Another country facing a severe water deficit is India, which occupies the fourth place with a total annual urban water use of 60 billion m^3. Other water-scarce countries such as Mexico, Spain, Egypt, Kazakhstan, Iran, Australia, and Pakistan are also on the list. The urban water supplies of the above countries with water shortage problems mainly depend on groundwater exploitation. After many years of groundwater abstraction, these countries are facing various problems in their urban water management systems, such as drawdown of groundwater tables, groundwater pollution, land subsidence, and seawater intrusion.

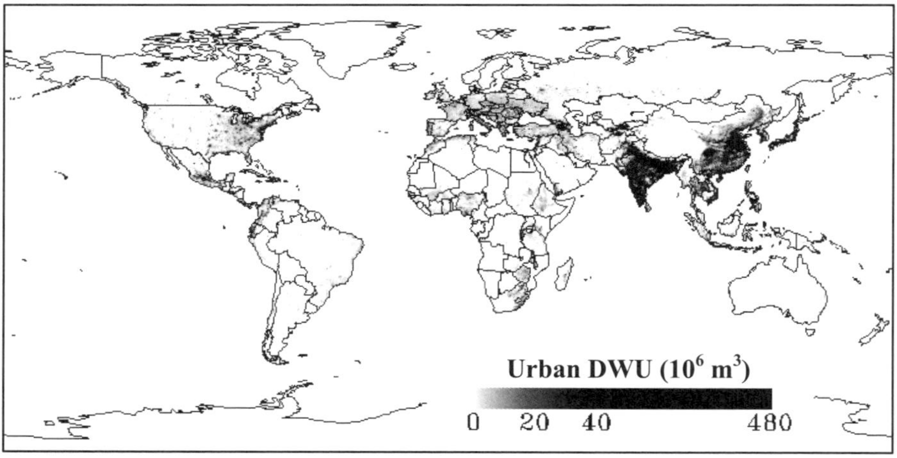

Fig. 3.3. Distribution of urban domestic water use in 2000 (spatial resolution is 0.5 degrees. Data source: Shen et al. 2007)

Table 3.2. World's top 30 countries of urban water use (UWU) in 2000 (Unit: 10^6 m^3/year)

Country	UWU	Country	UWU	Country	UWU
USA	271,657	Vietnam	20,177	Turkey	8,484
China	191,217	Ukraine	17,289	Argentina	7,448
Russia	61,542	Mexico	15,718	Egypt	7,224
India	60,605	Poland	14,218	Philippines	6,423
Canada	39,313	UK	13,261	Kazakhstan	6,312
Germany	37,316	Spain	10,566	Iran	6,087
France	35,049	Romania	9,667	Australia	5,691
Japan	29,128	Belgium	8,840	Netherlands	5,691
Italy	22,637	Korea	8,809	Azerbaijan	5,403
Brazil	22,028	Bulgaria	8,500	Pakistan	5,381

3.3 Groundwater Resources in Urban Water Management

3.3.1 Urban Water Supply and Groundwater

Groundwater plays a fundamental role in supporting the economic and social wealth of many urban areas in the developing world. Until now, no comprehensive statistics existed on the proportion of urban water supply worldwide that is derived from groundwater. Due to its relatively low cost and generally high quality, groundwater has often been the preferred source for reticulated public water supplies and is widely exploited for private, domestic and industrial uses.

Urbanisation and industrialisation have a profound effect on urban groundwater resources. On one hand, groundwater provides an economical and clean water source to the cities; on the other hand, intensive groundwater exploitation and massive effluent and waste disposal from urban areas deteriorate the urban groundwater system, including the local groundwater recharge/discharge regime and the groundwater quality.

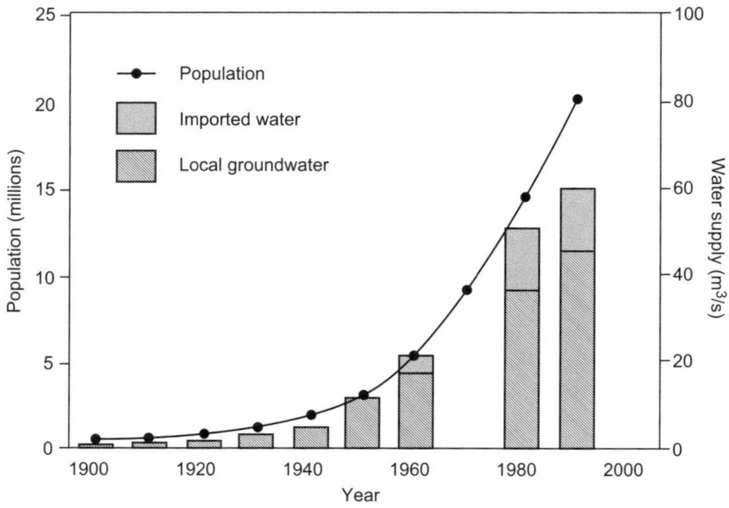

Fig. 3.4. Population growth and urban water demand in the Mexico City metropolitan area (Mazari and Mackay 1993)

The population growth of a city is largely dependent on the urban water supply. Figure 3.4 illustrates the population growth and urban water demand in the Mexico City metropolitan area. Before 1960, the urban water supply mostly could be met by abstracting groundwater; however, after 1960, the local water resource system could not meet the urban water demand and a large amount of water was imported from other basins. But even after 1960, groundwater still occupies the largest share of water supply.

In developing countries, the major problems in urban groundwater management come from the large population in the urban areas, and the lack of technologies for water management. As a result, groundwater depletion and pollution took place simultaneously in these cities. Lazaro (1979) pointed out in the World Bank's report, *Groundwater and Urban Development,* that

> *'a common feature of many developing cities has been the appearance of informal settlements located on marginal land or in burgeoning periurban districts... And, if cities are to provide adequate water supply, sanitation, drainage, and waste disposal to all their residents, municipal authorities need to evaluate critically how to manage the subsurface more sustainably, because these communities increasingly depend upon it, both for water supply and as a waste receptor'* (Lazaro 1979).

In the following sub-section, the geographic distribution of the world cities that depend highly on groundwater is presented in detail.

3.3.2 Distribution of Cities Highly Dependent on Groundwater

Groundwater Dependence Index

In order to calculate the grid-based dependence of urban water supplies on groundwater, an urban groundwater dependence index (UGDI) was designed according to the balance between urban domestic and industrial water use and the surface water flow in river channels in each of the 0.5-degree grid cells. The UGDI is calculated by following equation,

$$UGDI_{(i,j)} = \frac{UWU_{(i,j)}}{Q_{(i,j)}} \tag{3.1}$$

where, $UGDI_{(i,\,j)}$ is the urban groundwater dependence index, $UWU_{(i\,,j)}$ is urban water use of grid *(i,j)*, and $Q_{(i,\,j)}$ is total water flow in the river channels of the grid *(i, j)*. If *UGDI* is greater than a threshold value, then the urban water supply of that grid is considered to be highly dependent on groundwater. In this study, we set the threshold value as 0.4. This means that when the urban domestic and industrial water use is less than 40% of the total surface water flow in the river channels, people can manage the urban water supply using mostly abstraction of surface water. If the urban water demand is greater than 40% of the river discharge, then the urban water supply has to be met by exploiting groundwater in some seasons or diverting water from other basins. When *UGDI* is greater than 1, the surface water discharge in river channels is absolutely less than urban water demand, and then the urban water supply is assumed to be very highly dependent on groundwater or the water supply of other basins.

Distribution of World Cities Highly Dependent on Groundwater

Globally, there are many cities whose water supply is largely dependent on groundwater. Using the *UGDI,* we can calculate the groundwater dependence of urban water supplies in the world. Most large cities located in the arid and semiarid regions are highly dependent on abstracting groundwater to meet the huge demand. Some mega-cities in the humid regions also use large amounts of groundwater as part of their water supply systems (Figs. 3.5–3.7).

Specifically, in Asia and Europe the groundwater-supported cities are distributed from the North China Plain extending through the Silk Road to Central Asia, from South India, Northwest India, and East Pakistan to Iran and the Arabic Peninsula, from the Costal region of the Mediterranean Sea to the northwest of Europe (Fig. 3.5). Not only in most cities located in semiarid to arid regions, but also in many cities in humid regions, such as Tokyo, Osaka, Taipei, Manila, and Jakarta, groundwater was estimated to be an important source of their water supply (Fig. 3.5). This is mainly be

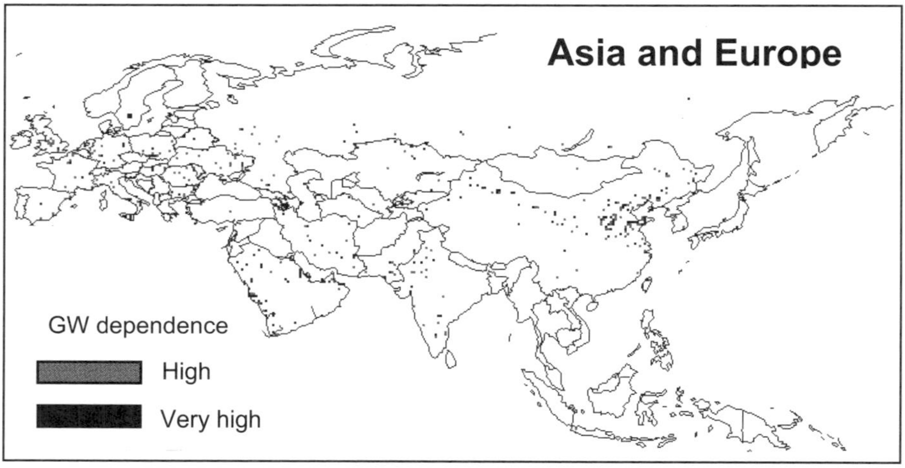

Fig.3.5. Distribution of Asian and European cities highly dependent on groundwater as classified by *UGDI* (2000)

cause these cities are located at small basins with relatively short river courses and the steep gradients of the riverbed from the river origin to the river mouth, which makes effective use of river water quite difficult. Hence, these cities have to abstract groundwater or bring water from other basins. The development of the water supply in Tokyo may be a good example of such cities. Today, the water supply for Tokyo is largely dependent on diversion water from the Ara-Tone River System. Because overexploitation of groundwater in the 1950s and 1960s resulted in severe land subsidence, the Japanese Government, as well as the Tokyo Metropolitan Government, implemented several laws and regulations around 1970 to limit the abstraction of groundwater. Concurrently, water resource development projects were implemented; one of the most important was the water diversion project from the Tone-Ara River System that was put into service in 1965. Thanks to three decades of regulation, the groundwater level of the Tokyo area has been rising. In many cities in developing countries, however, it is difficult to finance such a large-scale water resources development project for urban water use, and hence they are still dependent on groundwater.

The cities using groundwater intensively in Africa are mainly distributed in northern Africa (the cities in Egypt, Libya, Algeria, and Morocco), southern Africa (the cities in South Africa, Zimbabwe, and Namibia), and eastern Africa (some cities in Sudan, Ethiopia, and Kenya). Most of the cities largely supported by groundwater in Africa are in desert to savannah regions where water resources have large seasonal variability as affected by monsoons (Fig. 3.6).

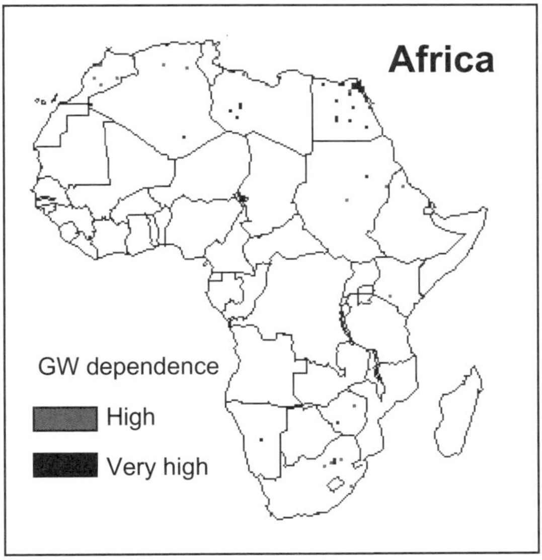

Fig. 3.6. Distribution of African cities highly dependent on groundwater as classified by *UGDI* (2000)

In North America, the large metropolitan areas of the USA, such as the urban areas in the western and eastern coastal regions, the western central states, and the south-western semiarid regions, and most large cities of Mexico (Fig. 3.7) are highly dependent on groundwater. In South America, the highly groundwater-dependent cities are distributed mainly along the western coastal region of Peru and Chile, where they have relatively low precipitation rates (Fig. 3.7).

The urban population worldwide living in cities highly dependent on groundwater amounts to 1.0 billion, of which around 40% living in the cities with *UGDI* between 0.4 and 1.0, and the other 60% living in cities very highly dependent on groundwater (*UGDI* > 1). Table 3.3 summarises the population of eight regions whose cities are highly dependent on groundwater. The population in cities highly dependent on groundwater in Asia (except for the Middle East and Central Asia) amounts to 400 million; the numbers for North America, Europe, Middle East and North Africa (M.E./N.A.), and South America, are estimated to be more than 100 millionin each of those regions.

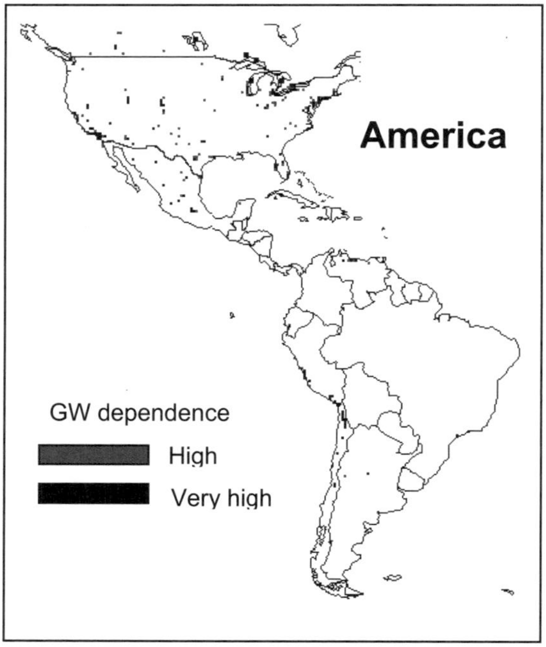

Fig. 3.7. Distribution of American cities highly dependent on groundwater as classified by *UGDI* (2000)

Table 3.3. Population in cities highly dependent on groundwater in eight regions of the world in 2000 (Unit: million)

Regions	Population in cities highly Dependent on Groundwater	Population in cities very Highly Dependent on Groundwater	Total
Asia	211	203	414
Europe	59	68	127
CIS	28	31	59
Middle East / North Africa	32	94	126
Sub-Saharan Africa	18	11	29
N. America	45	114	159
S. America	14	103	117
Oceania	8	3	11
World	415	627	1042

3.4 Development of Urban Groundwater Resource in Chinese Cities

3.4.1 Urban Development and Groundwater Exploitation in the North China Plain

China has experienced rapid economic development during the last 3 decades. Economic development and social reform have brought about rapid urbanisation. According to the *China Statistical Year Book*, the urban population of China in 1949 was 57.6 million, which was only about 10.6 of the total population; this figure grew to 191.4 million in 1980 (19.4% of total population), and 403 million in 2000 with an urban population ratio of 31%. The urbanisation of China has accelerated very much in recent years. The urban ratio in 2003 amounted up to 40.5% and the urban population exceeded 520 million. The latest estimation released by the China National Statistical Bureau in October 2007 states that the urban population ratio of China has reached 44% at the end of 2006 with a total urban population of 577 million. Fig. 3.8 shows the process of urbanisation in China during the last 50 years.

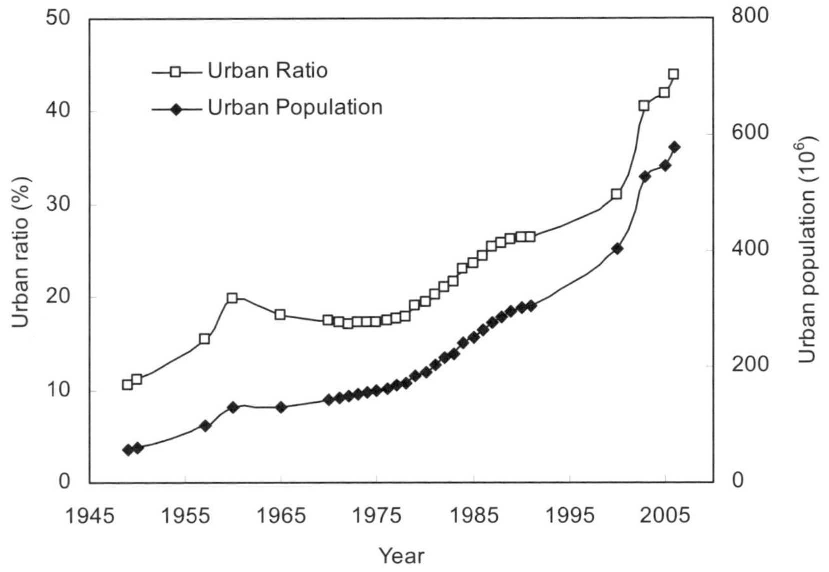

Fig. 3.8. Development of urbanisation in China 1949-2006 (Data source: China Statistical Year Book)

The North China Plain (NCP) is one of the economic centres of China, comprising Beijing, Tianjin, and Hebei Provinces. The population of the NCP now exceeds 100 million and the urban population is more than 40 million. Ranging from the semiarid to semi-humid climate, the annual precipitation of the NCP ranges from 400 to 600 mm. Most cities in the NCP rely on groundwater to support their water supply. During the last several decades, the shallow groundwater tables have declined in a very wide area of NCP. Fig. 3.9 illustrates the drawdown of the groundwater table in the cities in Hebei Province along the Beijing-Guangzhou railway corridor in the last decades. This region belongs to the piedmont plain of the NCP with rich groundwater resources. In 1964, the groundwater table from the north to south ranged from 2 to 7m. After the beginning of economic development, the groundwater level of this line dropped to around 10m beneath the land surface and some groundwater depression cones were observed. In 1993, the groundwater table beneath big cities such as Baoding, Shijiazhuang, and Handan dropped very fast to about 30m in depth. Fig. 3.10 shows the declination trend of the water level under the whole plain between 1959 and 1992. The groundwater table was around 3m below the land surface in the plain in 1959, when there were still artesian wells in some areas; however, the groundwater table beneath the whole plain had largely declined by 1993 and many groundwater depression cones appeared in big cities. In the piedmont plain, the decreasing rate of water table reached 1m annually, according to observations at Luancheng Ecological Research Station of Chinese Academy of Sciences (Shen, 2004).

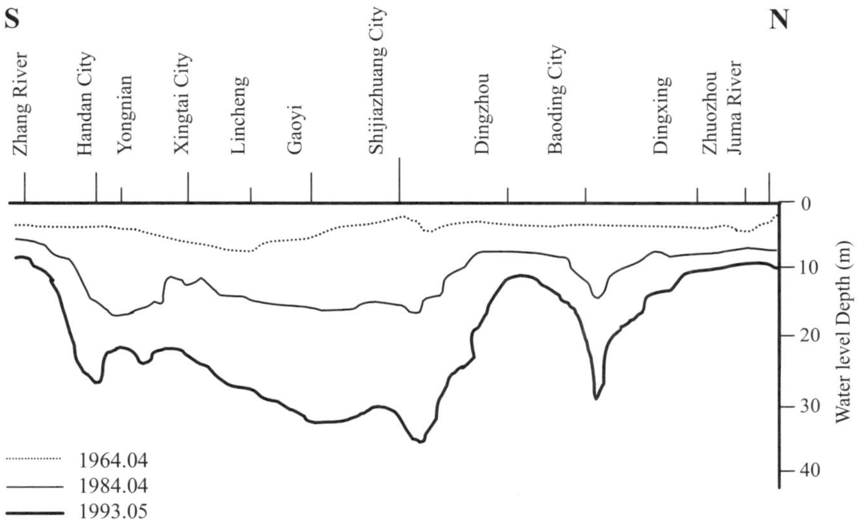

Fig. 3.9 Changes in the groundwater table along the Jing-Guang Line in piedmont plain between 1964 and 1993. (Jing-Guang Line is the railway between Beijing and Guangzhou. Shen 2004)

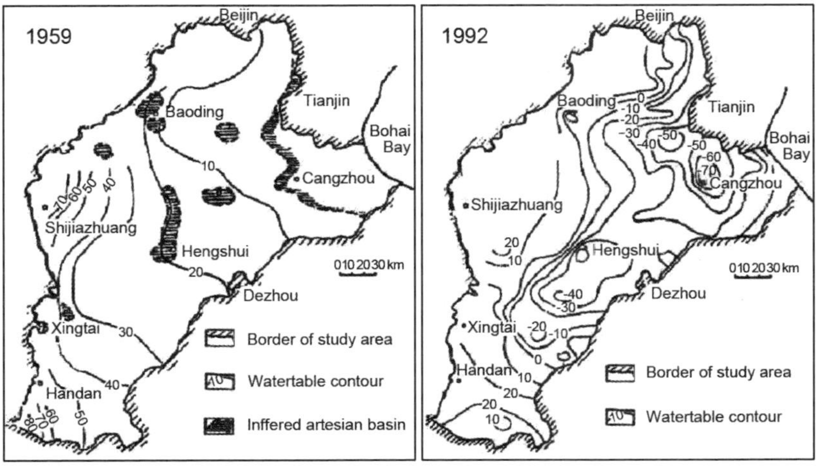

Fig. 3.10. Shallow groundwater potential in the North China Plain in 1959 and 1992 (Sun et al. 1999)

As a result of extensive groundwater over-abstraction, large cities in NCP such as Beijing, Tianjin, and Shijiazhuang are known today as water-scarce cities. In the past several years, Beijing diverted water from Shanxi and Hebei Provinces several times to augment the water deficit. Tianjin has been supported by transferring water from the Yellow River in Shandong Province for more than 10 years. To meet the chronic water deficit due to increasing urban water demand in the NCP, the Chinese government is now launching a huge water transfer project, i.e. the South-North Water Transfer Project, in which the government will try to divert water from the Yangtze River Basin to the NCP. The Middle and Eastern Routes of the Project are now under construction. According to the water diversion plan, the total water transferred across the Yellow River into the NCP will be 17 billion m^3 per year in 2020.

In addition to the drawdown of groundwater tables, groundwater over-abstraction causes land subsidence. There are many places undergoing land subsidence in the NCP. The most depressed area has subsided by more than 3m in Tianjin during the past several decades. Figure 3.11 shows the land subsidence in the NCP measured by using a combination of In-SAR remote sensing technology and ground observation. The most de-

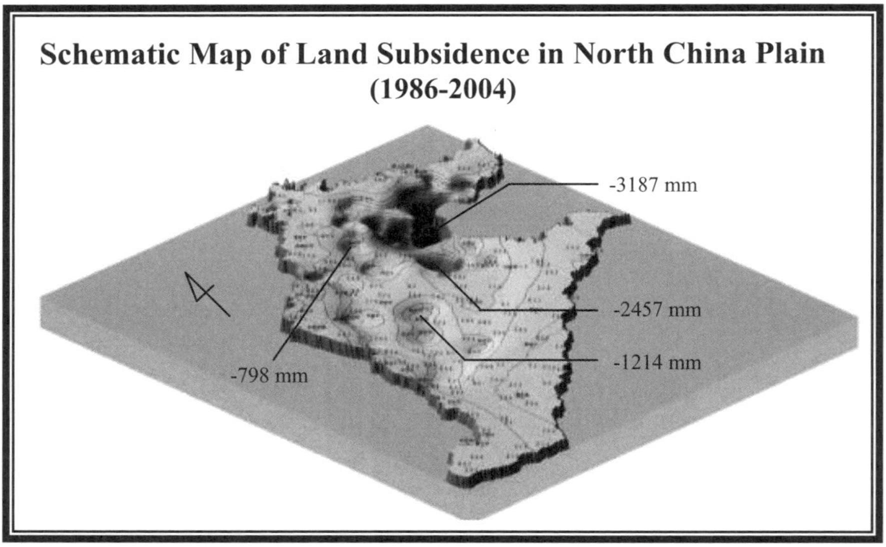

Fig. 3.11. Land subsidence in the North China Plain. The numbers on the map show accumulated subsidence of land surface from 1986-2004. (reproduced from: China Geo-Environmental Monitoring Institute, 2006)

pressed places are located in Tianjian and Cangzhou where saline groundwater in the shallow aquifer cannot be used for municipal and agricultural purposes. Intensive abstraction of deep groundwater over a long period has resulted in serious land subsidence. Consequently, seawater intrusion into the groundwater system undermines the use of groundwater in the coastal region. Other cities such as Beijing, Shijiazhuang, Hengshui, and Handan are also undergoing serious subsidence.

3.4.2 Urbanisation and Its Problems in Shijiazhuang City, Hebei Province

Shijiazhuang is a city located in the North China Plain. It has developed very fast in the past decades, and experienced problems associated with rapid urban expansion such as groundwater depression cones, land subsidence, and water pollution. These problems have become big issues in urban water management. In the following sub-section, the problems Shijiazhuang has experienced are elucidated.

Urban Expansion and Water Supply Change

Shijiazhuang is the capital and the largest city of Hebei Province and is important as the centre of administration, culture, economy and transportation. The current population is around 2.03 million. In the early 1900s, it was just a small village with less than 600 inhabitants. The construction of the Jing-Guang railway line (Beijing to Guangzhou) and the Shi-Tai line (Shijiazhuang to Taiyuan) provided opportunities for development. During World War II, the city expanded fast because of its important position in military operations. After becoming the capital of Hebei Province in 1968, it experienced rapid development. Especially in the last 30 years of economic reform, urban expansion in Shijiazhuang has been very significant. According to the Economic Annals of 2001, its population and built-up area are 1.6 million and 165.5 km^2. Today, the urban population amounts up to 2.03 million and the urban area exceeds 220 km^2. As a result, the landscape of Shijiazhuang region has changed over time. The impacts on the hydrological environment due to urbanisation are very significant (Shen et al. 2005). Fig. 3.12 shows the urban area of Shijiazhuang city in different historical periods (Xiao et al. 2006).

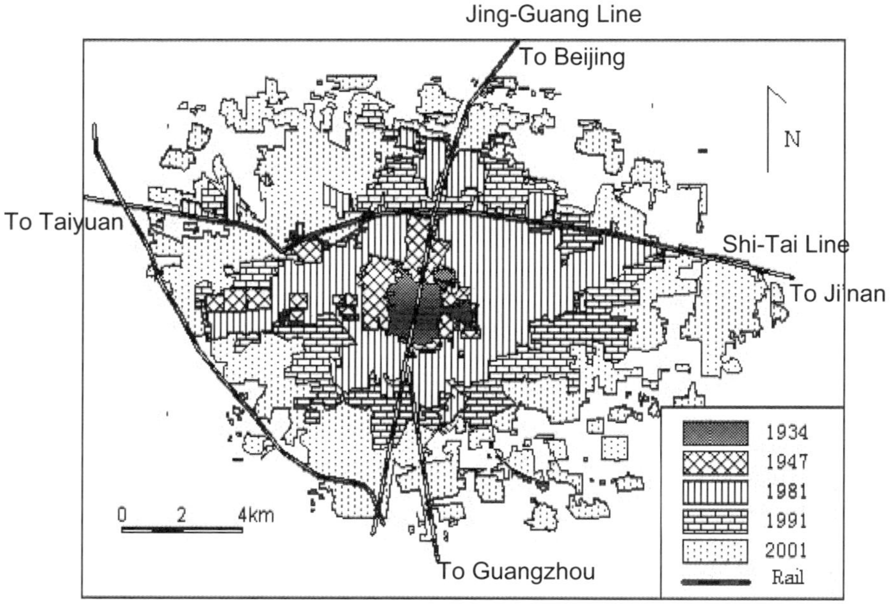

Fig. 3.12. Growth of urban area of Shijiazhuang City, China. The railway is that of 2001. (Xiao et al. 2006)

Associated with urbanisation, the water supply volume has increased significantly (Fig. 3.13) and groundwater has became the most important source of water supply in this city. The number of wells in the water supply system increased every year to meet the increasing water demand. At the same time, many wells have gone out of service every year due to fast drawdown of groundwater table. Shen et al. (2005) have divided the urbanisation and water resource changes of Shijiazhuang city into four periods.

Small Town Period (before 1950). Shijiazhuang was a small town and one of the important transportation sites in the North China Plain. The water supply was based on surface water and groundwater. Groundwater development dates back to ancient times. However, until the beginning of the twentieth century, only shallow wells (up to 10m in depth) were constructed. The water table was as shallow as 3 to 5m. As of 1950, there were two pumping wells which supplied 390,000 m^3/year for domestic use.

Urban Growth Period (1950–1955). Several factories were built and the urban area expanded to 2000 ha. In this period, there were five pumping wells to supply 1.7 million m^3/year for the domestic use; the water table fell by around 1m. The Hutuo River flowed all year around. It was reported in the literature that inland navigation through the major river channels of the North China Plain was very prosperous and played an important role in trade and business in the 1950s. The exploitation of water resources still maintained the balance between recharge and discharge of the natural local water cycle system.

Urban Improvement Period (1955–1981). The pumping wells were developed on a large scale after 1955, particularly since 1958. Towards the end of the 1960s, over a thousand hand-pump wells and wells with motor pumps, were put into service. The well depth ranged from 50 to 150m. During this period, urbanisation and industrialisation relied largely on groundwater, and the water table declined at a rate of 0.65m per year. In fact, after the flood in 1963, which resulted in the worst damage across the whole North China Plain, several reservoirs were constructed on the upper reaches of the Hutuo River in order to mitigate the flood damages. Fig. 3.13 shows the yearly changes of water use in Shijiazhuang from 1948 to 2000. The number of pumping wells was 44 in 1973, and then increased to 77 in 1980. The water for industrial and domestic use was 32% and 55% of the total water consumption, respectively.

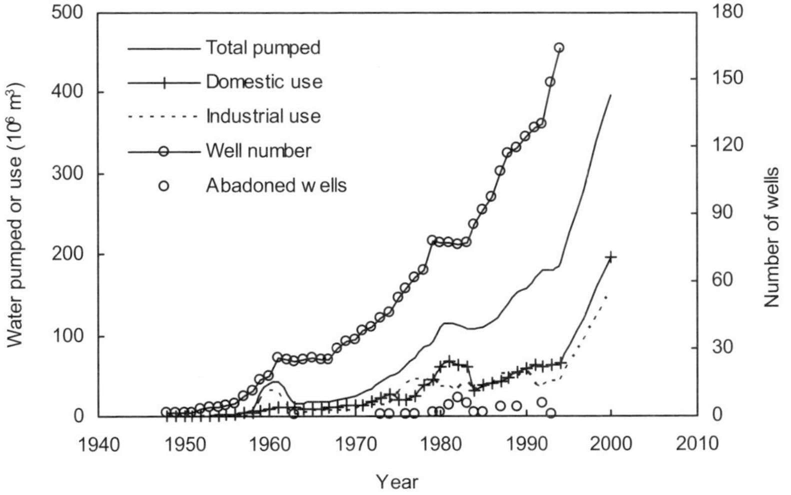

Fig. 3.13. Change of water use structure and groundwater pumping wells of Shijiazhuang over the past 50 years (Shen et al. 2005)

Rapid Growth of Urbanisation Period (1981–2007). Along with the urban re-development, the urban area has expanded rapidly. Today, most of the branches of the Hutuo River are dry and water flows only when water is discharged from the upstream reservoirs during large flood events. The domestic water supply only relies on groundwater. For example, the number of pumping wells increased to 164 with a total abstraction volume of 1.89×10^8 m^3 in 1994. As a result, the water table declined at a rate of 1.1m per year. In the late 1990s, a pipeline was built to transport water from the Huangbizhuang Reservoir to meet the rapidly increasing demand for domestic water. At the same time, deep groundwater was also exploited for use. The total water supply in Shijiazhuang reached 3.97×10^8 m^3 in 2000, of which surface water, shallow groundwater and deep groundwater were 14.4%, 82.5% and 3.1%, respectively.

Groundwater Depletion

Along with urban expansion and increasing municipal water demand, the water table has fallen across a very wide area to deeper than 40m. Fig. 3.14 shows the declining trend of the water table in the city area between 1960 and 1989. The groundwater table was at around 3 to 5m below the surface in 1965, and in some places artesian wells were observed. However, the water table has fallen significantly since 1970. The water table depression cone has appeared since 1973 with the deepest water table depth at 15.7m. In 2000, the water table at the centre of the depression cone reached 42.3m in depth. The recharge area became a pumping area from which much of the groundwater was used for municipal and industrial use.

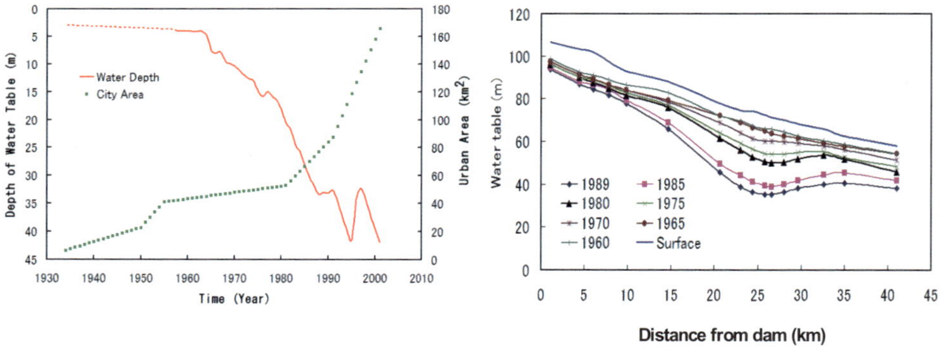

Fig. 3.14. The changes in groundwater depth and the built-up area (left) and the depression cone formation in the urban areas of Shijiazhuang city (right)

Before the 1970s, the amount of groundwater pumping was small, and neither the unconfined nor confined groundwater was affected by human activities. The water table kept fluctuating within a specific range in accordance with the annual precipitation. After the 1970s, the increase of groundwater pumping, agriculture production and urbanisation brought down the groundwater level and brought about other environmental problems.

Deterioration of Groundwater Quality

At the same time as urbanisation, huge quantities of wastewater were disposed from Shijiazhuang City and used as an alternative source of irrigation water and fertilizer for agriculture in the peri-urban areas. Wastewater irrigation practices began in the mid-1960s, and 80% of the irrigation water used in the croplands near the city was municipal sewage. It is reported that irrigated wastewater infiltrated into the local groundwater and contaminated the groundwater (Shen et al. 2005; Tang et al. 2003). Fig. 3.15 shows the variations of chloride (Cl^-) in groundwater from 1959 to 2000 in the urban area. In the early 1960s, chloride concentration was below 50 mg L^{-1}, which increased to 140 mg L^{-1} and 70 mg L^{-1} in the southeast and northwest areas of the city, respectively, in 2000. In the croplands irrigated with wastewater, nitrate concentrations range from 50 to 130 mg L^{-1}, but in the croplands irrigated with groundwater, which is far from the canal, nitrate concentrations were less than 35 mg m^{-1} (Tang et al. 2003). Comparing the areas irrigated with wastewater and groundwater irrigation in the NCP, it was found that more than 50% of groundwater samples from wastewater-irrigated areas had nitrate concentrations exceeding the WHO Drinking Water Guideline values, but less than 15% of samples from groundwater-irrigated areas exceeded this value.

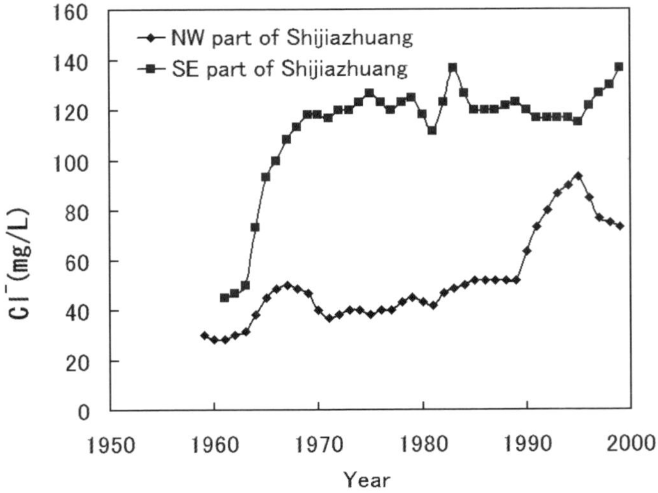

Fig. 3.15. The change of chloride concentration in groundwater in Shijiazhuang

Effects of Overexploitation on Local Water Cycle

Because of rapid economic development, the drying climate, and deterioration of the water resources, the Shijiazhuang region encountered serious water management problems. Therefore, it is important to study the main hydrological processes for sustainable water management in this region in the future. The water issues of the study area can be categorized as follows:

- Drying and shrinking surface water bodies. The drying of the Hutuo River has become more severe in recent years and it has become completely dry except for the occasional flood release.
- Over-exploited and exhausted groundwater. Because less surface water is available, large amounts of groundwater are extracted for urban use and irrigation in the surrounding areas.
- Water pollution. Another big environmental problem in the study area is the pollution of surface water and groundwater. To some extent, it can be said that the cities in the NCP have become the sources of the ephemeral rivers due to the large quantities of wastewater drained into the dried river channels. This not only produces polluted water but also contaminates the groundwater.
- Poor water management and water use inefficiency. Due to out-of-date water use methods, e.g. flood irrigation, and poor water management, e.g. high rates of water leakage from water mains, water use efficiency is still at a low level.

The rapid urban and industrial developments and extensions of irrigated farmland have doubled water demand. The consequence has been over-use of surface water resources and excessive exploitation of groundwater in the area. For example, in the late 1980s, the average grain yield of Shijiazhuang reached 12,000 kg/ha, and water consumption was about 850mm/year. The average excessive groundwater exploitation amounted from 80 to 100mm, which resulted in an annual decline of the shallow groundwater table of more than 0.8m/year in the 1970s and the 1980s, and 1.0 to 1.2m/year in the 1990s. Under the natural conditions, groundwater of the study area is recharged from three sources:

- direct rainfall infiltration,
- infiltration of surface water (rivers)
- the inflow from older rocks surrounding and underlying the study area.

To investigate the status of groundwater recharge and discharge, a water sampling survey was conducted along the Hutuo River in July and August 2000, and the tritium concentration of the water samples was measured.

Through a simple tritium model (Kitaoka. 1988) the mean residence time of water along the Hutuo River was estimated to be between 40 to 50 years (Fig. 3.16, Shen et al. 2005).

The infiltration of rainfall and stream water used to be the most importance sources for groundwater recharge. The infiltration of water to the underground occurred most readily in areas of outcropping piedmont alluvial fans. The thin cover of the semi-permeable sub-series sediments had no definite hindering effect on infiltration. The tritium distribution along Hutuo River, shown in Fig. 3.16, appears to be relatively uniform in the flow direction as a whole, suggesting that recharge to the groundwater may occur along the river banks. However, even if the river is important for lateral recharge, the groundwater has been seldom recharged from the river since the 1970s when the Hutuo River dried up in the study area. At the same time, huge amounts of groundwater have been pumped for regional economic development, which means that the pumping has become the most important type of discharge in the local water cycle.

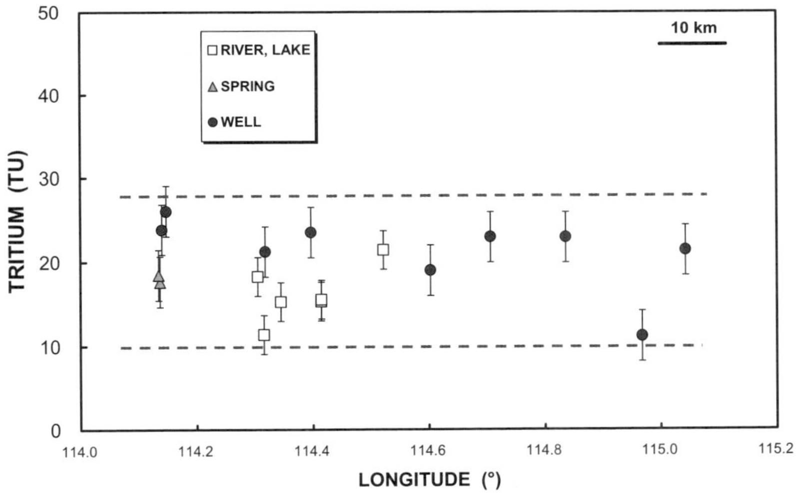

Fig. 3.16. Tritium distribution along Hutuo River running through the north side of the Shijiazhuang City (Shen et al. 2005)

3.5 Conclusion

Groundwater is an important source for urban water supplies worldwide. The global-scale analysis of groundwater dependence of urban water supply systems revealed that there are currently about one billion people in the world who are highly dependent on groundwater sources. At present groundwater plays an important role in urban water supply systems, and this will continue to be true in the future. However, in the developing world, the extensive exploitation of groundwater and poor management have brought about low water-use efficiency and widespread water contamination. Improved management of urban groundwater resources is urgently needed to mitigate actual and potential derogation caused by excessive exploitation and inadequate pollution control in large cities in the developing countries, such as Shijizhuang in China. Unless groundwater is protected in terms of both quantity and quality, there will be increased scarcity of water supply sources, which would push up the costs of the water supply. The high cost of water and limited availability may lead to an increase in populations who have difficulty in getting clean water, which would cause adverse health impacts on the city dwellers. An integrated approach for better groundwater management, including technology, legislation, and management, is urgently needed for the cities in the developing countries.

References

Bengtsson M, Shen Y, Aramaki T (2006) A global view of domestic water use: grid-simulation of the current situation. In: Proceedings of the 3rd APHW, ST1-04-A23-065-1156831475 (CD-ROM)

China Geo-Environmental Monitoring Institute (2006) Important advances in survey and monitoring of land subsidence over North China Plain (2003-2005) (in Chinese). Geo-technical report

Foster S, Lawrence A, Morris B (1998) Groundwater in urban development: Assessing management needs and formulating policy strategies. World Bank technical paper No. 390

Hagget P (1972) Geography: a modern synthesis. Harper & Row Publishers Inc, New York

Johnson JH (1967) Urban geography. Pergamon Press, Oxford

Kitaoka K (1988) A model of quality distribution in groundwater with reference to natural tritium concentration (in Japanese). J Groundwater Hydrology 30:77-93

Lazaro TR (1979) Urban hydrology: a multidisciplinary perspective. Ann Arbor Science Publishers Inc, Michigan

Mazari M, Mackay DM (1993) Potential for groundwater contamination in Mexico City. Environmental Science & Technology 27:749-802

Rugg DS (1972) Spatial foundations of urbanism. William C Brown

Shen, Y (2004) Study on hydrological processes of land and atmospheric system in semiarid agricultural region. Ph.D. Thesis, Chiba University, Japan

Shen Y, Oki T, Utsumi N, Kanae S, Hanasaki N (2007) Projection of future world water resources under SRES scenarios: 1, Water withdrawal. Hydrological Sciences Journal, in press

Shen Y, Tang C, Xiao J., Oki T, Kanae S (2005) Effects of urbanization on water resource development and its problems in Shijiazhuang, China. In: Sustainable water management solutions for large cities. IAHS Publ No. 293, pp 280-288

Sun J, Qi J, Yin Z, Zhang S, Jia X (1999) Environmental isotope study on groundwater circulation affected by global change and human activities in Quaternary aquifers of North China Plain. In: Proceedings of International Symposium on groundwater in environmental problems, Chiba University, Japan, pp 141-143

Tang C, Chen J, Shen Y (2003) Long term effect of wastewater irrigation on nitrate in groundwater in the North China Plain. In: Steenvoorden J, Endreny T (eds) Wastewater re-use and groundwater quality. IAHS Publ No 285, pp 34-40

United Nations, Department of Economic and Social Affairs (2004) Urban and rural areas 2003 (table), population studies, No. 231 (http://www.un.org/esa/population/publications/wup2003/2003UrbanRural2003_Web.xls)

Xiao J, Shen Y, Ge J, Tateishi R, Tang C, Liang Y, Huang Z (2006) Evaluating urban expansion and land use change in Shijiazhuang, China, by using GIS and remote sensing. Landscape and Urban Planning 75:69-80

4. Groundwater Potential in the Central District of Tokyo

Tomochika Tokunaga

Department of Environment Systems, the University of Tokyo

4.1 Introduction

Tokyo, the capital of Japan, is situated in the south-western part of the Kanto Plain, a large, flat plain with an area of approximately 16,000 km^2 (Fig. 4.1). The underground environment beneath the central district of Tokyo has been changing dramatically in accordance with the continuous increase and heavy use of underground space. Because of the complex interaction between the change of the groundwater environment and human activities underground, society has been affected in various ways. This chapter describes the temporal change of the groundwater environment and associated problems in the central district of Tokyo. Then, new techniques that are considered to be usable for the strategic management of groundwater resources are briefly presented. We believe that transferring our experience regarding the central districts of Tokyo to presently developing and expanding urban areas in the similar geological/hydrogeological settings, i.e., Bangkok, Hanoi and so on, is crucial to achieve sustainable urban development.

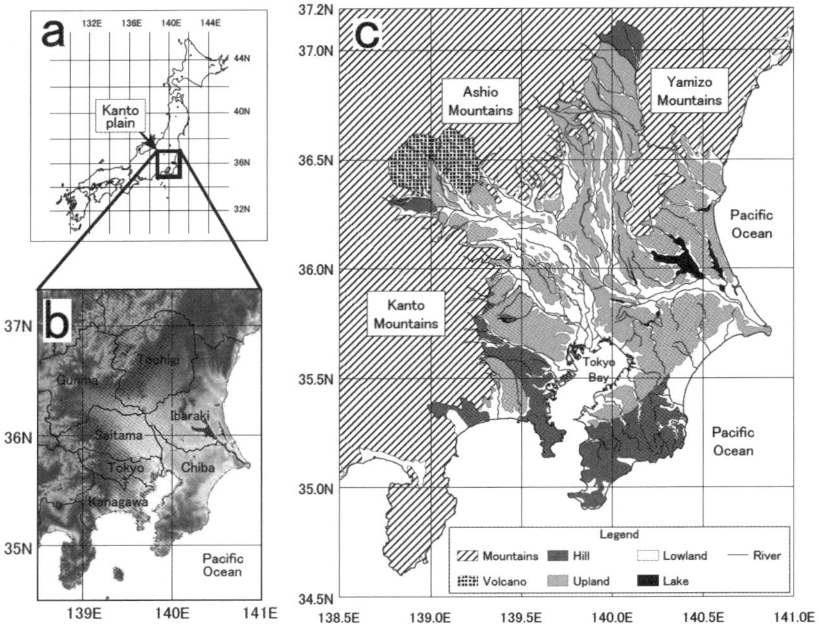

Fig. 4.1. Locality map of the Kanto Plain, Japan (Hayashi et al., submitted). a: Location of the Kanto Plain, b: Prefecture boundaries in the Kanto Plain, c: Simplified geomorphological map.

4.2 Hydrogeological System in the Central Districts of Tokyo and its Surroundings

4.2.1 Geological and Hydrogeological Settings

The Kanto Plain is surrounded by the Kanto, Ashio and Yamizo Mountains to the west and north and by the Pacific Ocean to the east and south. The Tokyo Bay, a shallow inner bay, is situated to the south of the plain (Fig. 4.1c). In the Tokyo area and its surroundings, Late Pliocene and younger sediments and sedimentary rocks unconformably overlay the Miocene sedimentary rocks. The shallow part of the Late Pliocene and Pleistocene formations constitute the confined aquifer-aquitard system. The Holocene deposits are distributed in the lowland area, and were deposited after the last glacial maximum. The upper part of the Holocene deposits is mainly composed of marine clay (Fig. 4.2).

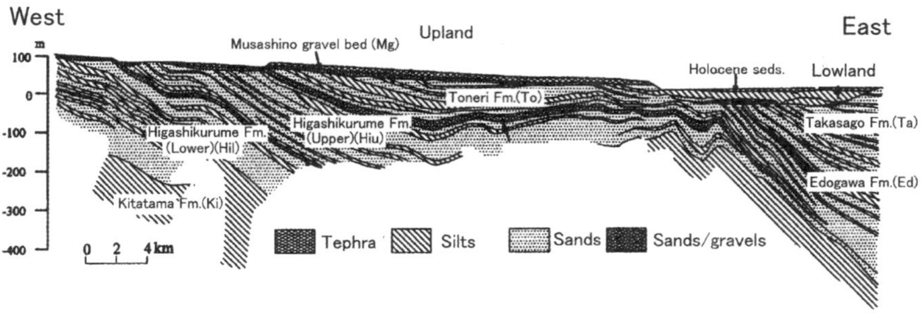

Fig. 4.2. Schematic east-west geological cross section of the shallow part of the Late Pliocene and upper formations of Tokyo (modified after Imaizumi et al. 2000). The Takasago, Edogawa, Toneri, Higashikurume Formations form the main confined aquifers in this area. The Kitatama Formation, thick mudstone layer, marks the bottom of the aquifer-aquitard system. Note that in some parts of western Tokyo, the unconfined aquifer which has developed in the Musashino gravel bed is considered to be hydrogeologically connected to the underlying confined aquifers.

According to the Institute of Civil Engineering of Tokyo (1977), this confined aquifer-aquitard system is bound at its bottom by a relatively thick mudstone layer, and the top of this mudstone becomes shallower toward the southwest. It is situated more than 600m deep in the northeast of Tokyo while it is about 100m deep in the southwestern part of Tokyo (Fig. 4.3). Below the aquifer bottom, the Pliocene sediments extend more than 2000m thick and mainly consist of alternating sandstones and mudstones, which constitute a reservoir system of methane gas dissolved in water and formation water.

In addition to the above mentioned confined aquifer system, there is also an unconfined aquifer in the area. Kawashima (2001) reported the temporal change of the depth to the water table of unconfined aquifers and showed that the water table has been more or less stable at least from the 1970s to the present (Fig. 4.4).

Fig. 4.5 shows examples of the change in groundwater potential of confined aquifers together with the extraction rate of fresh groundwater (white bars) and formation water (black bars). The formation water had been extracted for mining methane gas dissolved in water. This activity was one of the main causes for severe land subsidence in this area (Endo et al. 2001).

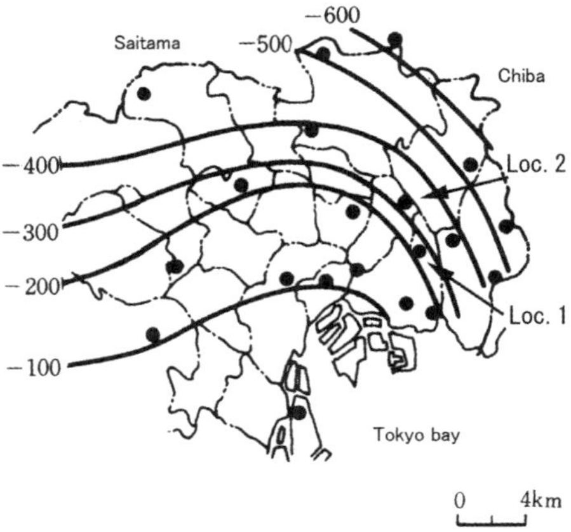

Fig. 4.3. Contour map showing the depth distribution of the bottom of the aquifer-aquitard system of Tokyo (modified after Institute of Civil Engineering of the Tokyo Metropolitan Government 1977)

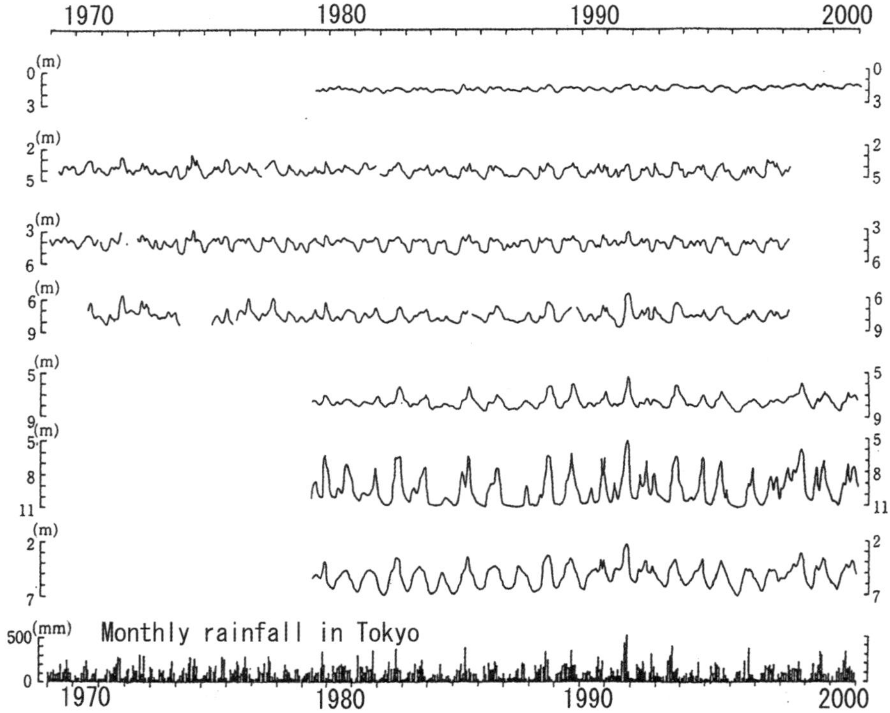

Fig. 4.4. Examples of the change in the unconfined water table (monthly mean data) at Tokyo (modified after Kawashima 2001). Water table depths are measured from the land surface.

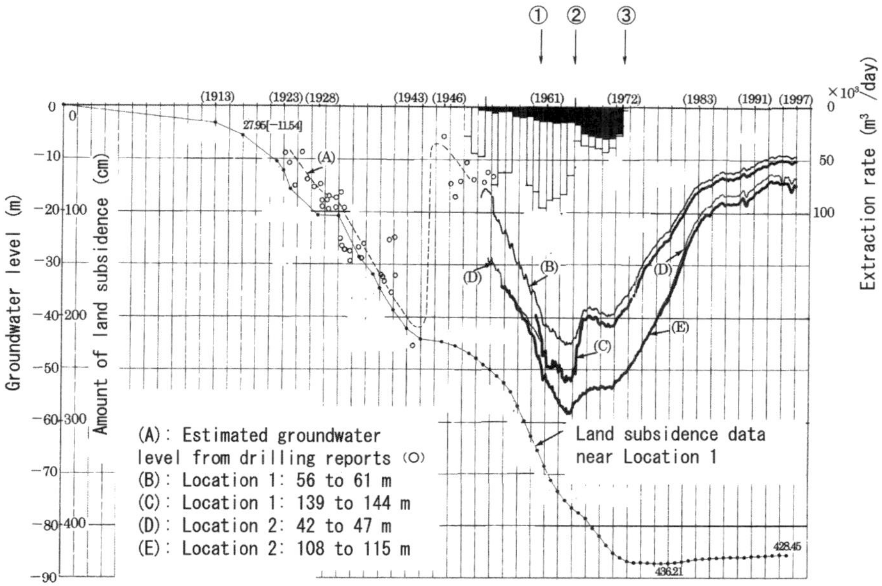

Fig. 4.5. Examples of the change of confined groundwater potential in Tokyo from 1881 until 1997 (modified after Endo et al. 2001). Localities of the wells are shown in Fig. 4.3. White bars indicate the extraction rate of groundwater at Koto and Sumida wards and black bars the extraction rate of formation water around the estuary of the Ara River. See Section 4.3.2 of this chapter for a discussion of the events shown at the top of this figure.

4.2.2 Unconfined vs. Confined Aquifers

In the previous section, we introduced the term unconfined and confined aquifers. Here, we follow the textbooks by Freeze and Cherry (1979) and Fetter (1994) to summarize the terms.

An aquifer is defined as a saturated permeable geological unit which can transmit significant quantities of water under ordinary hydraulic gradients. An alternative definition that is also widely used is that an aquifer is permeable enough to yield economic quantities of water to wells. The term aquitard is used to describe the less-permeable beds in a stratigraphic sequence. The definitions are somewhat arbitrary. This is because in areas of clay rich formations, silt with slightly higher hydraulic conductivity might be used to supply water to wells. On the other hand, the same silt might be considered as an aquitard if it were found in an area of sandy formations.

Aquifers can be close to the land surface with continuous layers of materials of high hydraulic conductivity extending from the land surface to the base of the aquifer. In this case, an unconfined aquifer, or water-table aquifer, can be defined as the zone from the bottom of the aquifer to the water table. A confined aquifer is an aquifer that is confined between two aquitards. This means that confined aquifers occur at depth while unconfined aquifers occur near the ground surface, as described above. It should be noted that the same aquifer can be defined as both unconfined and confined, depending on the relative position of the base of the top aquitard and the potentiometric surface (Fig. 4.6). The change from unconfined to confined aquifer conditions, or the hydraulic connection between unconfined and confined aquifers, is considered to exist in some parts of western Tokyo (Kayane, 1994).

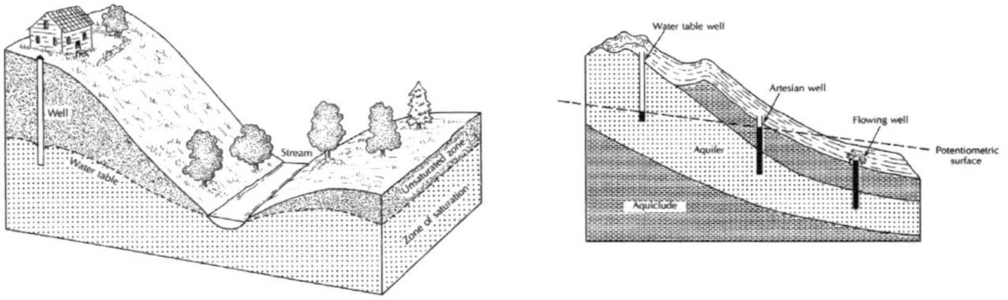

Fig. 4.6. Schematic figures showing an unconfined, or water-table aquifer (left) and a transition from unconfined to confined aquifer conditions (right) (Fetter, C.W., APPLIED HYDROGEOLOGY, 3rd Edition, © 1994, pp.111, 113. Reprinted by permission of Pearson Education, Inc., Upper Saddle River, NJ.

One consequence derived from the different conditions between unconfined and confined aquifers is the large difference in water volumes that can be drained from the aquifers. Fig. 4.7 illustrates the volume of water drained from the aquifers by unit drop of the hydraulic head. In the case of an unconfined aquifer, the unit drop of the hydraulic head corresponds to the drop of water table. Thus, the drained water volume is very close to the porosity of the aquifer. This is called specific yield. On the other hand, the drained water volume only comes from the elastic deformation of both water and the porous formation in the case of a confined aquifer, and hence, the water volume is rather small. This volume is called storativity for the confined aquifer. According to Freeze and Cherry (1979), storativities are at around 5×10^{-5} to 5×10^{-3}, while specific yields are between 1×10^{-2} to 3×10^{-1} for normal aquifers.

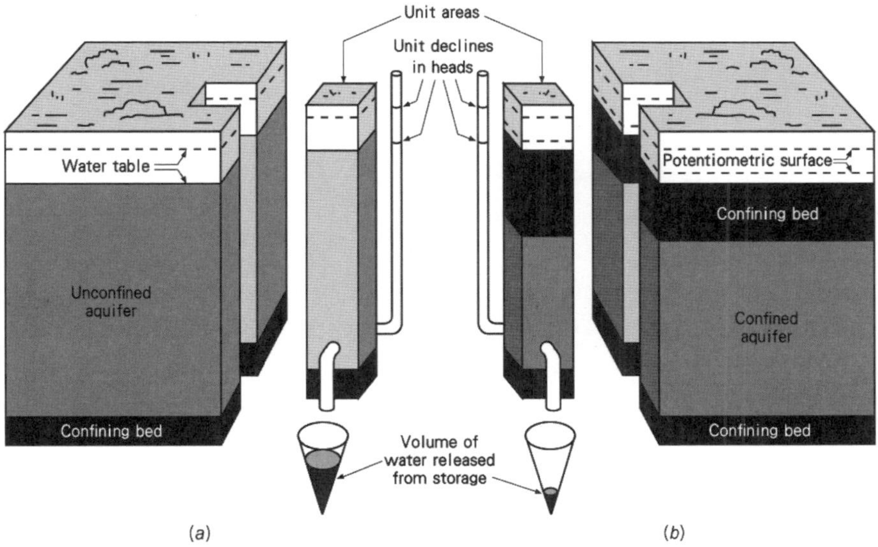

Fig. 4.7. Concept of specific yield for an unconfined aquifer (a) and storativity for a confined aquifer (b) (Heath 1983). Note that volume of water released from storage from an unconfined aquifer is much larger than that from a confined aquifer.

4.3 Historical Change of the Groundwater Situation in the Central Part of Tokyo

4.3.1 Heavy Extraction of Groundwater and Formation Water and Problems Induced by These Activities

In the Kanto Plain, rotary drilling technologies were introduced in the early 1920s (Shindo 1972). Thereafter, the number of wells significantly increased due to groundwater development and a considerable amount of groundwater was extracted. The rate of extraction increased drastically with economic growth in the 1950s and 1960s. As shown in Fig. 4.5, the groundwater potential of the confined aquifer dropped to about 50m below ground surface in the early 1970s. Because of the significant decline of groundwater potential in confined aquifers, severe land subsidence appeared as a direct result of over-abstraction of groundwater (Fig. 4.8).

Several shallow confined aquifers also changed and became unconfined (Fig. 4.9) and thus created oxygen-deficient air masses under the ground.

The oxygen-deficient air migrated along the aquifers during underground construction by the pneumatic caisson method and killed some people who stayed in basement floors in downtown Tokyo (Fig. 4.10).

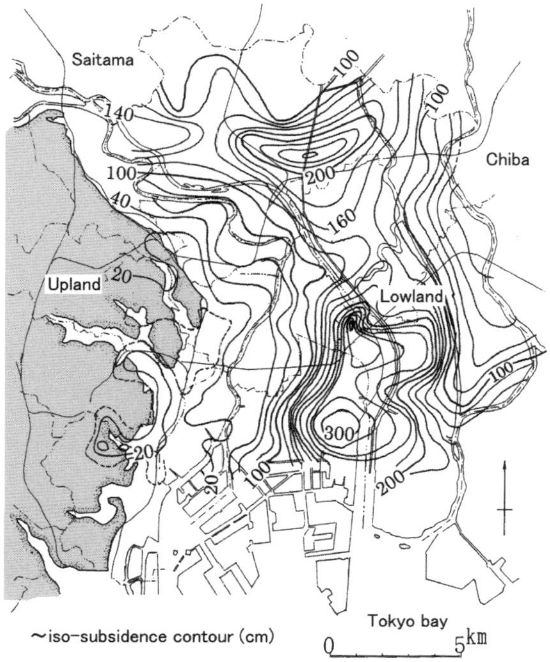

Fig. 4.8. Contour map showing the amount of subsidence between 1938 and 1977 in lowland Tokyo (modified after Endo et al. 2001).

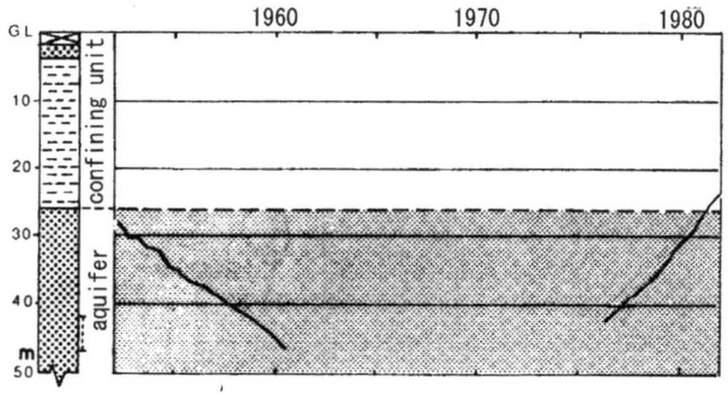

Fig. 4.9. Change of groundwater level at Loc. 2 (42 to 47 m) (modified after Endo and Ishii 1984). Location of the well is shown in Fig. 4.3. Note the unconfined condition in the aquifer between 1950s and 1970s.

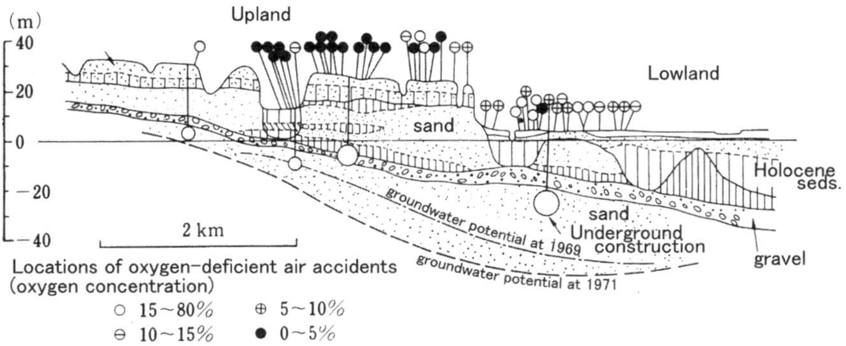

Fig. 4.10. Schematic east-west cross section showing the distribution of oxygen-deficient air accidents (modified after Mitamura and Takahashi 1993)

4.3.2 Regulation of Groundwater Usage as a Countermeasure to the Observed Problems

Because of the serious problems related to the over-abstraction of groundwater, local governments and the Japanese national government decided to regulate groundwater extraction in the central districts of Tokyo to the absolute minimum. From January 1961 until April 1974, the national government and the Tokyo Metropolitan Government with its surrounding three prefecture governments implemented the following groundwater regulation laws (also see Chapters 2 and 12). The national government restricted groundwater withdrawal for industrial use in 1961 through the Industrial Water Law. Under the regulation of this Law, no new wells were allowed to be installed for industrial use in the southern part of the alluvial lowland (Fig. 4.5, event 1). In 1966, the pumping of groundwater for industrial use in the southern part of the alluvial lowland was restricted (Fig. 4.5, event 2). In 1971, the pumping of groundwater for industrial use in the northern part of the alluvial lowland was restricted. The national government has also restricted groundwater withdrawal for air conditioning use since 1963 through the Law Controlling Pumping of Groundwater for Use in Building. In 1972, the extraction of methane gas dissolved in water was suspended in the Tokyo area by means of the purchase of mining rights by the Tokyo Metropolitan Government (Fig. 4.5, event 3).

After implementing these regulations, groundwater potential has recovered quickly, as shown in Fig. 4.5. The rapid recovery of groundwater potential has been considered to be because of the rather small volume of the storativities of the confined aquifers (see Section 4.2.2) and because of the relatively high recharge rate (2 to 3mm per day) in this region (Shimada 1988; Shimada et al. 2002).

4.3.3 Rapid Recovery of Groundwater Potential and New Problems

Even though the land subsidence in Tokyo area has ceased and groundwater potential has recovered, it has caused new types of damage to the underground infrastructure, which was constructed during the drawdown period of the groundwater potential in the region. The following shows an example of the problem.

The Tokyo Underground Station was designed in 1965 and has been in operation since 1972. At the time of its design, the groundwater potential at the location was 35m below the land surface. However, since then it has gradually recovered and reached 15m below the land surface in 1998. The detailed investigation for the possible damage to the station revealed that the buoyant force was quite high, and the critical groundwater potential for possibly severe damage was estimated to be 14.3m below the land surface (Shimizu 2004). The East Japan Railway Company decided to conduct countermeasure construction work to the station by applying the ground anchor technique (Fig. 4.11), which makes it possible to support the underground station until the groundwater potential is 12.8m below the land surface (Kurasawa 2001). Similar problems have been reported elsewhere in Japan, such as the Ueno Underground Station in Tokyo (Kurasawa 2001) and the rebuilding operation close to Osaka Station in Osaka (Hashimoto 2004). Hirose et al. (2004) summarized the published data on infrastructure damage caused by the recovery of the groundwater potential in the Tokyo area (Fig. 4.12). According to this study, the majority of the reported groundwater-related problems were situated in the upper buried terraces at depths shallower than -10m. There were no reports of damage in places with thick alluvial deposits. This occurrence suggests that the recovery of groundwater potential in the gravel/sand beds affects the stability of the structures because the gravel and sand beds on top of the buried terrace are used as foundations for structures such as stations.

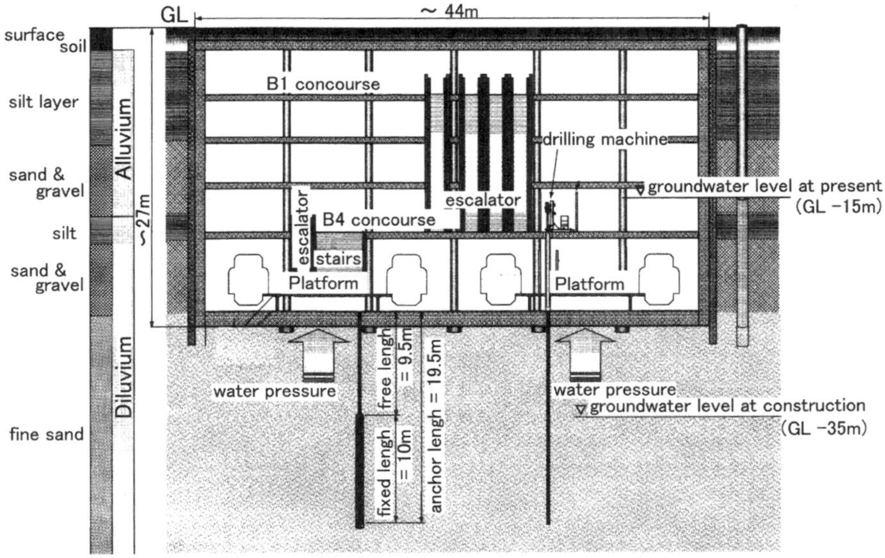

Fig. 4.11. Schematic diagram showing the situation at the Tokyo Underground Station and the image of the countermeasure construction work to the recovery of groundwater potential (modified after Kurasawa 2001).

Groundwater potential in the region's confined aquifers is still recovering (see Fig. 4.5), and accurate prediction of the rate of recovery and of the final groundwater level is necessary to plan maintenance operations for underground infrastructure.

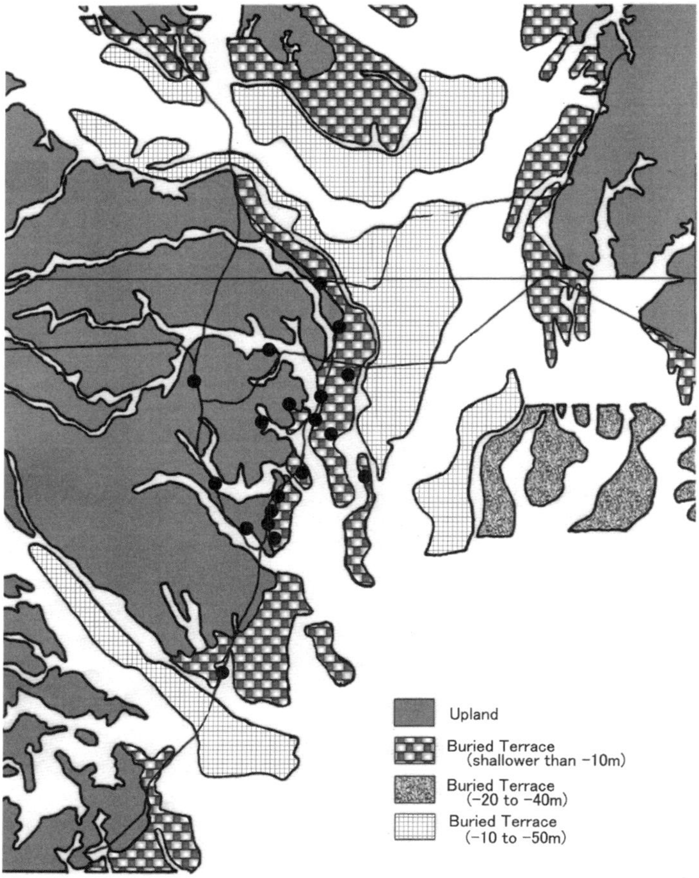

Fig. 4.12. Distribution of the reported infrastructure damages (black dots) caused by the recovery of groundwater potential (modified after Hirose et al. 2004). See text for details.

4.3.4 Some Examples of the Usage of Discharged Groundwater

Because of the recovery of groundwater potential, the amount of leakage into underground structures has increased. Usually, leaked water is drained to the sewer system. Several plans for efficient use of leakage water in underground structures have been presented to improve the surface water environment and some of them have been implemented.

One example comes from a project of the East Japan Railway Company, the Tokyo Metropolitan Government, and Kokubunji City (Shimizu 2004).

The Kokubunji tunnel of the Musashino line was constructed in 1973. This tunnel is oriented perpendicular to the general groundwater flow direction of the unconfined aquifer in the area, and experienced groundwater-related problems in 1974 and 1991 due to the rapid increase of the water table and seepage by heavy rain. To prevent the rapid increase of the water table of the upstream side, twenty-four drainpipes were set and have controlled the water table. From 2002, the drained groundwater has been used to restore a local small pond and to increase the flow rate of the river to improve the local surface water environment (Fig. 4.13). Similar approaches for confined aquifers have been under operation at the Tokyo tunnel and Ueno Station (Shimizu 2004).

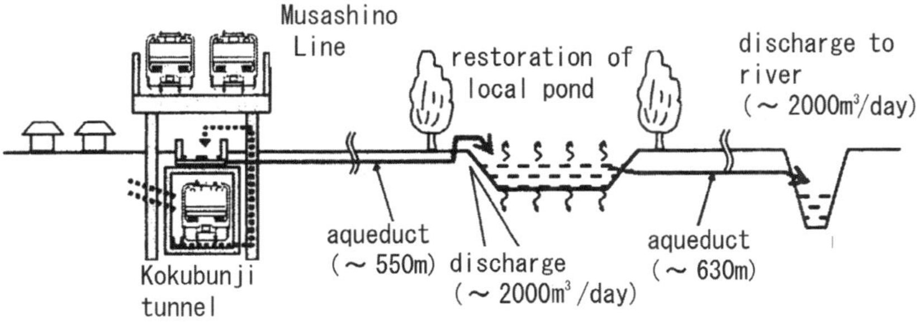

Fig. 4.13. An example of efficient usage of leaked water into a tunnel (modified after Shimizu 2004).

Kajino et al. (2004) presented an idea to use discharged groundwater for cooling the pavement in the urban area to mitigate the heat-island phenomenon. Their idea came from the fact that the groundwater temperature usually is very close to the annual mean temperature of the area and is much lower than the pavement temperature in summer. Considering that the groundwater potential has recovered in the Tokyo area and heat exchange can be done by applying cooler groundwater to the permeable pavement, it might be possible to control and improve summer urban environments.

4.4 New Methods for the Strategic Management of Groundwater Resources

4.4.1 Setting the Appropriate Groundwater Potential - Is it Possible?

Considering the changes of groundwater potential over time and related problems to underground infrastructure, it is desirable to set the appropriate groundwater potential of the confined aquifers and to develop and apply techniques for managing groundwater resources by keeping the groundwater potential around the appropriate level. Then, it might be possible to achieve sustainable groundwater usage both in terms of the environment and human society. The fundamental problem posed to us, then, is whether it is possible to set an "appropriate groundwater potential" or not, and if yes, how we can determine it.

According to a detailed analysis of the temporal change of pore water pressures of both aquifers and an aquitards in Tokyo (Hirose et al. 2004), it was found that the pore water in the aquitards had not completely been drained by the drawdown of the groundwater potential in the aquifer. This means that the consolidation of the aquitard had not yet reached its final stage. Thus, it is not appropriate to state that groundwater potential can be lowered to the values during the area's drawdown period. Once one starts extracting groundwater again and the groundwater potential drops to the level during the severe drawdown period, it might cause dewatering from the aquitard and larger land subsidence. This suggests that it is necessary to have a solid understanding of the spatio-temporal changes of pore water pressure in aquitards from the period of the drawdown to the present.

In addition to the necessity of obtaining knowledge of previous changes of pore water pressures in aquitards as described above, we need to predict the changes of groundwater potential in aquifers and aquitards, and the amount of subsidence/rebound of the formations with sufficient accuracy and spatial resolution in order to evaluate the possible environmental consequences by re-starting groundwater usage in Tokyo. These requirements can be achieved through a numerical modelling approach that takes into account both regional groundwater flow in the basin scale and local groundwater flow/deformation coupling processes. Recently, we have successfully developed this method and have confirmed its applicability to the central district of Tokyo and its surroundings (Aichi and Tokunaga 2006; Aichi et al. 2007). Details of this technique are presented in Chapter 6 of this book.

4.4.2 Satellite-Based Monitoring as High-Resolution Land Deformation Monitoring

Recent developments of satellite-based technologies can provide new tools for deformation monitoring. For monitoring land-surface movement, the technique called InSAR (Interferometric Synthetic Aperture Radar) has been used extensively. For example, Peltzer et al. (1996) used the InSAR technique to determine post-seismic surface displacement in the region of the 1992 Landers earthquake in the United States. Using the obtained surface displacement map, they made line plots of vertical displacement along roughly 10km segments spanning three different time intervals and estimated the contribution of pore pressure changes and groundwater flow to the post-seismic surface deformation. Stancliffe and van der Kooij (2001) applied this technique to monitor sub-surface steam movement at the Cold Lake heavy-oil production field, Canada. Hoffmann et al. (2001) analysed the spatial variations in subsidence and rebound in the Las Vegas Valley, U.S. using this technique and discussed the relationship between land subsidence and hydraulic head fluctuations in the basin.

Even though the InSAR technique has great potential for regional monitoring of land surface deformation, there are limitations in the InSAR method. These include de-correlation of surface scatterers due to vegetation or other surface changes, incoherence caused by large satellite orbit separations between the two image acquisitions used to make an interferogram, and noise from signal delay in the Earth's atmosphere (Ferretti et al. 2004).

To overcome these limitations, a new approach called the permanent scatterer (PS) InSAR method (Ferretti et al. 2000; 2001) has been introduced. As long as a significant number and density of independent radar-bright and radar-phase stable points, i.e., permanent scatterers, exist within a radar scene and more than fifteen SAR acquisitions have been collected, displacement time series and range change rates can be calculated (Colesanti et al. 2003). This method is said to be able to measure surface motions at a level of less than 1mm/year with sufficient resolution. Fig. 4.14 shows an example of the PSInSAR result of the San Francisco Bay, U.S. (Ferretti et al. 2004). Considering that there are many possible permanent scatterers in urban areas, the PSInSAR technique is very promising for the detailed spatial monitoring of land subsidence/rebound. Also, combining the PSInSAR technique with a numerical modelling approach to analyze deformation-fluid flow coupling processes, it will be possible to implement an efficient monitoring method for groundwater resource management.

Fig. 4.14. An example of PSInSAR analysis, San Francisco Bay Area, U.S. (Ferretti et al. 2004).

4.5 Conclusion

The temporal changes of groundwater conditions in the central districts of Tokyo and its surroundings were presented in this Chapter. The changes can be divided into three stages, i.e., a first stage in which there is deterioration of underground and surface environments due to over-extraction of groundwater; a second stage in which the regulation of groundwater extraction to the absolute minimum causes the recovery of groundwater potential of confined aquifers; and a third stage in which underground infrastructure can be damaged by buoyant forces and the increase of groundwater seepage due to the recovery of groundwater potential.

Underground environments in the central districts of Tokyo have dramatically changed over time and are being changed even now. Under these circumstances, it is necessary to have techniques and methods to monitor and analyze such changes with higher spatio-temporal resolution and accuracy in order to manage groundwater resources in a sustainable way.

Necessary and fundamental techniques and methods are being developed now, as this chapter has shown. Of course, further development and detailed studies of the geological/hydrogeological situations of a particular area should be continued to make our models more appropriate and produce more reliable results. Through these efforts, we believe that we can provide accurate information regarding the management of groundwater in developed urban areas such as Tokyo.

References

Aichi M, Tokunaga T (2006) A nested modeling scheme for high-resolution simulation of the aquitard compaction in a regional groundwater extraction field. H41B-0399, AGU Fall Meeting, San Francisco, CA, December 11-15

Aichi M, Hayashi T, Tokunaga T (2007) A nested modeling scheme to integrate regional groundwater flow and local groundwater flow/land subsidence; an application to the Tokyo metropolitan area, Japan. H53E-03, AGU Joint Assembly, Acapulco, Mexico, May 21-25

Colesanti C, Ferretti A, Novali E, Prati C, Rocca E (2003) SAR monitoring of progressive and seasonal ground deformation using the permanent scatterers technique. IEEE Trans. Geosci Remote Sens 40:1685-1701

Endo T, Ishii M (1984) Hydrogeology and subjects caused by recovering of groundwater level in Tokyo (in Japanese with English abstract). J Japan Soc Eng Geol 25:111-20

Endo T, Kawashima S, Kawai M (2001) Historical review of development of land subsidence and its cease in Shitamachi Lowland, Tokyo. J Japan Soc Eng Geol 42:74-84

Ferretti A, Novali F, Bürgmann R, Hilley G, Prati C (2004) InSAR permanent scatterer analysis reveals ups and downs in San Francisco Bay area. EOS 85:317-24

Ferretti A, Prati C, Rocca E (2000) Nonlinear subsidence rate estimation using permanent scatterers in differential SAR interferometry. IEEE Trans Geosci Remote Sens 38:2202-12

Ferretti A, Prati C, Rocca E (2001) Permanent scatterers in SAR interferometry. IEEE Trans Geosci Remote Sens 30:8-20

Fetter CW (1994) Applied hydrogeology, 3rd edn. Prentice Hall, p 691

Freeze RA and Cherry JA (1979) Groundwater. Prentice Hall, p 604

Hashimoto T (2004) Groundwater problems in Osaka plain (in Japanese). In: Proceedings of the 2004 Symposium, Japan Soc Eng Geol, pp 1-7

Hayashi T, Tokunaga T, Aichi M, Shimada J, Taniguchi M, Effects of human activities and urbanization on groundwater environments: An example from the aquifer system of Tokyo and the surrounding area. Sci Total Env, submitted

Heath RC (1987) Basic ground-water hydrology. Technical report. US Geological Survey (http://pubs.usgs.gov/wsp/wsp2220/pdf/wsp_2220.pdf)

Hirose M, Kawagoe T, Kiya H (2004). A relation of a groundwater level change for construction and underground structures (in Japanese). In: Proceedings of Ann. Meet., Japan Soc Eng Geol, pp 419-422

Hoffman J, Zebker HA, Galloway DL, Amelung F (2001) Seasonal subsidence and rebound in Las Vegas Valley, Nevada, observed by synthetic aperture radar interferometry. Water Resour Res 37:1551-1566

Imaizumi M, Komae T, Nishira S(2000) Evaluation of recharge mechanism of confined groundwater using long term tritium concentration records in the Metropolis of Tokyo (in Japanese with English abstract). J Japan Soc Eng Geol 41:81-102.

Institute of Civil Engineering of Tokyo Metropolitan Government (1977) Synthetic ground chart in Tokyo Metropolis (in Japanese). Gihodo Publ

Kajino N, Tokunaga T, Mogi G, Mogi K, Yasumochi S, Bando K (2004). Research on the pavement cooling system using groundwater (in Japanese). In: Proc. Ann. Meet., Japan Soc Eng Geol, pp 109-112

Kawashima S (2001) Temporal change of groundwater level in Tokyo. Found. Eng Equip 29(11):77-79

Kayane I (1994) Study on interaction between surface and groundwater in Tamagawa River system (in Japanese). (Report submitted to Tokyu Foundation for Better Environment)

Kurasawa N (2001) Examples of countermeasure constructions to overcome groundwater pressure due to the recovery of groundwater level –Ueno underground station and Tokyo underground station (in Japanese). Found Eng Equip 29(11):72-76

Mitamura M, Takahashi H (1993) Obstacles caused by groundwater development (in Japanese). In: Research Group on Water Budget (ed). Groundwater Resources and Environments. Kyoritsu Publ, pp 28-38

Peltzer G, Rosen P, Rogez F, Hudnut K (2004) Postseismic rebound in fault stepovers caused by pore fluid flow. Science 273:1202-1204

Shimada J (2002) The mechanism of unsaturated flow through a volcanic ash layer under humid climatic conditions. Hydrol Proc 2:43-59

Shimada J, Tang C, Tanaka T, Yang Y, Sakura Y, Song X, Liu C (2002) Irrigation caused groundwater drawdown beneath the North China Plain. In: Proceedings of Int. Groundwater Conf., Darwin, Australia, pp 1-7

Shimizu M (2004) Groundwater problems related to construction of underground structures (in Japanese). In: Proceedings of 2004 Symp., Japan Soc Eng Geol, pp 26-33.

Shindo S (1972) Groundwater in Southern part of the Kanto Plain (in Japanese). Soil Mech Found Eng 20(5):25-11.

Stancliffe RPW, van der Kooij MWA (2001) The use of satellite-based radar interferometry to monitor production activity at the Cold Lake heavy oil field, Alberta, Canada. Amer Assoc Petrol Geol Bull 85:781-93

Part II Technologies for the Analysis of Urban Groundwater

5. The Shallow Groundwater Environment and the Risk of Soil Liquefaction

Takeshi Hayashi

Faculty of Education and Human Studies, Akita University

5.1 Introduction

Various human activities connected with urbanisation and industrialisation impact the shallow groundwater environment. For example, shallow groundwater can be contaminated by domestic wastewater, industrial effluents and waste directly as seepage from the surface and on-site sanitation systems, such as septic tanks and cesspits, or by leakage from sewer pipes. The impermeabilisation of the land surface decreases the volume of infiltration. On the other hand, water main leakage and wastewater seepage recharge shallow groundwater. These anthropogenic groundwater sources strongly affect the subsurface water balance in urban areas.

Changes in the shallow groundwater environment may increase the risk of urban disasters. For instance, changes in the land use pattern increase surface runoff and induce rapid and extensive river discharge that can cause flooding. In East Asia, the occurrence of torrential rains over 50 mm/day has increased since the 1950s (Meteorological Agency of Japan 2005). Such torrential downpours flooded subsurface structures such as subway stations in urban areas because the watercourses and drainage systems cannot handle a large quantity of surface runoff. Furthermore, the depth of the groundwater table from the land surface is closely related with the occurrence of soil liquefaction. A rising groundwater table enhances the risk of soil liquefaction because soil liquefaction tend to occur in the well-sorted unconsolidated sands in saturated zone. Soil liquefaction destroys surface and subsurface structures, such as buildings, bridges and riverbanks, and damages lifelines such as water supply systems. Thus, the prediction of soil liquefaction is one of the major issues for preventing or limiting urban disasters.

In this chapter, characteristics and changes of the shallow groundwater environment in urban areas are summarized and the mechanisms and countermeasures of soil liquefaction are reviewed. Also, a case study in the eastern Tokyo metropolis is presented.

5.2 Characteristics of the Shallow Groundwater Environment in Urban Areas

5.2.1 Hydrological Process

Surface Impermeabilisation and Provision of Drainage

The impact of the increase of the impervious surface area on watersheds is thought to become detectable once impervious surfaces exceed 10% of the surface area (Lamptey et al. 2005). Surface impermeabilisation processes include the construction of roofs and of paved areas, such as roads, parking lots, industrial premises and airport aprons. On the other hand, some types of urban pavement such as tile, brick and porous asphalt are quite permeable, and conversely, some unpaved surfaces can become highly compacted with reduced infiltration capacity (Foster et al. 1998). Lerner (2002) estimates that roughly 50% of the impervious cover should be treated as permeable, particularly in residential areas.

Land surface impermeabilisation tends to lower evaporation. According to a study by the Bureau of the Environment of the Tokyo Metropolitan Government (1980), 2mm of rainwater evaporates at each precipitation event in some major cities in the Tokyo metropolis and Aichi prefecture. Given that all precipitation evaporates when the daily precipitation is lower than 2mm, and 2mm evaporates when the daily precipitation is higher than 2mm, they calculated the annual evaporation in the Tokyo metropolis to be approximately 15% of annual precipitation.

In many developing cities, the runoff from impermeable surfaces directly discharge into unlined drainage ditches, part of which seeps into the groundwater. In more developed or mature cities, surface runoff flows into storm sewers and discharges to surface water bodies such as rivers and canals (a separate sewerage system), or may be connected to a treatment plant (a combined sewerage system) (Barrett 2004).

Water Supply and Sewerage

With the development of urban areas, water usage per capita has been increasing in recent years. The world mean water usage per capita in 1995 was 1,756 litres/day: 174 litres for domestic uses, 352 litres for industrial uses and 1,231 litres for agricultural uses. However, these values vary depending on the region of the world. In Africa, for example, water usage is only 593 litres/day per capita: 63 litres for domestic, 494 litres for industrial and 37 litres for agricultural, while the usage in North America is 3,924 litres/day per capita: 428 litres for domestic, 1,602 litres for industrial and 1,897 litres for agricultural (Ministry of Land, Infrastructure and Transportation Japan 2004).

Water mains are pressurised to avoid infiltration of contaminants and to ensure distribution throughout the water distribution system. Thus, leakage from the systems occurs with the destruction of water mains by deterioration, compaction by traffic and land subsidence. The values of water loss from water mains in many cities are around 20 to 30%. Some cities with efficient and well maintained system report losses around 10%, but values of 30 to 60% are common in the less developed countries (Garcia-Fresca 2007; Foster et al. 1998). These values show that leakage is often greater than rainfall in arid climates (Foster et al. 1998). Note that, the mean value of losses for Japan as a whole was 9.7% in 1995, and loss in the Tokyo metropolis in 2005 was only 4.2% (Bureau of Waterworks, Tokyo Metropolitan Government 2006).

The installation of a sewerage system lags quite often behind population growth and the construction of water supply systems. On-site wastewater disposal via septic tanks, cesspits or pit latrines for high-density settlements may be semi-permanent, especially in low-income districts. Under these situations, most of the supplied water is recharged to the subsurface. When sewer pipes are located below the water table, groundwater may infiltrate into the pipes, and when located above the water table, the pipe may leak.

5.2.2 Change of the Groundwater Environment with Urbanisation

In general, urbanisation reduces recharge by impermeabilisation of ground surface (Lerner & Barrett 1996). However, a portion of what is considered impervious cover is permeable (Lerner 2002), and urbanisation introduces new sources of recharge, e.g. leakage from water mains and sewerage systems, leaks from storm sewers and recharge from lawns, parks and golf courses (Foster et al. 1998). Therefore, it is now acknowledged that total recharge to groundwater could increase with urbanisation (Fig.5.1). The rate of "urban enhanced" recharge is especially significant in arid climates and in cities in developing countries. For instance, the groundwater table in Jeddah, in Saudi Arabia has been rising approximately 0.1 m/year in recent years (Al-sefry and Şen 2006). The excess recharge in Riyadh has caused groundwater flooding in road underpasses and caused the death of many ornamental plants (Lerner 2002).

Foster et al. (1998) divided the change of shallow groundwater environment in regards to urbanisation into four stages. As groundwater is generally of high quality and is a more constant source than seasonally affected surface water sources, it is often the first resource to be abstracted by overlying cities (Foster et al. 1998). Therefore, the groundwater tables beneath cities have declined with the development of cities in the initial stage. As cities grow they reach the second stage, meaning that water supplies that were obtained from shallow groundwater may no longer be sufficient, either because the available resources are too limited by depletion of groundwater tables or because pollution has caused its quality to deteriorate. Next, extra water may be drawn from aquifers or surface water bodies in the city's hinterland or other river basins (the third stage). The fourth and final stage occurs when leakage from water mains and sewerages recharges the groundwater, causing the groundwater table to rise along with the development of cities.

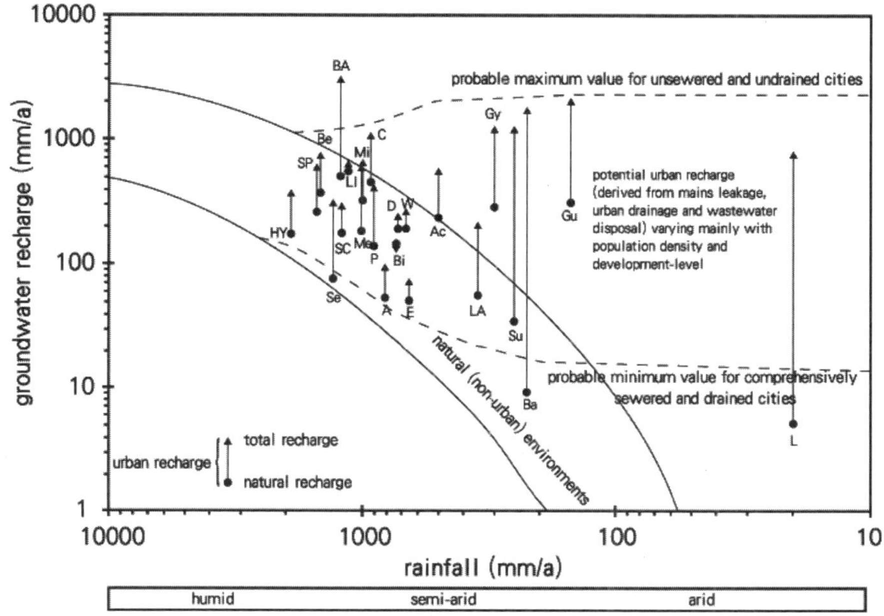

Fig. 5.1. Urban-enhanced groundwater recharge rates in twenty-three cities around the world (Garcia-Fresca 2007). **HY**: hat Yai, Thailand; **SP**: São Paulo, Brazil; **Be**: Bermuda, UK; **Se**: Seoul, Korea; **BA**: Buenos Aires, Argentina; **SC**: Santa Cruz, Bolivia; **LI**: Long Island (New York), USA; **Mi**: Milan, Italy; **Me**: Merida, Mexico; **C**: Caracas, Venezuela; **P**: Perth, Australia; **A**: Austin (Texas), USA; **Bi**: Birmingham, UK; **D**: Dresden, Germany; **W**: Wolverhampton, UK; **E**: Evora, Portugal; **Ac**: Aguascalientes, Mexico; **LA**: Los Angeles (California), USA; **Ba**: Baku, Azerbaijan; **Su**: Sumgayit, Azerbaijan; **Gy**: Gyandja, Azerbaijan; **Gu**: Gulistan, Uzbekistan; **L**: Lima, Peru.

5.3 Shallow groundwater and soil liquefaction

5.3.1 Occurrence of soil liquefaction in Japan and around the world

Wakamatsu (1991a, 1991b) recorded the distribution of the sites where liquefaction had occurred by recent and historical earthquakes in Japan. According to these studies, there are many liquefied sites all over Japan (Fig.5.2). Liquefaction has also occurred all over the world: the 1964 Alaska earthquake (USA, M8.3), the 1976 Guatemala earthquake (M7.5), the 1977 Romania earthquake (M7.2), the 1989 Loma Prieta earthquake (USA, M7.1), the 1994 Northridge earthquake (USA, M6.8), the 1995 North Sakhalin earthquake (Russia, M7.6), the 1999 Kocaeli earthquake (Turkey, M7.4) and the 2005 North Chile earthquake (M7.8). Liquefaction induced by earthquakes is one of the most harmful urban disasters, because liquefaction cuts off lifelines such as water mains, power cables, and roads.

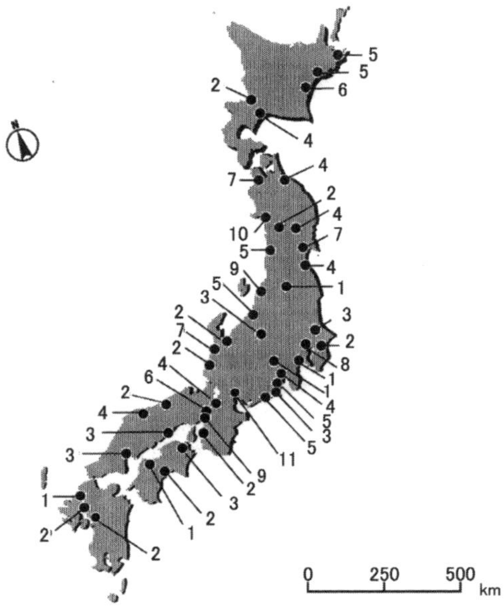

Fig. 5.2. Map of historic liquefaction sites in Japan (modified from Yoshimi and Fukutake 2005). Numbers show the frequency of liquefaction events.

5.3.2 Mechanism of soil liquefaction

Engineering Characteristics of the soil

Soil is a three-phase system comprised of solids, liquid and gas, but becomes a two-phase system (solids and liquid) beneath groundwater table. A schematic model of saturated soil is shown in Fig.5.3. In this figure, the solid phase is simplified to be composed of sand grains. Stresses in saturated soil are expressed by the following equation:

$$\sigma' = \sigma - p_w \tag{5.1}$$

where σ' is effective stress, σ is total stress and p_w is pore water pressure (Terzaghi and Peck 1948). Effective stress is stress acting on the soil skeleton while total stress is equal to the overburden pressure which is made up of the weight of soil vertically above the plane under consideration and any forces acting on the soil surface such as the weight of structures.

Increase of volume caused by shearing deformation is called dilatancy (Fig.5.4). Dilatancy is closely related to the change of pore water pressure and liquefaction.

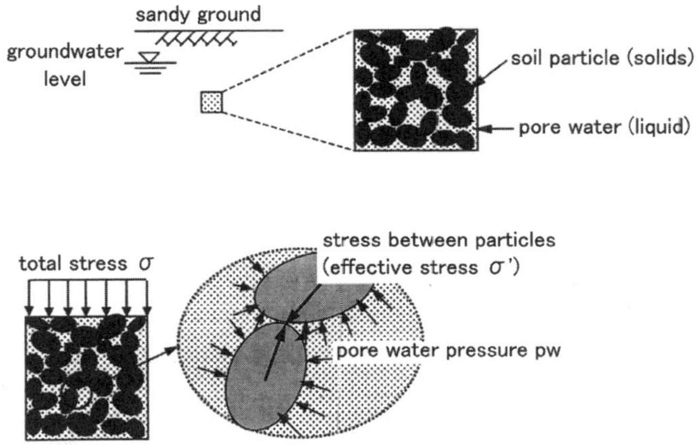

Fig. 5.3. Schematic model of sandy soil beneath groundwater table (modified from Yoshimi and Fukutake 2005)

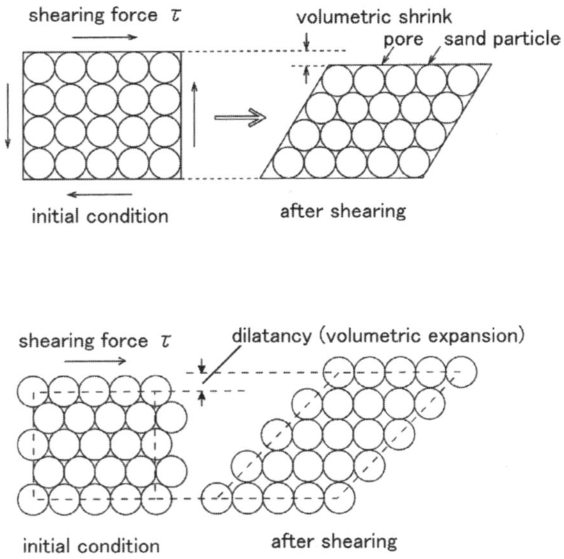

Fig.5.4. Schematic model of volumetric change caused by shearing deformation Top: loose sand, bottom: dense sand. (modified from Oka 2001)

Mechanism of liquefaction

When cyclic shearing load is applied to the saturated sandy soil, the movement of soil particles induced by cyclic loading can increase the pore water pressure and reduce the effective stress. If effective stress becomes zero, shearing stress also becomes zero, and the soil becomes a liquefied state. However, soil particles cannot stay afloat because of the higher specific gravity than water. Thus, soil particles settle again. Usually, liquefaction induces land subsidence because the porosity of re-settled soil is usually lower than that before liquefaction.

Liquefaction resistance of sandy soil increases significantly with decrease in the degree of saturation (Yoshimi et al. 1989). Yoshimi et al. (1989) studied the effects of the degree of saturation on the undrained cyclic shear strength of sand at a relative density of 60%. According to this study, liquefaction resistance of sand at 70 % saturation was approximately three times that at full saturation (Fig.5.5). This result suggests that liquefaction is hard to occur in unsaturated condition.

It was also demonstrated that grain size and degree of sorting of soil particles affect the occurrence of liquefaction significantly. Fig.5.6 shows the grain size distribution of three liquefied soils in Japan (Yoshimi and Fukutake 2005). According to this diagram, liquefied soils in Niigata City and Hachinohe Coast are mainly composed of well-sorted soil particles with grain size from sand to silty sand. On the other hand, the grain size of liquefied soil in Kobe Port Island has a wide range. This area is an artificial land reclaimed by soil that includes gravels (decomposed granite soil). This diagram suggests that a risk of liquefaction is relatively high in well sorted sandy soil under natural condition and an artificial reclaimed soil is easy to liquefy even if the diameter of soil is large.

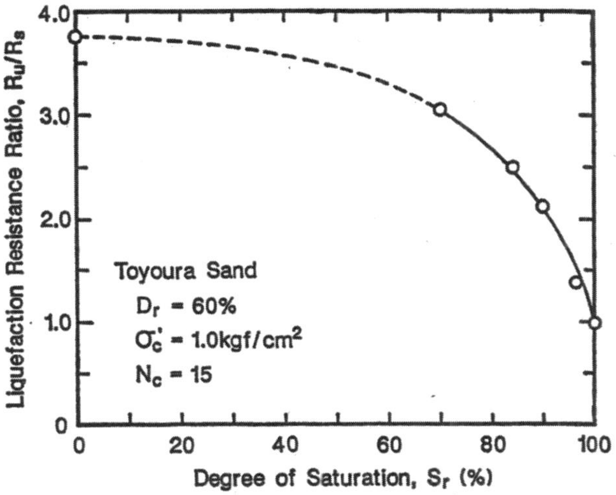

Fig. 5.5. Correlation between the degree of saturation of soil and liquefaction resistance ratio (Yoshimi et al. 1989)

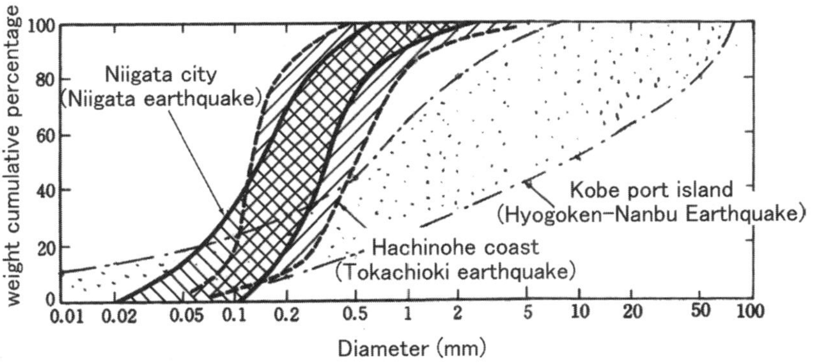

Fig.5.6. Grain size distribution of three liquefied soils in Japan (Yoshimi and Fukutake 2005)

5.3.3 Damage to Structures and Countermeasures against Liquefaction

Behaviour and Destruction of Construction

Liquefaction usually induces land subsidence because of the re-settlement of soil particles. For example, subsidence of about 20 to 50cm occurred by liquefaction at the 1989 Loma Prieta earthquake, the 1990 Luzon earthquake (Philippines, M7.8) and the 1995 South Hyogo Prefecture Earthquake (Yoshimi and Fukutake 2005). Also, ground flowed laterally by several meters, which significantly damaged surface and subsurface structures.

There are many studies on the subsidence or inclination of heavy structures, such as reinforced concrete buildings, caused by liquefaction. When the pile foundations are totally submerged in the liquefied layer, structures built on top of the piles are apt to subside or incline by liquefaction. Settlement for a built structure depends on its weight and width, and is not dependent on the number of piles, the height (the number of floors) of the building or the existence of a basement. When the pile foundation is penetrated into the soils situated below the liquefied layer, piles are mostly destroyed by liquefaction. In the 1964 Niigata Earthquake and the 1995 South Hyogo Prefecture Earthquake, pile foundations were destroyed although no ground structures were built on these foundations (Yoshimi and Fukutake 2005).

Underground installations, such as septic tanks, fuel tanks, manholes and sewerage, often rise to the surface as a result of liquefaction. Considering the density of liquefied sand (approximately 1.9 g/cm^3), underground installations that have lower densities than sand can be uplifted by liquefaction. Sewer networks are easily destroyed by liquefaction because the apparent density of manholes and sewers is much less than the density of the liquefied sand layer (Oka 2001).

As for earthen structures, such as earth dams, banks of roads and railways, the sandy part of structures and the sandy layer in foundations can liquefy and induce the damages (i.e. collapse, cracking and subsidence) to these structures.

Effect of aseismic base isolation

The liquefied layer is considered to have the effect of aseismic base isolation. That is to say, liquefied layer can reduce the damage by earthquakes to the structures. In the 1995 South Hyogo Prefecture Earthquake, damages to structures that had been built on unimproved ground or on ground where only the upper layers was compacted were relatively small (Fukutake 2001). In the 1999 Kocaeli Earthquake (Turkey), many structures built on hard ground collapsed, but those built on the liquefied soft ground did not collapse though subsidence occurred. These phenomena have been evaluated by experiments and numerical simulations (Yoshimi and Fukutake 2005). Fig.5.7 shows the outline and the results of a vibration experiment. It was revealed that acceleration of the foundation was lowest in the unimproved condition, and acceleration increased with the thickness of the improved soils.

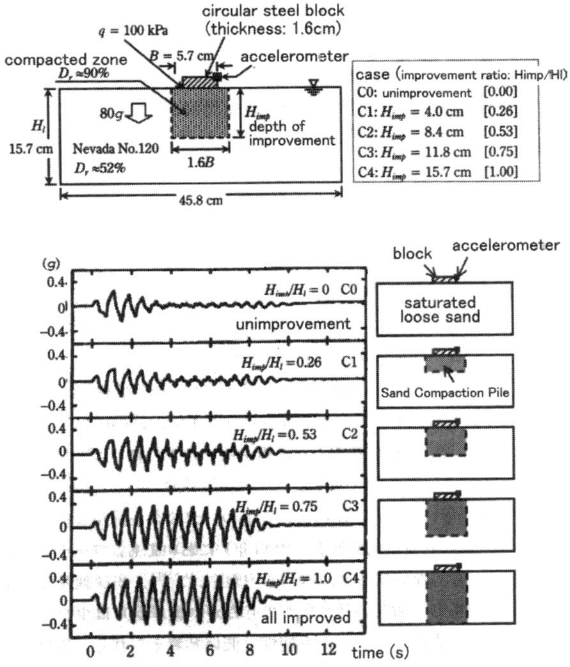

Fig. 5.7. Vibration experiment with varied thickness of compacted soils (Yoshimi and Fukutake 2005)

Countermeasures

There are many countermeasures against liquefaction. However, so far there has been no economically feasible countermeasure proven to be effective. Recent approaches for the countermeasures against liquefaction can be classified into two: improving the ground to reduce liquefaction or ensuring the safety by the improved pile foundation.

Measures to increase the liquefaction resistance are divided into two: improvement of the soil and change of the stress conditions by dewatering. In the former type, soil is improved by consolidation and chemical grouting. In the latter type, construction of artificial drainage routes for pore water (gravel drain) and cut-off walls are used for dewatering of soils. As we saw in the previous section, the degree of saturation in the soil is closely related to the likelihood of liquefaction. Dewatering of the shallower part of the ground increases the effective stress on the saturated part of the ground because of the reduction of pore water pressure. When saturated sandy soil becomes unsaturated, it can remain unsaturated for a long time even beneath the groundwater table. According to Okamura et al. (2003), the unsaturated condition of compacted soil beneath the groundwater table (5 to 7m in depth) has been maintained for 26 years. This fact suggests the efficiency of the dewatering method. Fig.5.8 shows the example of the dewatering by constructing cut-off walls and groundwater pumping in a tank yard in Japan.

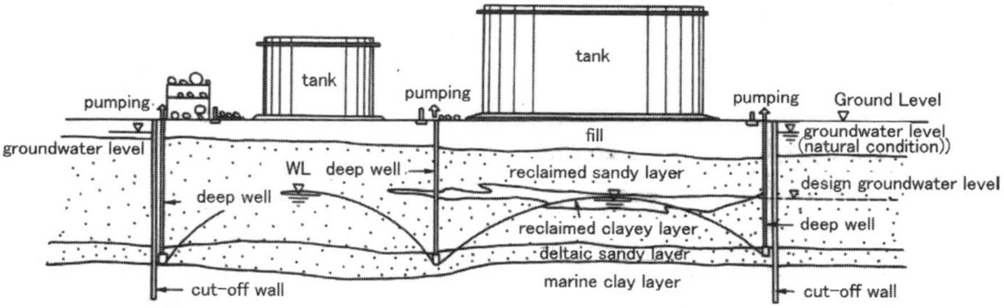

Fig. 5.8. Schematic diagram showing the method of artificial groundwater table depression (Ohmori 1988)

Soil improvements such as consolidation and dewatering are effective approaches to increase the liquefaction resistance of soil. However, it is difficult to prevent liquefaction for every earthquake. As mentioned earlier, liquefaction has an effect of aseismic base isolation (Fig. 5.7). Recently, artificial two-layered structures of ground that is made by improvement of only upper part of the soil are designed (Fukutake 2001). In these approaches, liquefaction is allowed to prevent structures against earthquakes. It is important to take an integrated approach combining characteristics of soil and design of pile foundations and constructions.

5.4 Case Study in Eastern Tokyo

The eastern Tokyo area, also called the Tokyo lowland area, is located in the southern part of the Kanto Plain, which is the largest sedimentary plain in Japan. Geomorphological and geological information of this area is described in Chapter 4. In this chapter, the correlation between the shallow groundwater environment and the risk of soil liquefaction in this area is discussed.

The surface and shallow subsurface conditions of this area has been extensively modified by urbanisation, specifically projects such as artificial landfills and river improvements and replacements. There are also artificial, reclaimed lands that have been constructed since the 17th century in the southern, coastal part of this region (Fig. 5.9). Thicknesses of these landfills and reclaimed soils are generally around several meters, however in some parts of the thickness exceeds 10m. These landfills, reclaimed soils and sandy layer having a thickness of several to 10m (upper Yurakucho Formation) comprise a shallow unconfined aquifer. The top of the Lower Yurakucho Formation, which is a clayey silt layer, is the boundary of this aquifer.

In this area, the public water supply system and sewerage system reached 100% in 1988 and 1995, respectively. Water mains are buried mainly in the artificial landfills and in the upper Yurakucho formation. On the other hand, some sewer mains are buried at depths below 30m to maintain gravity flow. Therefore, sewer mains are built not only in the upper Yurakucho Formation but also in the lower formations.

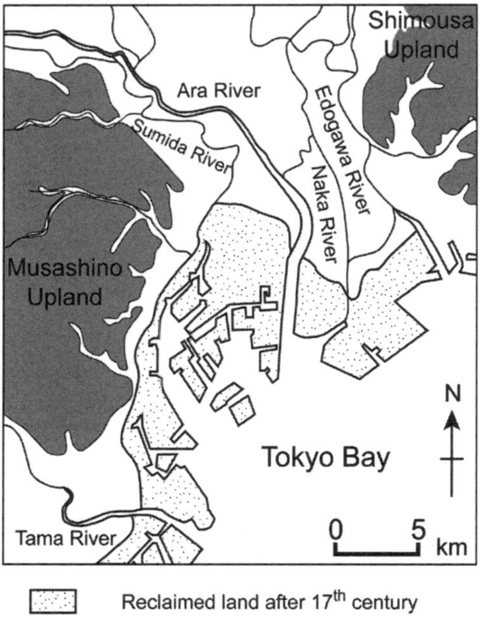

Fig.5.9. Location map of the Tokyo lowland area

5.4.1 Changes in the shallow groundwater environment

Long-term change of groundwater level and groundwater table

A large amount of deep confined groundwater was abstracted from the 1920s until the 1970s, contributing to land subsidence in the eastern area of Tokyo. However, some studies reported that shallow groundwater levels had not been depressed during this time, with the exception of seasonal change (Kawashima 2001; Institute of Civil Engineering, Tokyo Metropolitan Government 1987). These facts suggest that the lower Yurakucho Formation has prevented deep groundwater abstraction (Chap. 4) from affecting the shallow groundwater.

The groundwater table in the western part of this region in 1929 is shown in Fig.5.10. At the time, this part was the most urbanized area in this region. The groundwater table is lower than 2m below ground around Tokyo station and in some other areas, but in most of eastern Tokyo, the groundwater table is shallower than 2m below the surface. On the other hand, the groundwater table was located near the surface on the eastern side of the Ara River, because there were many ill-drained paddy fields in this part. Following urbanisation after the 1920s, the number of shallow open wells decreased drastically. The Institute of Civil Engineering of the Tokyo Metropolitan Government compiled groundwater level data from more than 7,000 boring logs and made a groundwater table map of this region (Fig.5.11; Institute of Civil Engineering, Tokyo Metropolitan Government 1987). By the time of the study, the eastern part of this region had already been completely urbanized. A comparison between Fig.5.10 and Fig.5.11 shows that the groundwater table is depressed by one meter or more.

Groundwater balance

The Tokyo Metropolitan Government has estimated the groundwater balance including both shallow groundwater and deep groundwater and the result is shown in Table 5.1 (Tokyo Metropolitan Government 1998; Bureau of Environment, Tokyo Metropolitan Government 1980). These estimates show that, since the mid-1970s, the leakage from water mains has been larger than the infiltration from precipitation. The rates of infiltration from precipitation and leakage from water mains are approximately 266,000m^3/day and 492,000m^3/day (average of 1985 to 1987), respectively. Groundwater discharge to sewerages and subway tunnels are 211,000m^3/day and 21,000m^3/day respectively. That is, groundwater corresponding to approximately 90% of the infiltration from precipitation discharges to the anthropogenic drainage system.

The groundwater balance shows a positive value, which means that a part of recharged groundwater naturally discharges to rivers or to Tokyo Bay. Because the areas above sea level are limited to the surrounding parts of this region (Fig.5.11), discharge of groundwater to the rivers or sea is possible only from those high-elevation lands. On the other hand, in areas below sea level, the groundwater table is also distributed beneath the ground surface (Fig. 5.11). Groundwater in such areas is considered to discharge to anthropogenic groundwater sinks such as sewerage and subsurface construction.

Anthropogenic groundwater sources and sinks have played an important role in groundwater balance in this region. Furthermore, human induced land subsidence restricts the area where groundwater can discharge naturally to the rivers or to the sea. Therefore, the future groundwater balance is strongly affected by the changes in rates of anthropogenic recharge and discharge, as happens with the modification of water mains and sewerage.

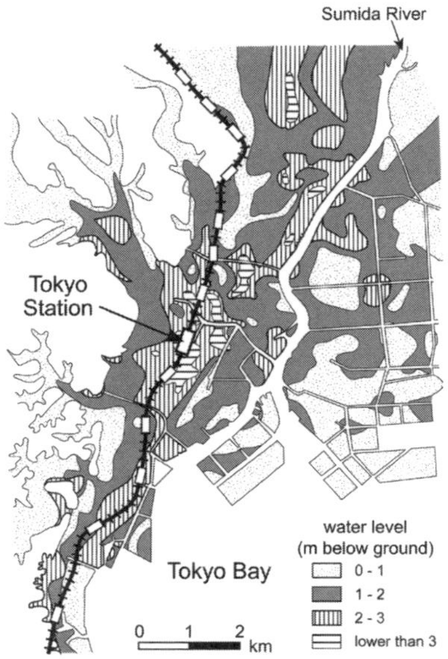

Fig. 5.10. Groundwater table in the southwestern part of the Tokyo lowland in 1929 (modified from Bureau of reconstruction, Ministry of Interior 1939)

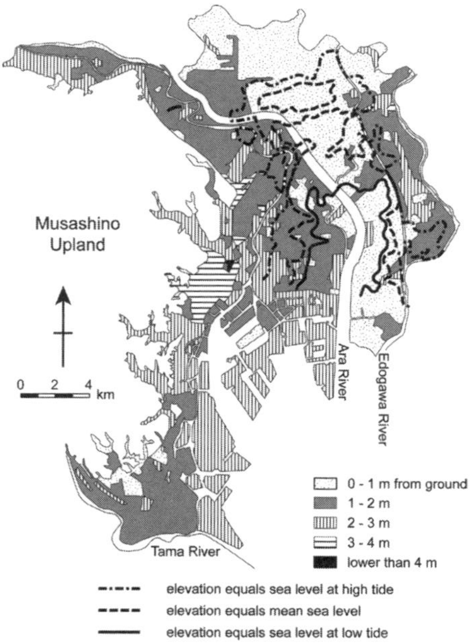

Fig. 5.11. Groundwater table in the Tokyo lowland area before 1987 (modified from Institute of Civil Engineering, Tokyo Metropolitan Government 1987).

Table 5.1. Groundwater balance in the eastern Tokyo including deep groundwater (Tokyo Metropolitan Government 1998; Bureau of Environment, Tokyo Metropolitan Government 1980)

Unit: ×10,000m^3/day		1985-1987	1994-1996	Difference
Recharge	Infiltration from precipitation	27	23	-4
	Leakage from water pipes	49	38	-11
Discharge	Discharge to sewerage	21	27	6
	Discharge to subsurface construction	2	3	1
	Groundwater abstraction	12	11	-1
Total		41	20	-21

5.4.2 Soil liquefaction

Soil liquefaction in the Tokyo lowland

Soil liquefaction has occurred repeatedly in the Tokyo lowland. However, liquefied areas are concentrated in the riversides and natural levees of present or former rivers (the Naka, Edo and Tama Rivers) and in the reclaimed land surrounding the Sumida and Ara Rivers (Fig.5.12; Institute of Civil Engineering, Tokyo Metropolitan Government 1999, 1987a, 1987b; Wakamatsu 1991a, 1991b). These facts suggest that the liquefiable ground has some particular geological and soil characteristics.

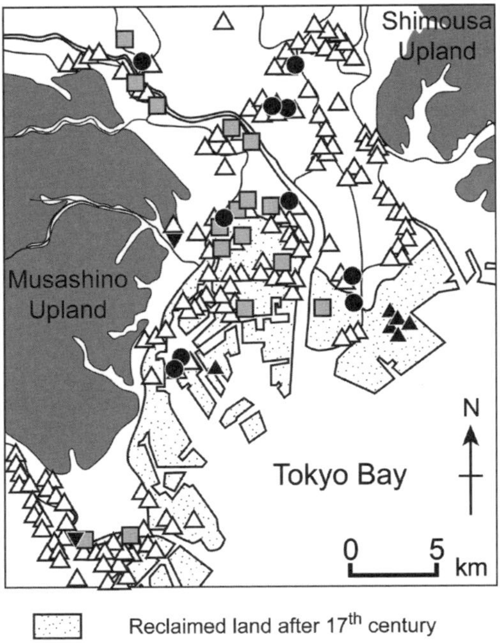

Fig. 5.12. Map of historic liquefaction sites in the Tokyo lowland area (data from Institute of Civil Engineering, Tokyo Metropolitan Government 1999, 1987a, 1987b; Wakamatsu 1991a, 1991b).

Geological and soil characteristics of liquefied ground

The Institute of Civil Engineering of the Tokyo Metropolitan Government has compiled a historical record of soil liquefaction events and has reported the geological and soil properties of liquefied ground (Institute of Civil Engineering, Tokyo Metropolitan Government 1999, 1987a, 1987b). Based on these reports, in liquefaction events, clear groundwater and sand has gushed into riverside and natural levee areas, while muddy groundwater had gushed into the reclaimed area.

According to the same studies, the areas that made up the liquefied ground in the riverside and natural levee were the sand layers that had an average diameter of 0.15 to 0.30mm and contained less than 20% fine-grained fraction. On the other hand, soil in the reclaimed area was made up of a silty sand layer with an average diameter of 0.15 to 0.25mm and contained between 20 and 50% fine-grained fraction.

Prediction of liquefaction and possible countermeasures

The Institute of Civil Engineering of the Tokyo Metropolitan Government made a prediction map for soil liquefaction in Tokyo based on the geological and soil characteristics of liquefiable ground, distribution of the groundwater table and subterranean geological structure estimated by boring logs (Institute of Civil Engineering, Tokyo Metropolitan Government 1999, 1987b). On the basis of this map, almost all parts of the Tokyo lowland area are susceptible to liquefaction (Fig. 5.13). The risk of liquefaction is especially high in the eastern and the northern part of the Tokyo lowland.

A groundwater table map (Fig. 5.11) was made based on information collected before 1987. Although the information on the current status of shallow aquifers is limited, it is estimated that there has been little change during the last couple of decades. Hence, the risk of liquefaction is likely to be similar to that in 1987. Countermeasures should be taken against liquefaction, especially in the area of high liquefaction risk. There are various technical methods involved in liquefaction countermeasures as mentioned in 5.3.3. Although soil improvement approaches such as compaction and solidification are proven countermeasures, it is difficult to apply these approaches to a wide area because of cost, noise, land cover and other reasons. Similarly, the artificial lowering of the groundwater table requires a large investment in construction and maintenance of facilities for groundwater pumping, treatment and drainage. Considering the groundwater balance shown in Table 5.1, the major groundwater source and sink are the leakage from the water mains and seepage into the sewer mains, respectively (Chap. 5.4.1). Therefore, decreasing the volume of the leakage or increasing groundwater drainage by sewer mains may possibly lower the groundwater table. That is to say, the repair of water mains will contribute not only to the reduction of non-revenue water for waterworks but also to the mitigation of liquefaction risks.

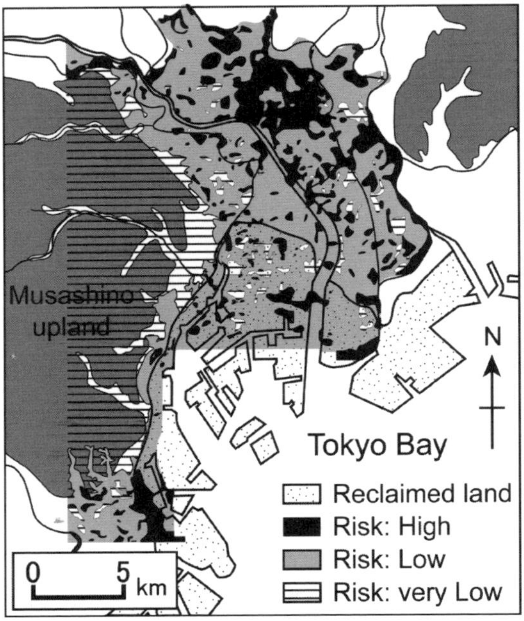

Fig. 5.13. Prediction map of liquefaction (modified from Institute of Civil Engineering Centre, Tokyo Metropolitan Government 1999)

5.5 Conclusions

In this chapter, changes of the shallow groundwater environment with urbanisation and characteristics of soil liquefaction have been presented, and changes in the shallow groundwater environment and the risk of soil liquefaction in the Tokyo lowland area have been discussed.

In the process of urbanisation in the Tokyo lowland area, the shallow groundwater environment has not been affected by deep groundwater abstraction and the shallow groundwater table has not been changed to a great degree. However, the main groundwater source presently is the leakage from water mains, and the groundwater balance of this region is dominated by anthropogenic groundwater sources and sinks.

As for the prediction of the risk of soil liquefaction, as well as the analysis of possible countermeasures, it is important to evaluate the situation of the present groundwater table in the Tokyo lowland. However, because the number of shallow open wells has been drastically decreased with urbanisation, the delineation of the present groundwater table is very difficult.

Hence, new techniques and methods for the measurement of the groundwater table beneath the covered urban area are required to be developed.

Sea level rise due to global warming (IPCC 2007) will induce the rising of the groundwater table in coastal urban areas such as the Tokyo lowland area, which will eventually exacerbate the risk of soil liquefaction. Thus, it is important to evaluate on and to take measures against the liquefaction based on long-term environmental change. Hence, it is essential to take an integrated approach combining hydrological/hydrogeological studies about the urban groundwater cycle and engineering studies about the evaluation of risk of and possible countermeasures against soil liquefaction.

References

Al-sefry SA, Şen Z (2006) Groundwater rise problem and risk evaluation in major cities of arid lands – Jedddah case in Kingdom of Saudi Arabia. Water Resources Management 20: 91–108

Arai T, Shindo S, Ichikawa A, Yoshikoshi A (1987) Hydrological environment in urban area (in Japanese). Kyoritsu Publ

Barrett MH (2004) Characteristics of urban groundwater. Urban Groundwater Pollution, pp 29-51

Bureau of Pollution, Tokyo Metropolitan Government (1980) Report on groundwater balance (in Japanese)

Bureau of Reconstruction, Ministry of Interior (1939) Report on geology in Tokyo and Yokohama district.

Bureau of Waterworks, Tokyo Metropolitan Government (2006) Waterworks of-Tokyo metropolis (in Japanese)

Foster S, Lawrence A, Moris B (1998) Groundwater in urban development assessing management needs and formulating policy strategies. World Bank Technical Paper No. 390

Foster SSD, Gale IN, Hespanhol I (1994) Impacts of wastewater use and disposal on groundwater. Technical report No. WD/94/55 by British Geological Survey. Nottingham, UK

Fukutake K (2001) Aseismic base isolated foundation using nonlinear characteristics of soft ground (in Japanese). MENSHIN 34:26-40

Garcia-Fresca B (2007) Urban-enhanced groundwater recharge: review and case study of Austin, Texas, USA. In: Urban Groundwater - Meeting the Challenge, pp 3-18

Institute of Civil Engineering, Tokyo Metropolitan Government (1999) Prediction map of liquefaction in Tokyo Metropolis (in Japanese)

Institute of Civil Engineering, Tokyo Metropolitan Government (1990) Land reliquefaction phenomena at the times of the past earthquake in Tokyo lowland (in Japanese). Annual report, pp 281-291

Institute of Civil Engineering, Tokyo Metropolitan Government (1987a) Characteristics of liquefied area during the Kanto earthquake and the method of evaluating liquefaction potential in Tokyo lowland (in Japanese). Annual report, pp 225-236

Institute of Civil Engineering, Tokyo Metropolitan Government (1987b) Prediction map of liquefaction in Tokyo Metropolis (in Japanese)

IPCC (2007) Intergovernmental Panel on Climate Change fourth assessment report climate change 2007: Synthesis report

Japan Meteorological Agency (2005) Report on anomalous weather 2005 (in Japanese)

Kansai branch of Japan Society of Civil Engineers (2002) Evaluation of effect of groundwater control on environment of ground (in Japanese)

Kawashima S (2001) Groundwater environment in Tokyo Metropolis (in Japanese). Journal of Ground Water Technology 43(3):6-19

Kjellén M, Mcgranahan G (1997) Comprehensive assessment of the freshwater resources of the world. Stockholm Environment Institute

Lamptey BL, Barron EJ, Pollard D (2005) Impacts of agriculture and urbanisation on the climate of the Northeastern United States. Global and Planetary Change 49:203- 221

Lerner DN (2002) Identifying and quantifying urban recharge: a review. Hydrogeology Journal 10:143-152

Ministry of Land, Infrastructure and Transport (2004) 2004 water resources in Japan (in Japanese)

Ohmori K (1988) Example of design and construction in soft soil foundation of tanks (soil improvement with decreasing of pore water pressure). Soil and Foundations 18(12):122-129

Oka F (2001) Science of ground liquefaction (in Japanese). Kin-mirai Publ

Okamura M, Ishihara M, Tamura K (2003) Degree of saturation of sand improved with SCP 27 years ago (in Japanese). In: Proceedings of annual conference of Japanese Geotechnical Society 38, pp 2027-2028

Section of Reconstruction, Ministry of Interior (1930) Report on geology in Tokyo and Yokohama (in Japanese)

Shiklomanov IA (1998) World water resources - A new appraisal and assessment for the 21st century. A summary of the monograph World Water Resources

Terzaghi K, Peck RB (1948) Soil mechanics in engineering practice. John Wiley & Sons Inc

Tokyo Metropolitan Government (1998) Project on conservation of water environment in Tokyo Metropolis (in Japanese)

Wakamatsu K (1991a) Maps for historic liquefaction sites in Japan (in Japanese)

Wakamatsu K (1991b) Regional maps for soil liquefaction sites in Japan (in Japanese)

Yasuhara M (1998) Leakage from water pipes and groundwater recharge in foreign countries. Journal of Groundwater Technology 40(10):1-7 (in Japanese)

Yoshimi Y, Fukutake K (2005) Physics, evaluation and measures of soil liquefaction (in Japanese). Gihodo Publ

Yoshimi Y, Tanaka K, Tokimatsu K (1988) Liquefaction resistance of a partially saturated sand. Soils and Foundations 29(3):157-162

6. Coupled Groundwater Flow/ Deformation Modelling for Predicting Land Subsidence

Masaatsu Aichi

Department of Geosystem Engineering, the University of Tokyo

6.1 Introduction

Land subsidence is one of the major problems to consider in the attempt to achieve sustainable groundwater usage in urban areas, as mentioned in Chapters 3, 4 and 12. For better management of groundwater-related environments, a combination of monitoring and predictive modelling is desirable (Fig. 6.1).

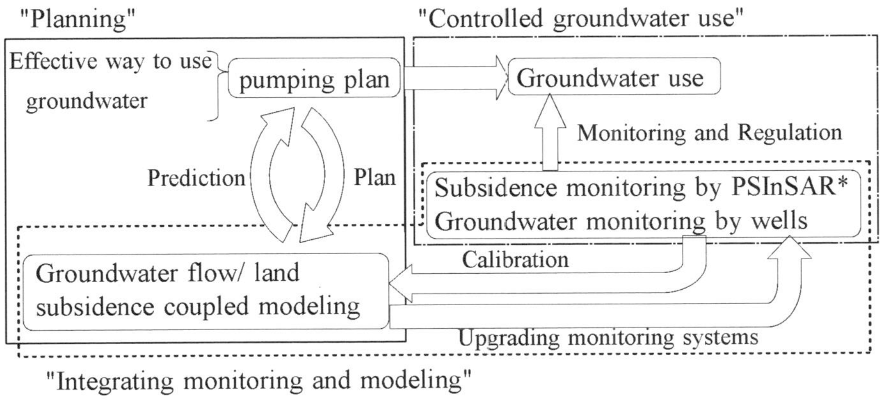

Fig. 6.1. A strategy for sustainable groundwater use in urban areas by integrating monitoring and modelling. *See Chapter 4 for the details of the PSInSAR technique

Numerical simulation is a prospective tool for quantitatively predicting the change in groundwater-related environments such as land subsidence. In this chapter, the present state and the potential of numerical modelling techniques are presented.

6.2 Theoretical background of Groundwater Flow/ Deformation Coupled Processes in variably saturated porous media

Land subsidence is a coupled process of both groundwater flow and land deformation.

Groundwater flow is governed by Darcy's law:

$$q_i = -K_{ij} \frac{\partial}{\partial x_j}\left(\frac{p_w}{\rho g} + x_3 \right) \tag{6.1}$$

where q_i is the volumetric flux of water, p_w is the pore water pressure, K_{ij} is the hydraulic conductivity, ρ is the density of water and g is the acceleration of gravity. It should be noted that the Darcy's law corresponds to a reduced form of the momentum balance equation.

To describe variably saturated groundwater flow, capillary pressure and saturation dependent hydraulic conductivity should be taken into account. A set of empirical equations developed by Mualem (1976) and Van Genuchten (1980) has been widely used (Japanese Geotechnical Society, 2004).

The empirical relation for capillary pressure in an unsaturated zone was developed by van Genuchten (1980):

$$p_w = -\frac{1}{\eta}\left(S_e^{-\frac{1}{m}} - 1 \right)^{\frac{1}{n}} \tag{6.2}$$

where S_e is the effective saturation $\left(= \frac{S_w - S_r}{1 - S_r} \right)$, S_w is the water saturation, S_r is the residual saturation and $m\left(= 1 - \frac{1}{n} \right), n, \eta$ are parameters which are dependent on soils. The empirical relation for unsaturated hydraulic conductivity was developed by Mualem (1976):

$$K_{ij} = K_{ij}^{ws} S_e^{0.5} \left\{ 1 - \left(1 - S_e^{\frac{1}{m}} \right)^m \right\}^2 \tag{6.3}$$

where K_{ij}^{ws} is the saturated hydraulic conductivity and m is as defined above.

Stress-strain relations for porous materials have been described in terms of effective stress. Effective stress is a stress which works in the solid phase of porous material. It is often assumed that the effective stress σ'_{ij} can be written as a function of water saturation as follows:

$$\sigma'_{ij} = \sigma_{ij} + S_w p_w \delta_{ij} \tag{6.4}$$

where σ_{ij} is the total stress (Japanese Geotechnical Society, 2004).

Soils and soft rocks are known to behave elastically before effective stress reaches the yield criterion of past maximum effective stress, which the rocks have experienced (Wood, 1991). When the effective stress exceeds the past maximum effective stress, soils and soft rocks start to behave plastically (Fig. 6.2). This is a reason why land subsidence does not recover even after pore pressure recovers and effective stress becomes smaller.

To describe elastic deformation and plastic deformation by different equations respectively, it is convenient to separate the strain into the elastic strain and the plastic strain as follows:

$$\varepsilon_{ij} = \varepsilon^e{}_{ij} + \varepsilon^p{}_{ij} \tag{6.5}$$

where ε_{ij} is the total strain, $\varepsilon^e{}_{ij}$ is the elastic strain and $\delta\varepsilon^p{}_{ij}$ is the plastic strain.

In this study, linear poroelasticity (Biot, 1941) was used for the stress-strain relation in the elastic state:

$$\varepsilon^e{}_{ij} = \frac{1}{2G}\left(\sigma'_{ij} - \frac{\nu}{1+\nu} \sigma'_{kk} \delta_{ij} \right) \tag{6.6}$$

where G is shear modulus and ν is drained Poisson's ratio. Yield criterion f and increment of plastic strain $\delta\varepsilon^p{}_{ij}$ were represented by the modified Cam-clay model (Roscoe and Burland, 1968) described as below:

$$f = \sigma_q{}^2 - M^2 \left[\sigma_p' \left(\sigma_{pc}' - \sigma_p' \right) \right] = 0 \tag{6.7a}$$

$$\delta\varepsilon^{p}{}_{ij} = -\frac{1}{2M^4\sigma_p'}\left[\frac{(1-\phi)\lambda}{\sigma_p'}+\frac{1}{K}\right]\frac{\partial f}{\partial \sigma_{ij}'}\left[\frac{\partial f}{\partial \sigma_{kl}'}\delta\sigma_{kl}'+\frac{\partial f}{\partial \sigma_{kk}'}\delta\sigma_{kk}'\right] \tag{6.7b}$$

where ϕ is porosity, σ_q is the deviator stress, σ_p' is the mean effective stress, σ_{pc}' is the past maximum mean effective stress, λ is the slope of the virgin isotropic compression line in the plastic deformation and M is the critical ratio. The deviator stress and the mean effective stress are written as:

$$\sigma_q = \left[(\sigma_{ii}')^2 + \frac{3}{2}\left(-\sigma_{ii}'\sigma_{jj}'+\sigma_{ij}'\sigma_{ji}'\right)\right]^{\frac{1}{2}} \tag{6.8a}$$

$$\sigma_p' = \frac{\sigma_{ii}'}{3} \tag{6.8b}$$

Porosity can be described in terms of strain:

$$\frac{1-\phi_0}{1-\phi} = 1+\varepsilon_{ii} \tag{6.9}$$

where ϕ_0 is initial porosity.

The governing equations applied for the coupled processes are force equilibrium of the porous material and mass conservation of groundwater.

Force equilibrium of continua is written as:

$$\frac{\partial \sigma_{ji}}{\partial x_j} + F_i = 0 \tag{6.10}$$

where F_i is the body force.

Mass conservation of groundwater is written as:

$$\frac{\partial(\rho\phi S_w)}{\partial t} + \frac{\partial}{\partial x_i}(\rho q_i) + Q = 0 \tag{6.11}$$

where q_i is the Darcy flux and Q is the source term of groundwater.

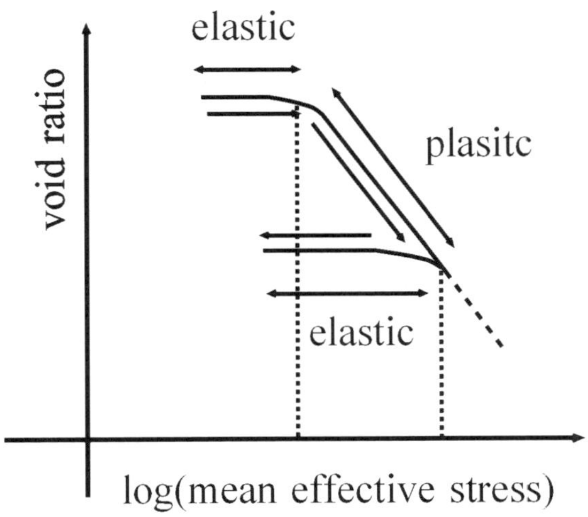

Fig. 6.2. General behaviour of soil: the void ratio e is the ratio between the pore volume and the solid volume

6.3 Nested Modelling Approach

6.3.1 Necessity of the Nested Modelling Approach

In the areas that have experienced both significant drawdown/land subsidence and the rapid recovery of hydraulic head of groundwater such as Tokyo, past maximum effective stress inside the clayey layer can be different from the one that is estimated from the analysis of hydraulic head in the aquifer (Fig. 6.3). This situation can happen because the pore pressure diffusion in the clayey layers tends to be delayed due to lower hydraulic diffusivity.

The amounts and rates of land subsidence are highly dependent on the spatio-temporal distribution of the past maximum effective stress inside clayey layers because the plastic deformation of clayey layers is larger than the elastic deformation. Furthermore, the possibility of land re-subsidence caused by the future groundwater extraction is dependent on the past maximum effective stress profile. For sustainable groundwater usage without significant land re-subsidence, high resolution modelling to simulate effective stress conditions in the clayey layers is needed.

Because hydraulic heads in aquifers are controlled by the regional groundwater flow system including the effects of groundwater pumping, we should analyze both small-scale processes within clayey layers and large-scale processes in the whole groundwater basin. A high-resolution modeling of coupled regional groundwater flow/deformation processes is quite computer-intensive and is difficult to apply to a real case. Here, the nested modelling approach is adopted to solve this difficulty.

The nested modelling approach is a domain decomposition technique. In this technique, both a regional scale model and a local scale model are prepared (Fig. 6.4). The spatial discretization for the regional scale model can be as large as that of usual large scale groundwater flow modelling, while that of the local model should be fine enough to be able to calculate the pore pressure and strain distribution in the clayey layers. A new approach to handle these types of problems has been developed by Aichi and Tokunaga (2006) and is briefly described in the next section.

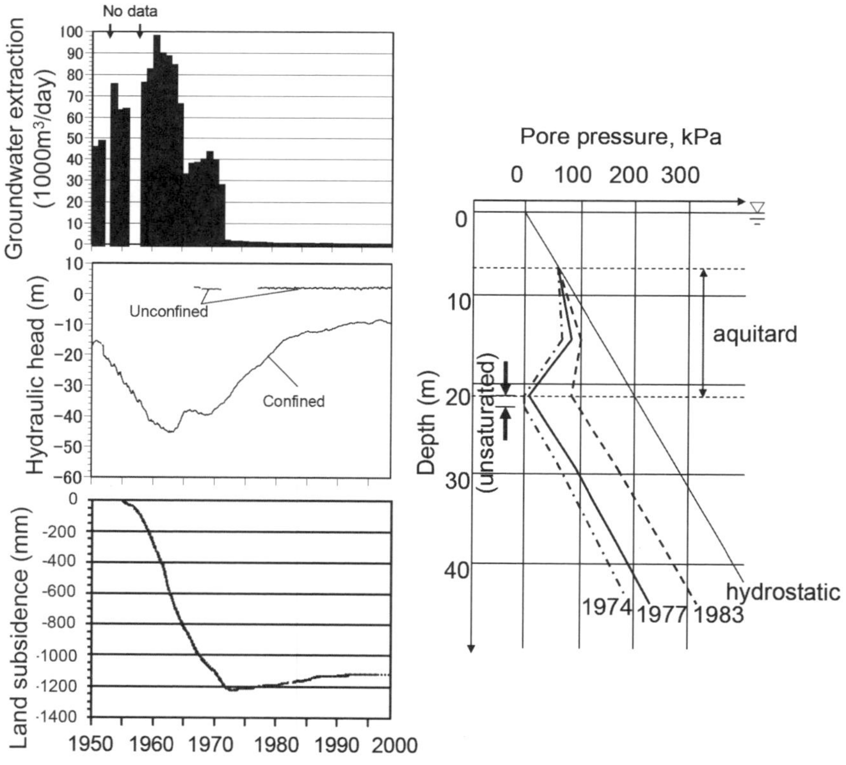

Fig. 6.3. An example of the temporal change in groundwater use, hydraulic head and land subsidence in the Tokyo lowland. The right hand side of the figure showing the pore pressure profile is drawn after Hirose et al (2004)

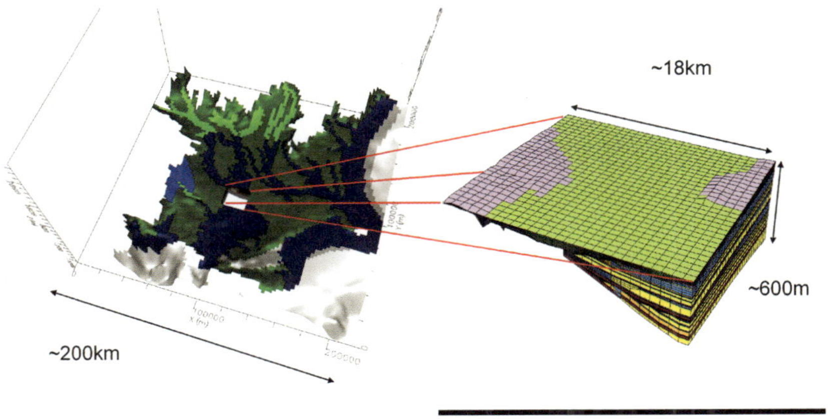

Fig. 6.4. An example of a nested modelling approach: overview of a regional groundwater flow model for the Kanto Plain and a local scale groundwater flow /land subsidence coupled model for the central district of the Tokyo.

6.3.2 Nesting Procedure

The nesting procedure to transfer boundary conditions at the interface between the regional and local scale models in this study is shown in Fig. 6.5. This procedure starts by running the regional model for the entire domain to obtain the first estimation of the hydraulic head at the boundary of the local domain.

The specified hydraulic head boundary condition for the aquifer of the local model is given from the calculated hydraulic head in the regional model (Fig. 6.6). The hydrostatic head profile is assumed in the aquifer when the condition is transferred from the regional to the local model. For the horizontal direction, the specified head is directly introduced at the shared node first and is linearly interpolated for the nodes between shared nodes. For the aquitards in the local model, the boundary condition is assumed to be no flow because the groundwater flow in the aquitards is usually vertically dominant due to the contrast of hydraulic conductivity. The groundwater flux at the inner boundary of the regional domain is given by summing up the flux obtained from the local model (Fig. 6.6). Next, the regional domain is calculated using this specified flux boundary condition. The local model is run using the results from the regional model again and the flux at the boundary for the regional model is obtained.

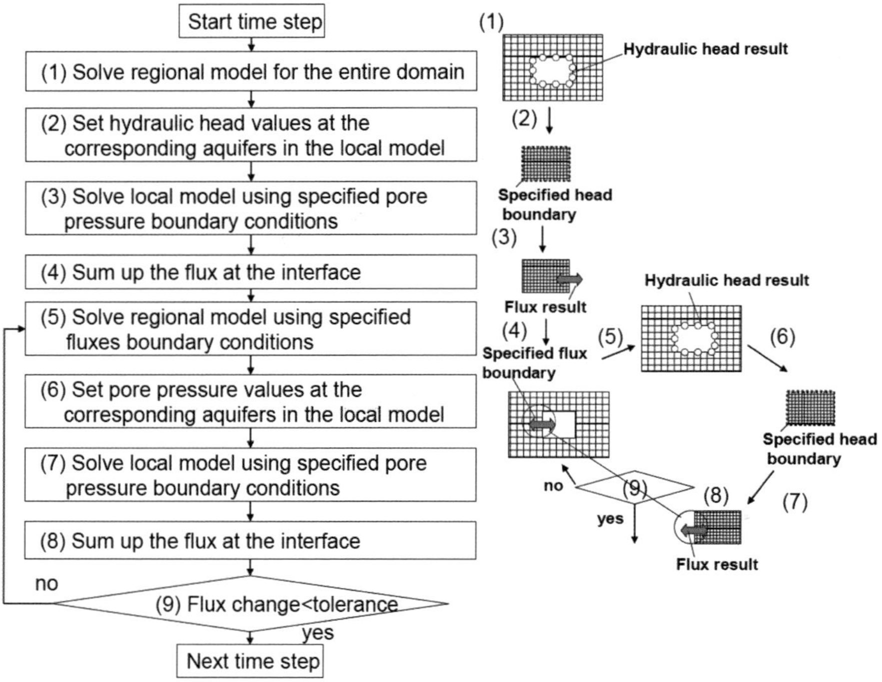

Fig. 6.5. Nesting scheme for transferring boundary condition between the regional and the local scale models

Flux continuity at the interface between the regional and the local domain is checked by comparing the fluxes calculated from the local model in this step and those in the previous step. If their difference is small enough, the iterative procedure finishes and the simulation proceeds to the next time step.

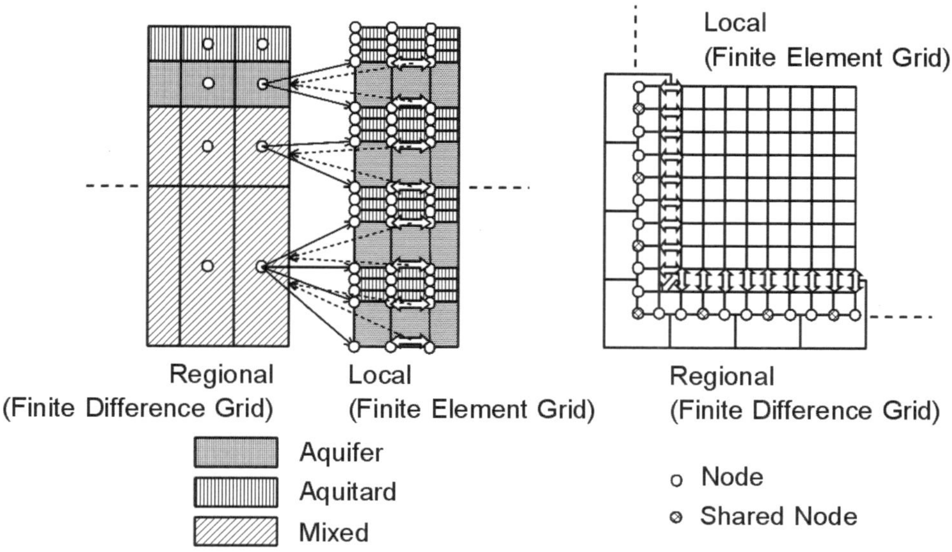

Fig. 6.6. An example of the transfer of the boundary conditions at the interface between regional and local scale models: The left figure shows the relation in a vertical cross section. In this case, the middle and deep aquifers in the regional model are modelled as a mixed layer of an aquitard and an aquifer. The right figure represents the relation in a horizontal section of the aquifer.

6.4 Application to the Central Districts of Tokyo

6.4.1 The Central Districts of Tokyo

As mentioned in Chapter 3, the Central Districts of Tokyo experienced land subsidence problems from the 1960s to the early 1970s. After the adoption of severe regulations regarding groundwater pumping, the hydraulic head recovered rapidly and the land subsidence stopped. The land surface uplift has been observed recently (Endo et al., 2001) and can be explained as elastic expansion due to the increase of pore pressure in the subsurface formations.

To examine the potential of numerical modelling for reproducing these processes, we conducted a preliminary study using this nesting procedure (Aichi et al., 2007).

Here, the regional model for this case study was taken as the whole Kanto Plain. Nakayama et al. (2007) also developed a Kanto Plain scale model to consider regional groundwater flow while assessing the effects from subsurface structures on groundwater flow.

The target period of calculation for the current analysis was from 1920 to 2000. This timeframe was selected because the pumping of groundwater by electrically driven pumps started in the 1920s (Shindo, 1972).

6.4.2 Constructing the Regional Groundwater Model

Hydrogeological Structures

The aquifer system in the regional model was modelled as consisting of seven layers. These layers are named I to VII, in descending order. Layer I mainly consists of clay-rich material in the lowland, while that in the upland mainly consists of loamy soils. Layers II through VII consist of gravel-silt-sand alternations. Below layer VII is a thick mudstone and was modelled as the hydrogeological basement in this study.

Pumping Rate

The time dependent pumping rate and its spatial distribution were modelled from the records reported by local governments.

For the periods when these records were absent, the pumping rate was estimated by the following procedure. In 1920, the pumping rate was assumed to be zero because Shindo (1972) stated that electric pumps were used after 1920s and hence, the amount of groundwater usage before then was not significant. In 1945, the pumping rate was also assumed to be zero because Shindo (1972) and Endo et al. (2001) have stated that hydraulic head recovered around 1945 after pumping was stopped due to the destruction of factories by air raid attacks during World War II. From 1920 to 1945, the pumping rate was calculated using the pumping rate from 1960 to 1970, weighed by the GDP ratio. The pumping area was assumed to be the same location as the urban area read from old maps. If there was qualitative information that the pumping rate was more or less constant, the constant value was used. Otherwise, linear interpolation was used. Fig. 6.7 shows an example of pumping rates used for this modelling.

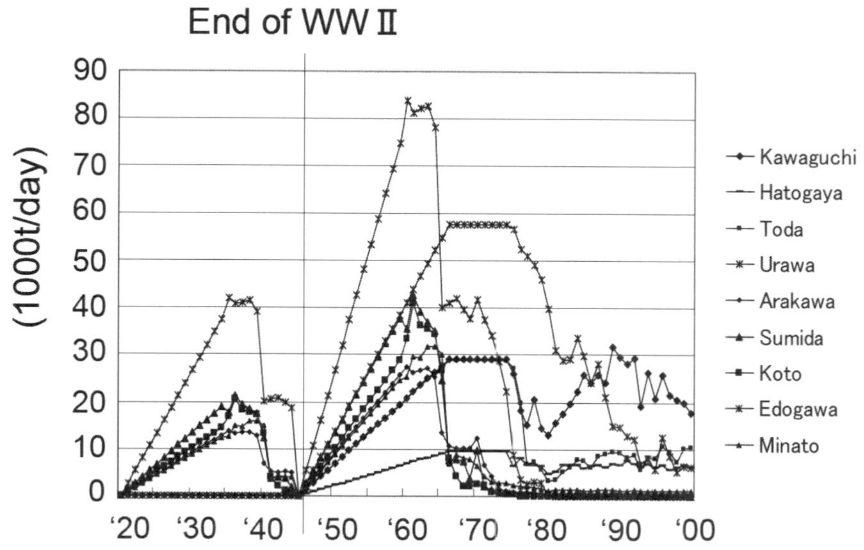

Fig. 6.7. An example of input of pumping rates used for this study

Because the distribution of the pumping intervals has not been well reported, it was assumed that the groundwater was equally extracted from the whole thickness of the aquifer and the pumping rates were proportional to the relative thicknesses of the main aquifers in layers V through VII. However, in the central part of the Kanto Plain, we followed the existing reports (Kino, 1965; Saitama Prefecture, 1973) on the depth of the pumping intervals and the screens were assumed to be situated mainly in the layers VI and VII.

Modelling the Distribution of Squeezed Water from the Aquitards

The regional model only simulates groundwater flow in order to reduce the computational efforts. If land subsidence has occurred in an area outside the local model, it is necessary to take into account the water squeezed from the aquitards and to introduce the estimated volume into the model as a virtual source of water. In the Kanto Plain, land subsidence in the northern part has occurred and it should be taken into account (Fig. 6.8).

The squeezed water is considered to be equal to the volume of land subsidence assuming that the solid particles are incompressible. That is, the distribution of the source term can be calculated from the levelling result.

In the lowland area where clayey alluvial sediments are distributed, consolidation of the alluvium (shallow layer) has been approximately half of the total deformation (Fig. 6.9a), while in the upland area, layers with pumped aquifers have mainly been contracted (Fig. 6.9b). This observation was used for setting up the depth distribution of virtual water sources.

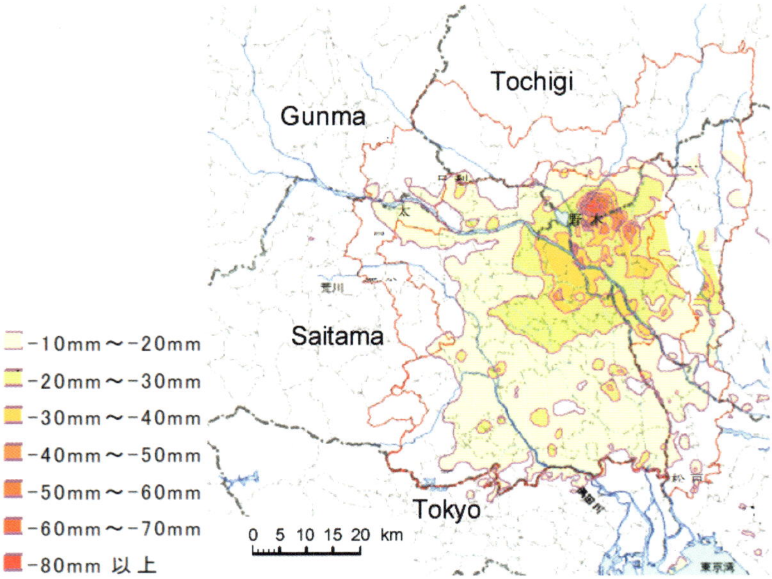

Fig. 6.8. Land subsidence in the northern part of the Kanto Plain (after Ministry of Land, Infrastructure and Transport, 2005)

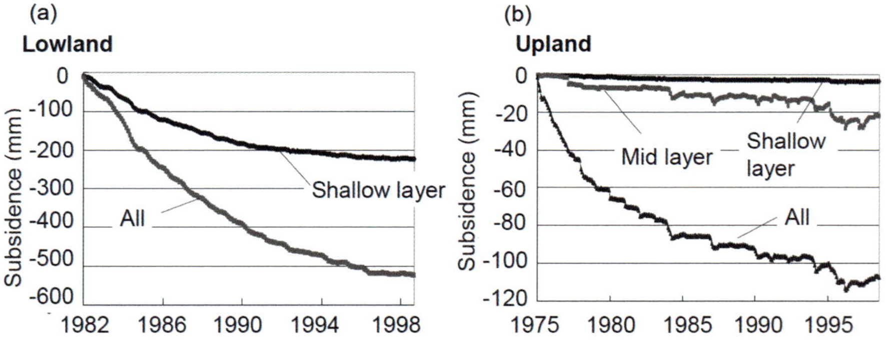

Fig. 6.9. Depth dependent contraction: a) example from lowland area and b) one from upland

Boundary Conditions and Initial Conditions

Boundary conditions for the regional model were set from the hydrogeological structures and previous studies (National Land Agency, 1998). The base of layer VII and lateral boundaries for the plain were modelled as no flow boundaries. From past observations (Kawashima, 2001; Ministry of Land, Infrastructure and Transport, 1974-1997), the water table of the unconfined aquifers was more or less constant through the observed period even though the fluctuation with shorter periods is apparent; thus, specified head boundary condition was applied for the top boundary. For simplicity, the sea bottom was also treated as a specified head boundary and density-driven flow was not considered.

The initial condition was established from steady state calculations with a pumping rate of zero in 1920.

Preliminary Tuning of Regional Model

Before going into the process of integration with the local model, a rough matching of the regional model with measured data was carried out to check the appropriateness of the regional model.

Fig. 6.10 shows a comparison of calculated results and observed data on the temporal change of hydraulic heads at the observation wells in the northern part of Tokyo. The calculated results showed slightly higher values but the patterns of the temporal changes were well reproduced. This model is used as the regional model for our nested simulation.

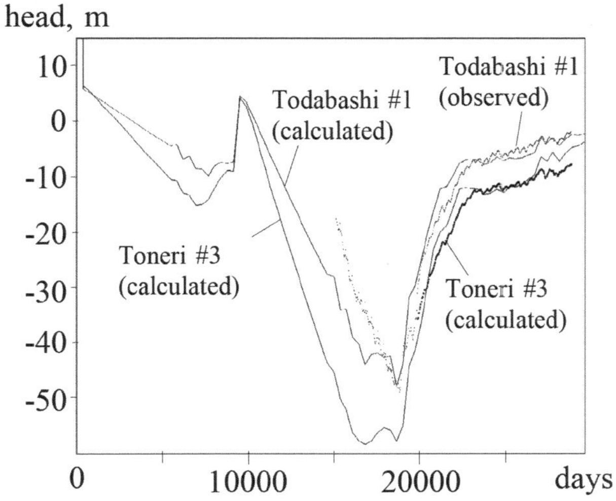

Fig. 6.10. A comparison of calculated and observed hydraulic head at observation wells in the northern part of Tokyo

6.4.3 Land Subsidence Modelling for the Local Area of Interest

Hydrogeological Structures

The local model in this study covers the eastern part of the Central Districts of Tokyo where land subsidence was significant (Fig. 6.11).

In the local model, twenty-four aquifers and twenty-three aquitards were modelled based on the existing geological and hydrological information (Institute of Civil Engineering, Tokyo Metropolitan Government, 1996) while seven layers were modelled in the regional model. Fig. 6.11 shows the domain of the local model. Generally, layers become thinner toward the southwestern area and many layers thin out completely in that direction, while the northeastern area has about a 600m thick aquifer-aquitard system.

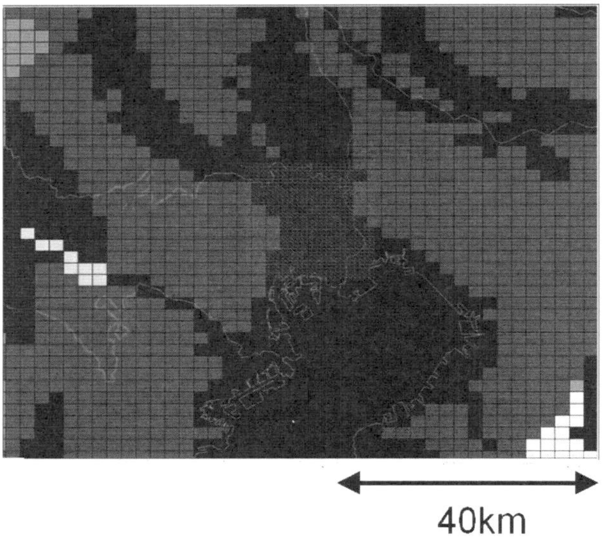

Fig. 6.11. The local model (fine mesh) constructed in this study: the black area is lowland or sea and the light coloured area is upland. Coarser mesh is for the Kanto Plain scale model.

The physical properties of aquifers and aquitards were estimated from the existing reports (Institute of Civil Engineering, Tokyo Metropolitan Government, 1996; Nakayama and Ohka, 1992; Isozaki and Sugawara, 1986; Saitama Prefecture, 1993; Endo et al. 1997). Hydraulic conductivity of the aquifers was set to be about 10^{-5}cm/s to 10^{-2}cm/s (Nakayama and Ohka, 1992) and that of aquitards was set from 10^{-10}cm/s to 10^{-8}cm/s (Isozaki and Sugawara, 1986). Other physical properties that are related to the deformation of the formations are summarized in Table 6.1.

Table 6.1. Physical properties used in this study

Properties	Range	Source
Drained bulk modulus	300MPa – 3GPa	Saitama Prefecture, 1993
Compression index	0.5 - 0.7	Institute of Civil Engineering, Tokyo Metropolitan Government, 1996
Critical ratio	0.3 - 0.8	
Wet density	1.7g/cm^3 – 2.5g/cm^3	
Initial void ratio	0.5 - 1.0 (aquifers) 1.3 - 1.9 (aquitards)	

Boundary Conditions and Initial Conditions

The boundary conditions for displacement were set as follows: the base of the model was fully constrained and the four side faces of the model domain were constrained so that there is no displacement perpendicular to these faces. The initial condition should demonstrate no displacement for the entire domain before groundwater extraction starts. To make this condition, the procedure explained below was used.

The procedure started from the steady state nested calculation with no deformation. Based on the calculated pore pressure distribution and density distribution of soils, the vertical effective stress distribution was calculated. Horizontal effective stresses can be calculated from the drained Poisson's ratio. These effective stress components gave the first estimate because shear components were omitted. All stress components, including shear stress, were obtained from iterative calculations under the steady state condition. When the calculated displacements became small enough, calculated stresses were used as initial conditions for the transient analysis.

6.4.4 Results and Discussion

Regional Pattern of Hydraulic Head

Fig. 6.12 shows a comparison between the observed and calculated equipotential maps in layer VI in 1995. The calculated hydraulic heads were generally higher than those observed. However, the location of the centre of the calculated drawdown was consistent with the observed location.

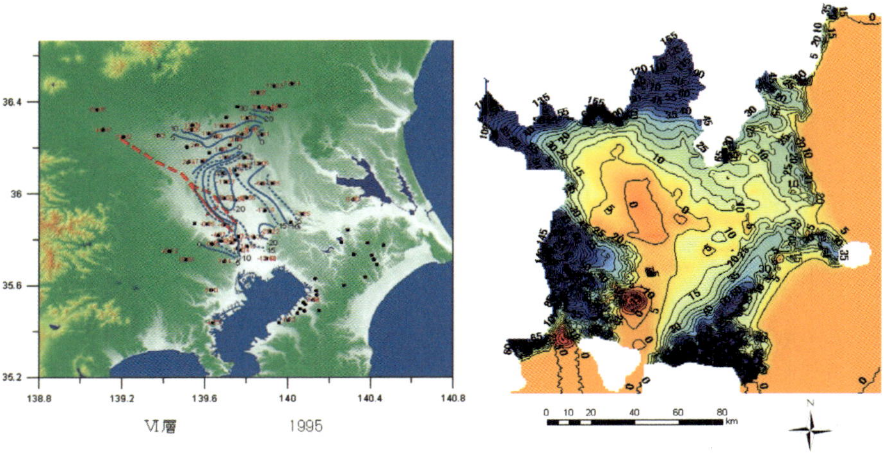

Fig. 6.12. Observed (left) and calculated (right) hydraulic heads in layer VI in 1995

Pore Pressure Profiles in Aquitards

An example of calculated pore pressure profile in aquitards is shown in Fig. 6.13. Calculated results showed a non-linear profile in the aquitard. This indicates that the nested modelling scheme can represent the spatio-temporal distribution of the past maximum effective stress, an important factor in the prediction of land subsidence.

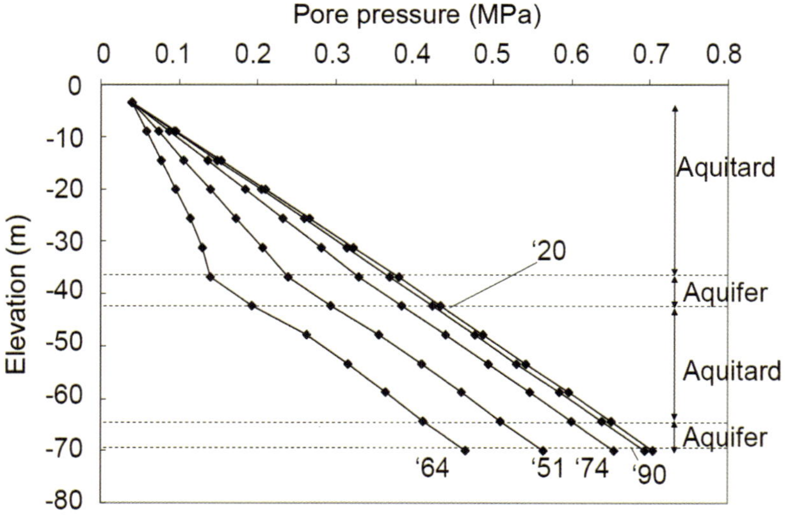

Fig. 6.13. An example of calculated pore pressure profile in aquitards

Displacement of Land Surface

As shown in Fig. 6.3, the land subsidence occurred mainly before the early 1970s and uplift occurred mainly after 1975. Fig. 6.14 shows the calculated and the observed land surface displacement.

The calculated result represented the overall tendency of land subsidence: large in the alluvial lowland and small in the non-alluvial western area. The order of the calculated land subsidence and uplift was also the similar with the observed changes. This implies that the nested modelling scheme worked properly.

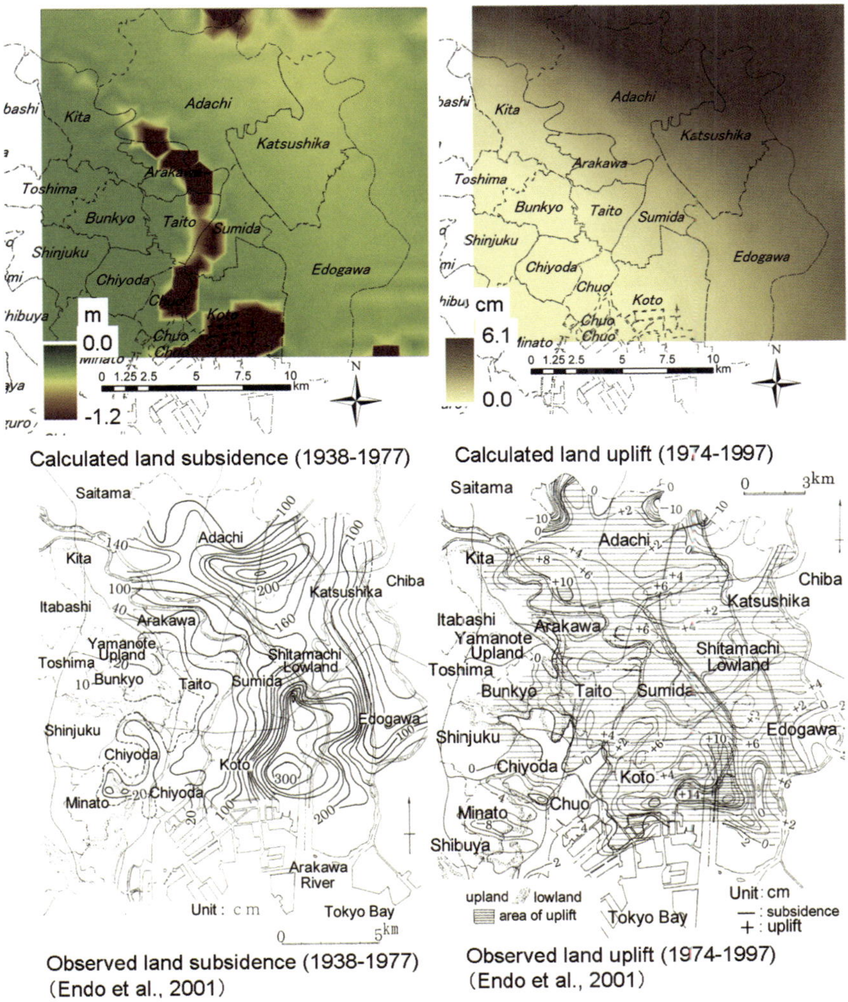

Fig. 6.14. Comparison of calculations and observations regarding land surface displacement in the eastern part of the Central Districts of Tokyo

However, it should be noted that the spatial distribution of calculated subsidence in the lowland area was different from the observed distribution. This is not only caused by the uncertainty of the input data but is also explained by land subsidence in the deep layers which occurred under layer VII due to the production of methane dissolved in groundwater (Endo et al., 2001), which was not taken into account in the simulation.

6.5 Conclusion

A nested modelling scheme was introduced to simulate the coupled land subsidence/groundwater flow processes in Tokyo. Preliminary results suggest that the proposed scheme could reproduce a major tendency in the distribution of hydraulic heads and the overall tendency of surface displacement. Further efforts are necessary to represent quantitatively the actual processes occurring in urban areas and to make this new scheme suitable for analysing practical problems.

References

Aichi M, Tokunaga T (2006) A nested modeling scheme for high-resolution simulation of the aquitard compaction in a regional groundwater extraction field. AGU Fall Meeting, San Francisco, CA, December 11-15, H41B-0399

Aichi M, Hayashi T, Tokunaga T (2007) A nested modeling scheme to integrate regional groundwater flow and local groundwater flow/land subsidence; an application to the Tokyo metropolitan area, Japan. AGU Fall Meeting, Acapulco, Mexico, May 22-25, H53E-03

Biot MA (1941) General theory of three-dimensional consolidation. J Appl Phys 12:155-164

Bishop AW, Blight GE (1961) Some aspects of effective stress in saturated and partly saturated soils. Géotechnique 13:177-196

Endo T, Kawashima S, Kawai M (2001) Historical review of development of land subsidence and its cessation in Shitamachi Lowland, Tokyo (in Japanese). J Japan Soc Eng Geol 42:74-87

Gambolati G, Ricceri G, Bertoni W, Brighenti G, Vuillermin E (1991) Mathematical simulation of the subsidence of Ravenna. Water Resour Res 27:2899-2918

Hirose M, Kawagoe T, Kitani H (2004) A relation of a groundwater level change for construction and underground structures (in Japanese). Annual Study Meeting of Japan Society of Engineering Geology, Niigata, Japan, Oct 28-29, p46

Institute of Civil Engineering, Tokyo Metropolitan Government (1996) On Subsurface structures in Tokyo (23-ku) (in Japanese). Intelligence Board of the Tokyo Metropolitan Government, Tokyo

Isozaki Y, Sugawara T (1986) Groundwater in the Kanto Plain (in Japanese). In: Research group on groundwater use in agriculture (eds) Groundwater in Japan. Chikyu Corp, Tokyo

Japanese Geotechnical Society (2004) Behaviour and evaluation of variably saturated soils. Japanese Geotechnical Society, Tokyo

Kawashima S (2001) Temporal change of groundwater level in Tokyo (in Japanese). The Foundation Engineering & Equipment, Monthly 340:77-83

Kino Y (1965) Subsurface geology and deep groundwater in Kuki city, Saitama, Japan -from Kuki Gs-1 borehole (in Japanese). Monthly report of Japan Geological Survey 16:243-257

Leake SA, Lawson PW, Lilly MR, Claar DV (1998) Assignment of boundary conditions in embedded ground water flow models. Ground Water 36:621-626

Ministry of Land, Infrastructure and Transport (2005) Outline of the meetings on the guidelines to prevent land subsidence (in Japanese) (http://www.mlit.go.jp/kisha/kisha05/03/030330/01.pdf)

Ministry of Land, Infrastructure and Transport (1974-1997) Groundwater level chronology (in Japanese). National Water Well Association of Japan, Tokyo

Mualem Y (1976) A new model predicting the hydraulic conductivity of unsaturated porous media. Water Resour Res 12:513-522

Nacul EC, Aziz K (1990) Efficient use of domain decomposition and local grid refinement in reservoir simulation. The 65th Annual Technical Conference and Exhibition of the SPE, New Orleans, LA, September 23-26, SPE20740

Nakayama T, Ohka H (1992) Classification of the deep underground condition in Yamanote Upland -Engineering geological study in Tokyo (in Japanese). Annual report of Institute of Civil Engineering of Tokyo Metropolitan Government, H4:193-202

Nakayama T, Watanabe M, Tanji K, Morita T (2007) Effect of underground urban structures on eutrophic coastal environment. Science of the Total Environment 373:270-288

National Land Agency (1998) Groundwater map (in Japanese). National Land Agency, Tokyo

Roscoe KH, Burland JB (1968) On the generalised stress-strain behaviour of 'wet' clay. In: Heyman J, Leckie FP (eds) Engineering plasticity. Cambridge Univ Press, New York, pp 535-609

Saitama Prefecture (1992) Report on the construction of the land subsidence observation well at Kasukabe (in Japanese). Oyo Corp, Tokyo

Saitama Prefecture (1993) Report on the construction of the land subsidence observation well at Yashio (in Japanese). Sumiko Consultants, Tokyo

Saito R, Nakayama A (1977) Accuracy of levelling of high precision in Tokyo (in Japanese). Annual Reports of the Institute of Civil Engineering of the Tokyo Metropolitan Government, S52:409-417

Shearer TR (1998) A numerical model to calculate land subsidence, applied at Hangu in China. Engineering Geology 49:85-93

Shindo S. Groundwater in southern part of the Kanto Plain (in Japanese). Soil Mechanics and Foundation Engineering 171:25-35

van Genuchten MTh (1980) A closed-form equation for predicting hydraulic conductivity of unsaturated soils. Soil Sci Soc Am J 44:892-896

Wood DM (1991) Soil behavior and critical state soil mechanics. Cambridge Univ Pr, New York Melbourne

7. Groundwater Contamination in Urban Areas

Keisuke Kuroda and Tetsuo Fukushi

Department of Urban Engineering, the University of Tokyo

7.1 Introduction

Groundwater has been used everywhere in the world for a long time because of its easy accessibility and good quality. In urban areas, groundwater as a source of domestic, commercial and industrial water has greatly contributed to the development of cities. Groundwater in urban areas is sometimes contaminated with multiple contaminants at higher concentrations than in rural areas. For example, one of the most prevalent contaminants in urban groundwater, nitrate, is commonly the product of agricultural runoff due to the use of fertilisers in rural areas. In urban areas, however, fertilisers in agricultural fields are rather minor sources, but leaky sewage, septic tanks, industrial spillages, landfill leachates, and fertilisers used in gardens and parks are other, more common sources of nitrate. Those sources of nitrate can also become sources of other hazardous substances or health-related microorganisms. Other than nitrogen, there are many contaminants in urban groundwater including arsenic, fluoride, heavy metals, and volatile organic carbons. Several kinds of pharmaceuticals, N-nitrosodimethylamine (NDMA), and perfluorinated surfactants (PFSs) can also be detected in urban groundwaters.

In this chapter, various contaminants of urban groundwater and the estimation methods of those sources are described. The method to estimate the contamination sources of nitrate in groundwater is also presented, using Tokyo as a case study.

7.2 Groundwater Contamination in Urban Areas

7.2.1 The Mechanisms of Groundwater Contamination in Urban Areas

Groundwater in urban areas can suffer from many kinds of contamination from both natural and anthropogenic sources. Among contaminants from natural sources, arsenic and fluoride are the most serious. Originally, these elements are contained in rocks, and then released into groundwater by the weathering of these rocks. As for anthropogenic contamination, industrial wastewater, domestic wastewater and leakage of fuels are sources of nitrogen, heavy metals, volatile organic compounds, pathogenic microorganisms and pharmaceuticals. These contaminants are either discharged to the ground by factories, warehouses and households, and then migrate to the subsurface together with rainwater infiltration, or they can be directly discharged into subsurface soil layers through leaking sewer pipes. Pollution by petroleum often takes place in storage tanks installed in subsurface. In suburban areas where agricultural land or livestock farms exist, pollution by nitrogen and pathogenic microorganisms by fertilisers and livestock wastewater can sometimes be found in significant levels in groundwater. The characteristics of each contaminant and its mechanism of pollution are described below.

7.2.2 Arsenic

Arsenic is one of the most prevalent contaminants of groundwater in the world. In Asia, arsenic contamination has brought about serious consequences in Bangladesh, Cambodia, India (West Bengal region), North China, Philippines and Vietnam (UNEP/WHO/UNICEF 2002). In Hanoi city, Vietnam, groundwater, which is a major source of the domestic water supply, is highly contaminated with arsenic (Nga et al. 2003).

Arsenic can cause severe and irreversible health effects, even in very small amounts. Early symptoms can range from the development of dark spots on the skin to a hardening of the skin into nodules, often on the palms and soles. Over time, the symptoms can become more pronounced and in some cases internal organs including the liver, kidneys and lungs can be affected. In the most severe cases, cancer can occur in the skin and internal organs and limbs can be affected by gangrene. The damage due to arsenicosis in humans is irreversible (UNEP/WHO/UNICEF 2002).

The elevated concentration of arsenic occurs due to the solution of arsenic by redox reaction of arsenic-containing minerals. Normally, arsenic is attached with iron oxides and immobilised in aquifers near river deltas. However, when dissolved oxygen in groundwater is depleted and aquifer becomes anaerobic, iron oxides are reduced. Then, arsenic becomes mobile and can be released into groundwater. High concentrations of arsenic are found in groundwater in a variety of environments. This includes both oxidising (under conditions of high pH) and reducing aquifers and in areas affected by geothermal, mining and industrial activities. In areas where geothermal activity is strong, arsenic pollution becomes very prevalent (Smedley and Kinniburgh 2002). (See Chapters 2 and 14)

7.2.3 Fluoride

A high concentration of fluoride in groundwater has caused serious health effects in Eastern Africa, the Middle East, Mexico and Asian countries including India, Thailand and China. For example, a maximum of 28mg/L of fluoride, which is extremely high, was detected in arid areas in North China, where evaporation is large (Liu and Zhu 1991). In the Chiang Mai basin, North China, membrane filtration is used to remove fluoride in groundwater (Matsui et al. 2006). Excessive exposure to fluoride causes dental fluorosis (discoloured, blackened or mottled teeth) and crippling skeletal fluorosis (permanent spine, bone and joint deformations). As with arsenicosis, the damage caused by fluorosis is irreversible (UNEP/WHO/UNICEF. 2002). Fluoride in groundwater is, like arsenic, mainly caused by the weathering of rocks and the elution of minerals. (See Chapters 2 and 15)

7.2.4 Nitrate

Nitrate pollution in groundwater is the most prevalent type of anthropogenic pollution and has been reported in many countries around the world. In the U.S., nitrate concentration exceeded 10mgN/L in 586 sources of groundwater which were used in the water supply system in 1986 (WHO 1998). In France, the United Kingdom and Denmark, nitrate concentration has increased by 30% on average from 1980 to 1990 (Wieringa 1995). In the annual groundwater quality survey in 2005 in Japan, 4.2% of groundwater samples exceeded the national standard of 10mgN/L, which was more frequent than any other parameter designated by the national environmental standard (Ministry of the Environment, Japan 2005).

Elevated nitrate concentrations in drinking water causes a human health hazard such as a disease known as methemoglobinemia or nitrate cyanosis. This disease is caused by the bacterial reduction of nitrate to nitrite in the intestinal tract. The nitrite then enters the blood stream and combines with haemoglobin to form methemoglobin, which reduces the blood's capacity to transport oxygen. In the worst case it leads to death. The reduction to nitrite occurs primarily in infants because their gastric juices are nearly neutral in pH, unlike those of adults, and provide a better environment for nitrogen-reducing bacteria. In its uncomplicated acute form, methemoglobinemia is generally easily diagnosed and its treatment is understood (National Research Council 1972).

In many cases nitrate pollution in groundwater is caused by fertiliser and livestock wastewater in agricultural areas. In Denmark and the Netherlands, nitrate concentration increased at a rate of 0.2 to 1.3mg/L per year (0.060.4mg/L as nitrate nitrogen) due to use of synthetic fertilisers, discharge of livestock waste and land use change (WHO, 1985). However, nitrate concentration in urban aquifers can be similar or even higher than aqifers in surrounding areas. Nitrogen in urban groundwater is caused by various sources including point sources (e.g. landfills, industrial spillages and coal gasification works), multipoint sources (e.g. leaky mains, leaky sewers, and soakaways) and diffuse sources (atmospheric deposition, house building and recreation areas) (Ford and Tellam 1994; Wakida and Lerner 2005).

7.2.5 Volatile Organic Compounds (VOCs)

Groundwater contamination by volatile organic compounds (VOCs) has been found in industrial sites in urban areas. VOCs, in particular the chlorinated solvents such as trichloroethene (TCE), probably represent the most troublesome class of organic contaminants found in urban groundwater to date (Pankow and Cherry 1996). They are dense non-aqueous phase liquids (DNAPLs) and are able to penetrate deep below the water table where they slowly dissolve over decades and form plumes. VOC plumes may extend several kilometres due to their low sorption, limited chemical reaction and varied, but often low, biodegradation potential (Jackson 1998; Rivett et al. 2001). Due to their widespread use, mobility and persistence, chlorinated VOCs are considered to be of the highest risk of all industrial contaminants (Ellis and Rivett 2007).

7.2.6 Heavy Metals

Groundwater contamination by heavy metals such as lead, mercury, cadmium and chromium can be derived from wastewaters used in industry and mining, leachates from landfill sites and the earth's crust. Heavy metals accumulate inside the human body. Once they are taken into the human body, their toxicity persists for a long time and causes significant harm. For example, mercury is a potent neurotoxin contaminant, which passes more easily into the brain of a foetus or a young child. The developing brain is particularly sensitive to methylmercury. Depending on the dose and exposure level, it can interfere with brain development and cause mental retardation, cerebral palsy and seizures. Mercury may cause cancer and damage the stomach, large intestine and lungs. It also may permanently harm unborn children (UNEP/WHO/UNICEF 2002).

7.2.7 Petroleum Oils

The increasing demand for oil products by industrial society is associated with increasing risks of accidental oil spills. Various authors (Schellekens 1974; Schwille 1975; Vanlooke et al. 1975) have reported incidents of accidental oil spills from corroded subsurface tanks, road tankers and pipelines (Kessler and Rubin 1987). Organic contaminations released by such oil spills pose a serious threat to groundwater in urban areas. These contaminants are often released as non-aqueous phase liquids (NAPLs) during spills of gasoline, crude oil, creosote, coal tar or chlorinated solvents. Once released, the liquids seep downward and dissolve into groundwater. In many cases, the impacted groundwater contains a mixture of contaminants, either due to the complexity of the NAPL (e.g. gasoline) or due to co-disposal/co-spillage (e.g. landfill leachates). Many organic contaminants are hazardous to human health and the environment, and therefore threaten our potable water sources and natural ecosystems (Schirmer 2004).

7.2.8 Pathogenic Microorganisms

Groundwater pollution by pathogenic microorganisms takes place with the migration of domestic wastewater into subsurface aquifers. Basically, pathogenic microorganisms diffuse into an aquifer together with groundwater flow, while some of them are retained or killed during diffusion. It was reported that the behaviour of pathogenic microorganisms in aquifers varied between species of marker bacteria and viruses (Meschke and Sobsey 2003). In the UK, sewage-derived bacteria (thermotolerant coliforms, faecal streptococci and sulphite-reducing clostridia) and viruses (enteroviruses, noroviruses and coliphages) were regularly detected at depths of 60m in unconfined sandstone and at a depth of 91m in confined sandstone. The detection of noroviruses and Coxsackievirus B4 in groundwater corresponded with seasonal variations in virus discharge to the sewer system. It was reported that sandstone aquifers, which allow rapid penetration of microbial contaminants, are vulnerable to microbial contamination (Powell et al. 2003).

7.2.9 Other and Emerging Contaminants

Other contaminants in urban groundwater include pesticides, PCBs and dioxins. In addition, pharmaceuticals, Perfluorinated surfactants (PFSs) and NDMA (N-nitrosodimethylamine) are emerging contaminants in groundwater in urban areas.

Pharmaceuticals are derived from domestic wastewater. Some kinds of pharmaceuticals have little sorbability to soils and diffuse into groundwater. Perfluorinated surfactants (PFSs) are contained in wax, detergents and emulsion breakers. They have little biodegradability and accumulate inside human body. N-nitrosodimethylamine (NDMA) is a contaminant of recent concern for drinking water regulators because of its discovery in contaminated groundwater supplies, in reclaimed water and treated drinking water from treatment plants where chlorination is the primary disinfection process. NDMA has been shown to cause cancer in laboratory animals such as rats and mice when the animals are exposed at high levels over their lifetime. It is identified as a carcinogen by California's Proposition 65 and as a probable human carcinogen by the United States Environmental Protection Agency (California Department of Health Services 2000).

7.3 Tools for Analysing Groundwater Contamination in Urban Areas

Groundwater contamination in urban areas is caused by a large variety of contaminant sources and the complexity of recharge pathways. Apart from point source contamination, such as industrial contamination, it is not easy to estimate dominant contaminant sources such as domestic wastewater.

Several tools have been proposed to estimate contaminant sources and pathways of contamination of urban groundwater. They include water balance analysis, mass balance analysis, solute balance analysis, stable isotopes, and markers such as pharmaceuticals and X-ray contrast media. In this section, those methods are introduced with regard to analysing groundwater contamination by domestic wastewater.

7.3.1 Sources and Pathways of Groundwater Contamination by Domestic Wastewater

In urban areas where sewerage coverage is nearly 100%, the primary route of groundwater contamination by domestic wastewater is leaking sewers. Sewage in sewer pipes and groundwater surrounding sewers interact with each other. That is, when the groundwater level is above a sewer pipe, groundwater infiltrates into sewers and dilutes sewerage water; on the other hand, when the groundwater level is below a sewer pipe, sewage exfiltrates into groundwater through joint defects and corrosion, and contaminates ambient groundwater (Fenner 1990; Reynolds 1994). Spillages from on-site wastewater treatment facilities, such as septic tanks, also cause groundwater contamination by domestic wastewater. In the following sections, methods to estimate pollution related to domestic wastewater will be described.

7.3.2 Mass Balance of Contaminants

When the hydrological information of an aquifer and the characteristics of water quality parameters are known, mass balance analysis of contaminants can be an effective tool to estimate the contaminant loads. The sum of estimated contaminant loads from known sources, divided by the flow rate of groundwater, will give an estimated concentration of contaminants. The difference between an estimated concentration and a measured concentration of contaminants may be due to an unknown source or sink of the pollution. Yang et al. (1999) estimated the quantity of leaking sewage into groundwater using a hydrological model and mass balance analysis of chloride, sulphate, and total nitrogen in Nottingham, England. After that, Wakida and Lerner (2005) estimated the pollution source of nitrogen in groundwater from the nitrogen concentration of each possible source in that region.

7.3.3 Stable Isotopes

Stable isotopes are elements that have a common atomic number (proton number) but a different mass number (sum of proton and neutron). Stable isotopes are some of the most popular tracers used for groundwater studies. The chemical behaviour of stable isotopes is the same as normal elements, but their physical and biological behaviour is slightly different due to differences in mass number. The ratio of stable isotopes in the environment differs in each country and in each region. The ratio also differs depending on the origins of the substance, e.g. natural or anthropogenic origins. Thus, these isotopic differences are useful for source estimation. Isotopes used for groundwater study include: oxygen (^{18}O), hydrogen (^{2}H), nitrogen (^{15}N), carbon (^{13}C), chrorine (^{17}Cl), boron (^{11}B), sulphur (^{34}S) and strontium (^{87}Sr).

In general, the isotope ratio is expressed as the ratio of isotopes of the sample to isotopes of its standard sample by the unit of ‰. For example, nitrogen has two stable isotopes: ^{14}N and ^{15}N. Thus, the nitrogen isotopes ratio $\delta^{15}N$ is expressed as follows (Barrett et al. 1999; Bath et al. 2006):

$$\delta^{15}N(‰) = \left[\frac{R_{sample}}{R_{s\tan dard}} - 1\right] \times 1000 \qquad (7.1)$$

where $R = {}^{15}N/({}^{14}N + {}^{15}N)$.

Stable isotopes are useful in the source estimation of nitrogen contamination in groundwater. Nitrogen isotopes differ depending on the origins such as the atmosphere, synthetic fertiliser, soil, and human and animal manures (Fig. 7.1). However, it may be better to combine isotope analysis with other methods to strengthen the estimation because the accuracy of estimation from stable isotopes decreases when more than two sources of nitrogen are mixing or when denitrification occurs. It has been reported that the use of the oxygen stable isotope together with the nitrogen stable isotope is a powerful tool to evaluate denitrification of nitrate (Fukada et al. 2004).

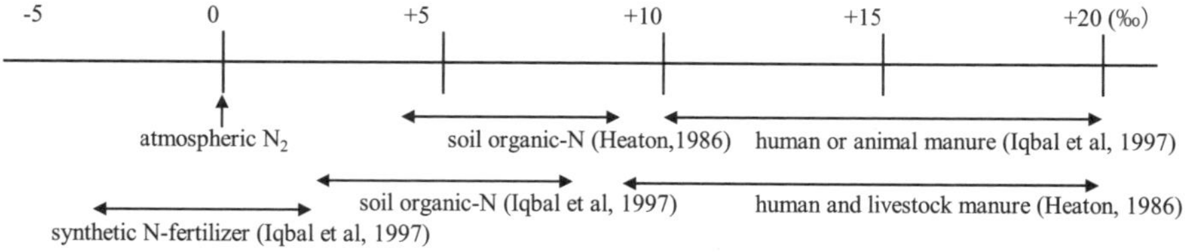

Fig. 7.1. Nitrogen stable isotope ratios in various sources (Heaton 1986; Iqbal et al. 1997)

7.3.4 Marker Species of Contamination

If a chemical substance is to be used as a marker species for the estimation of contamination source in groundwater, it should have the following characteristics: little variance in concentration in the pollution source, little adsorptive ability and little degradability in soil. Some kinds of pharmaceuticals and X-ray contrast media can satisfy these requirements (Fenz et al. 2005; Wolf 2003). Fenz et al. (2005) used the anti-epileptic drug carbamazepine as a marker species of sewerage and estimated the percentage of leaking sewage to be less than 1%.

7.3.5 Statistical Approach

Statistical approaches have also been applied to the analysis of groundwater contamination. A Geographic Information System (GIS) helps to integrate groundwater quality data with geographical data such as land use and elevation. For example, Lee (2003) analyzed nitrate concentration in South Korea using land use information within a certain diameter of the groundwater called the ROI (radius of influence). LaMotte and Greene (2006) attempted to determine the natural or anthropogenic factors that best explain the occurrence and distribution of elevated concentrations of nitrate in shallow groundwater in Maryland and Virginia, USA. They proposed two types of statistical models: logistic regression (logit) and geostatistics. The logit statistical model was developed so that it could predict nitrate concentration at existing wells by relating statistically significant geographic variables to 529 nitrate samples. Geostatistics is the analysis of data that are distributed in space and/or time (Olea 1991). It can be used to predict the spatial patterns of contamination and interpolate values of the attribute of interest at unsampled locations. Geostatistical analysis offers a set of tools and methods aimed at understanding and modelling the spatial variability of water quality data (de Marsily 1986; Dagan 1989; Goovaerts 1997; Deutsch and Journal 1998).

7.4 Nitrogen Contamination of Groundwater in Tokyo

7.4.1 Groundwater Use and Recharge in Tokyo

Groundwater has been used in Tokyo for a long time. With the rapid economic growth of Japan in the 1960s, the groundwater abstraction volume in Tokyo increased significantly, resulting in serious land subsidence. In response, groundwater abstraction was banned in the 1960s and 1970s, which brought about a rising of the groundwater level in the decades since. Recently, the rising groundwater level has brought about various problems on underground infrastructure including railway stations and subway stations. It still requires more time to find out if the problems of rising groundwater can be mitigated in Tokyo by controlled abstraction of groundwater. On the other hand, a growing number of large consumers of water, such as hospitals and department stores, have switched from public water supply to private water supply systems using groundwater in order to cut the cost of water supply and to prepare for the possibility of an emergency.

Groundwater in the urban area has an important role as an alternative source of water in case of emergency. Municipalities have drilled boreholes for emergency use. In Tokyo, there were 621 boreholes as of 2006, of which 570 were for domestic use and 51 were for potable use. Private wells are also registered as emergency wells, which can supply water to the general public in case of earthquake, for example. In 2006, there were 5,267 and 168 registered private wells for domestic and potable use, respectively (Tokyo Metropolitan Government 2006).

As a result of increasing urbanisation in the last couple of decades, artesian wells have reportedly been diminishing because of decreasing rainwater infiltration due to the increase of impermeable area (Tokyo Metropolitan Government 2006). Hence, to enhance infiltration, rainwater infiltration facilities, such as permeable pavements and rainwater infiltration trenches, have been installed in some wards in Tokyo. In such areas, groundwater can be utilised for various purposes including domestic water, environmental restoration and water for landscape. Groundwater can also be used for spraying onto pavements or building walls to mitigate the urban heat island effect.

In this way, the situation surrounding groundwater use in urban areas has considerably changed from that of the 1970s, when the ban on groundwater abstraction was brought into force. Up-to-date management of urban groundwater is now needed with the aim of securing safety of urban infrastructure, maintaining water supplies in an emergency and achieving environmental soundness. In order to support such a policy change, quantitative and qualitative data on urban groundwater are needed. As groundwater abstraction has been banned for thirty years, scientific data on the groundwater quality in Tokyo are insufficient to evaluate groundwater as a water resource. Hence, our investigation was carried out to elucidate the present status of groundwater quality in Tokyo and to estimate the pollution sources and the availability of groundwater for beneficial uses.

7.4.2 Geology of Tokyo

The surface geology, as well as the sampling well distribution, in the central Tokyo district is shown in Fig. 7.2. The central Tokyo district is mainly characterized by the Musashino Terrace in the west and the Tokyo lowland in the east. The elevation of the Musashino Terrace is about 8–50m above mean sea level and that of the Tokyo lowland is less than 8m.

The Musashino Terrace covers most of the western half of the central Tokyo district. A loamy layer of 5–8m thickness covers the surface. Sometimes a mud layer exists in the lower part of the loam layer. Below this loam layer there is a gravel layer, which is the main aquifer for unconfined groundwater in this area.

In the Tokyo lowland that covers the eastern part of Tokyo, there are several alluvial layers. Among these, the Yurakucho Layer is the dominant alluvial layer. The upper part consists of sand layer with 5–10m thickness. This is the main aquifer for unconfined groundwater in this area.

The Pleistocene Edogawa layer lies underneath both the western part and eastern areas of the central Tokyo district. Its thickness gradually changes from the west to the east, and reaches more than 100m in the eastern lowland. The upper part of this layer is often a confined aquifer of good quality.

The Tertiary Kazusa Group lies below the Edogawa layer. The depth of the Tertiary Kazusa Group increases from the southwest of the central Tokyo to the northeast, as the thickness of overlying Edogawa layer gradually increase in this direction.

7.4.3 Sampling Methodology

Groundwater samples were collected from public boreholes, such as emergency wells, and private boreholes, such as those for public baths, between October 2005 and December 2006. In total, 106 groundwater samples were obtained from both unconfined (51 boreholes) and confined (55 boreholes) aquifers. 3 out of 51 unconfined groundwater sources were artesian (see Fig. 7.2). Dissolved oxygen (DO), pH, Oxidation Reduction Potential (ORP), electric conductivity (EC), ammonia nitrogen, ferrous iron and alkalinity were measured immediately after sampling in the field. Major cations, major anions, total organic carbon, turbidity, *E. coli* and total coliforms were measured in the laboratory. Borehole information such as depth was obtained by interviewing the local caretakers.

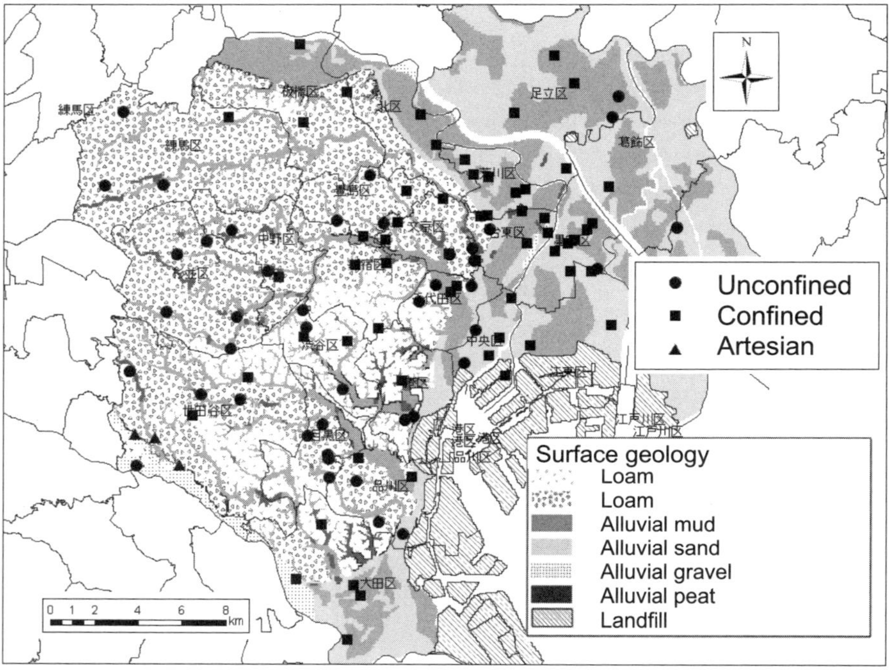

Fig. 7.2. Distribution of sampled boreholes and surface geology

7.4.4 Nitrogen Distribution in the Central Tokyo Area

Nitrogen was the most prevalent contaminant detected from both confined and unconfined aquifers. The distributions of nitrogen concentration are shown in Figs. 7.3 and 7.4 for unconfined and confined aquifers, respectively. 10% (5 out of 51) of the unconfined groundwater sources and 7% (4 out of 55) of confined groundwater sources contained nitrate nitrogen at higher concentration than the Japanese Standard of 10mgN/L.

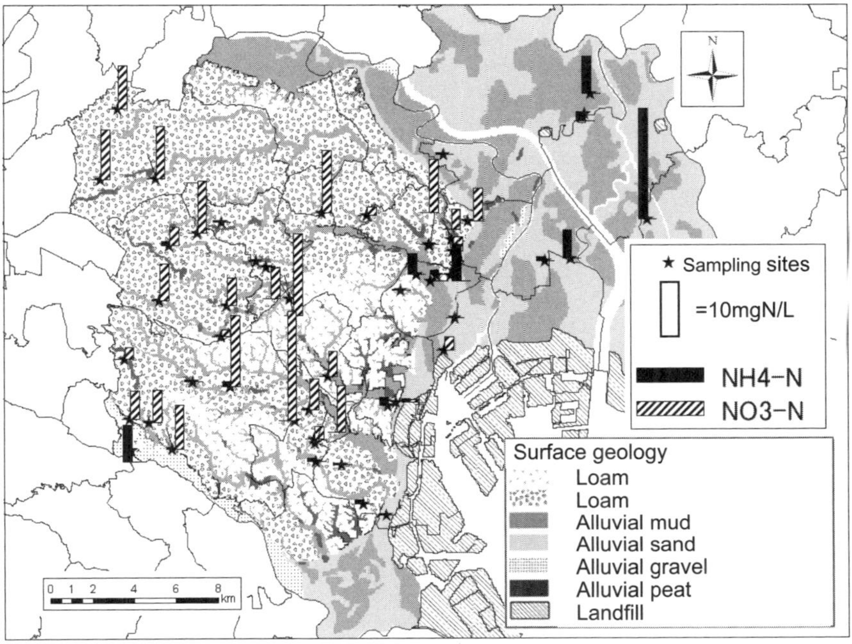

Fig. 7.3. Distribution of nitrogen concentration in unconfined aquifers

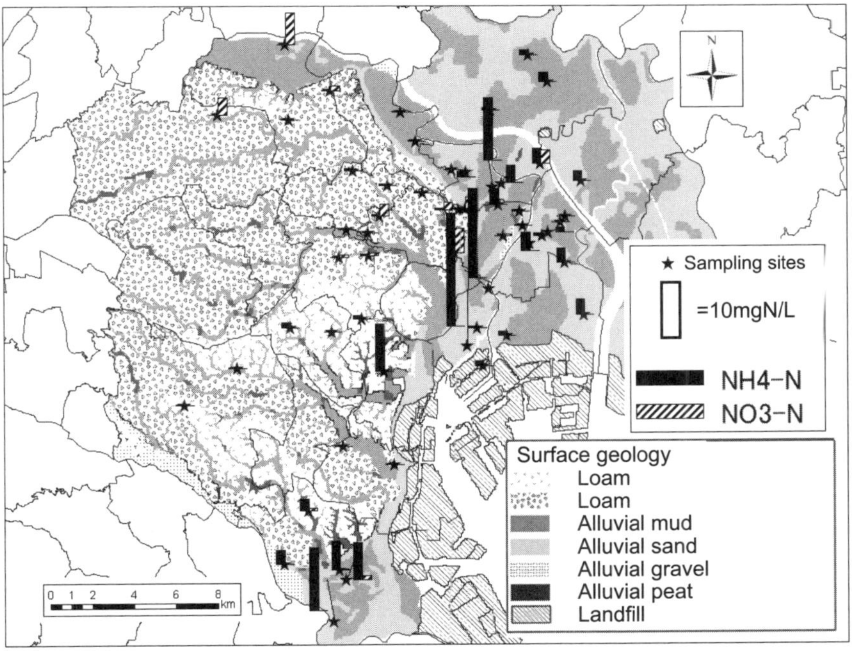

Fig. 7.4. Distribution of nitrogen concentration in confined aquifers

As for unconfined aquifers, geological conditions determined the predominant nitrogen species found in groundwater. In the western terrace area of Tokyo, only nitrate was detected, whereas in the eastern delta area, ammonia nitrogen was predominant in the alluvial sandy aquifer.

The difference of nitrogen species between the western and eastern areas of Tokyo may be due to soil permeability and the existence of oxygen in the aquifers. Fig. 7.5 shows the relationship between dissolved oxygen and nitrogen species in the groundwater. In groundwater with low dissolved oxygen, ammonia nitrogen was dominant, whereas in groundwater with high dissolved oxygen, nitrate nitrogen was the dominant species. In the western terrace of Tokyo, dissolved oxygen was quite high due to the high permeability of the subsurface layer. There, ammonia nitrogen, which may originate from domestic wastewater, fertilisers or soil, is oxidized to nitrate nitrogen. On the other hand, in the eastern lowland of Tokyo, dissolved oxygen was low due to low permeability of the soil. In this area, nitrate, which may originate from leaky mains and precipitation, could be denitrified and only ammonia nitrogen could remain in the groundwater.

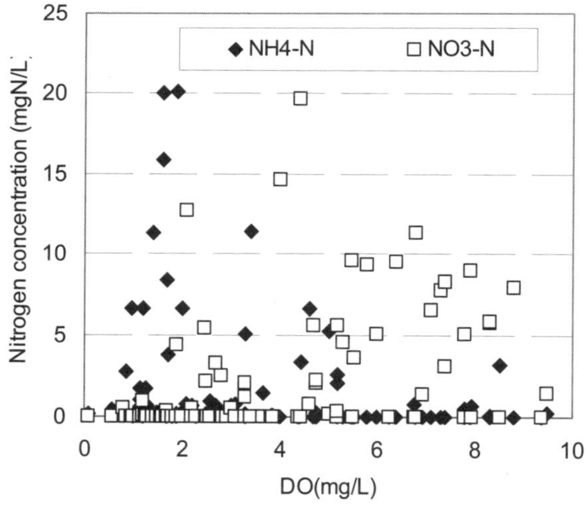

Fig. 7.5. Dissolved oxygen and nitrogen species

7.4.5 Estimation of Nitrogen Contamination Sources by Taking Water and Nitrogen Mass Balance

In unconfined aquifers, infiltration of rainwater is the main source of recharge. Hence, groundwater quality, including nitrogen, is largely affected by the land use in the vicinity of a borehole. This implies that the amount of nitrogen loading into the unconfined aquifer could be estimated from land use data. The estimation procedure had two steps: 1) estimation of groundwater recharge by rainwater infiltration and leakage of water mains, and 2) estimation of nitrogen loads.

Water Balance Study

Water balance at each unconfined groundwater source was calculated by the following equation:

$$\int R_a dt - \int F_{riv}^{(out)} dt - \int E_a dt = \int F_G^{(in)} dt \tag{7.2}$$

where R_a is precipitation, $F_{riv}^{(out)}$ is surface runoff, E_a is evapotranspiration, and $F_G^{(in)}$ is infiltration into the ground. The unit is mm/year for each parameter.

It was assumed that precipitation onto an impermeable area would not infiltrate into the ground and would be directly discharged into sewers or rivers. In other areas, the infiltration was calculated by subtracting surface runoff and evapotranspiration from precipitation.

There are different reports regarding surface runoff on permeable surfaces. After comparing several studies (Ogawa 2000; Miyake 1978; Tsukada 1978), it was assumed for this analysis that 20% of precipitation becomes surface runoff on permeable surfaces in Tokyo, where the topography is rather flat and the surface soil has low permeability (Kuroda et al. 2007).

The land use data within the "recharge area," which is assumed to be within a radius of 300m from a borehole for gravel layers and 50m for sand layers, were obtained by GIS. Then, permeable and impermeable areas in each recharge area were calculated using land use data and their impermeability ratios (Table 7.1).

Table 7.1. Land use classification in the recharge area* and the percentage of impermeable area in each land use

Land use	Percentage to total area	Percentage of impermeable area**
Forest and trees	2.5%	0%
Paddy fields	0.0%	100%
Fields	2.3%	0%
Developed land	0.0%	1%
Open space	3.7%	1%
Industrial area	0.8%	70%
Normal low-rise residential area	23.7%	75%
Dense low-rise residential area	5.6%	90%
Middle- and High-rise residential area	3.5%	75%
Commercial area	15.5%	90%
Roads	15.0%	90%
Parks	7.6%	15%
Other public area	19.5%	85%
Schoolyard with pavement	-	85%
Schoolyard with soil	-	40%
Rivers and lakes	0.1%	100%
Others	0.0%	0%
Ocean	0.0%	100%

*: 50m or 300m radius from boreholes
**: modified from Corbitt (1998)

In Nerima ward, rainwater infiltration facilities, such as permeable pavements and rainwater infiltration trenches, have been installed (Fujita 1986). Based on the researches by Fujita (1986) and Murakami (2005) on the effects of rainwater infiltration facilities for rainwater runoff in the region, it was supposed that 70% of precipitation on impermeable surface infiltrates into the ground through those infiltration facilities.

Evapotranspiration on permeable surfaces was estimated using the Makkink equation as shown below:

$$E_p = a\frac{\Delta}{\Delta+\gamma}\frac{R_s}{\lambda}+b \tag{7.3}$$

where E_p is possible evapotranspiration [mm/day], Δ is the slope of saturated evaporation curve at relevant temperature [hPa/K], γ is the psychrometry constant [hPa/K], R_s is insolation [J/(m^2*day)], λ is vaporization heat [J/Kg]. a [-] and b [mm/day] are constants depending on the region. Adopting the constants in Tokyo, the following approximation was used in calculating Eq. 7.3.

$$\frac{\Delta}{\Delta+\gamma}=\frac{1}{1.05+1.4\,exp(-0.0604T)} \tag{7.4}$$

where T is mean daily temperature [°C].

Evapotranspiration, E_a, was calculated by multiplying possible evapotranspiration, E_p, with the empirical seasonal coefficient (C).

$$E_a = C \cdot E_p \tag{7.5}$$

As for insolation, temperature, and precipitation, daily data between 2004 and 2006 were collected using data from the Japan Meteorological Agency.

The estimated water balance in this study is shown in Table 7.2. Only 9% of precipitation infiltrates into the subsurface, whereas 72% is lost as surface runoff. In the sampling sites, the ratio of impermeable surface to total recharge area is 71.4%. This result is consistent with the past report by Arai (1987), who estimated that the infiltration rate is 140mm/year and surface runoff is 1,020mm/year with an impermeable ratio of 80% in the Musashino Terrace of Tokyo.

Table 7.2. Water balance at sampling points

	Rate (mm/y)	Ratio
Precipitation	1,367	100%
Infiltration	127	9%
Evapotranspiration	261	19%
Surface runoff	979	72%

Nitrogen Balance Study

Nitrogen loads from known sources, i.e. precipitation, leaky mains and fertilisers in fields, were taken into consideration as known sources of nitrogen. Nitrogen leaching into groundwater in agricultural areas is known to be in proportion to the infiltration rate of precipitation (Ogawa 2000; Kunimatsu 1989). Kunimatsu (1989) described the relationship between infiltration rate and nitrogen leaching as shown in Fig. 7.6.

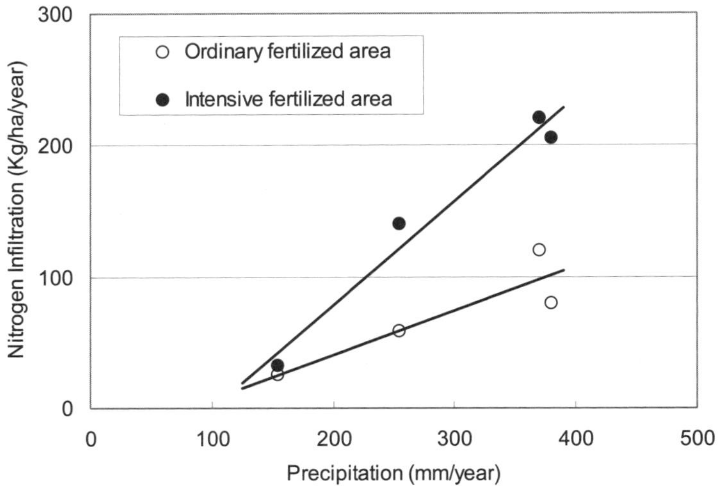

Fig. 7.6. Relationship between infiltration rate and nitrogen leaching into groundwater in agricultural area in Japan (Kunimatsu 1989)

Using this relationship, it was estimated that the nitrogen load into groundwater is 90kgN/ha/year with an infiltration rate of 370mm/year in agricultural fields in this study.

On the other hand, Kunimatsu (1989) proposed the following equation that expresses the relationship between nitrogen in fertilisers and nitrogen leaching:

$$P_N = 0.32 \cdot F_N + 9.6 \tag{7.6}$$

where P_N is leaching nitrogen into groundwater [KgN/ha/year] and F_N is nitrogen in fertilisers [KgN/ha/year]. Using Eq. 7.6, it was estimated that the average nitrogen leaching in the Tokyo metropolitan area in 2004 was 70kgN/ha/year, considering the field area of each crop and the standard amount of fertilisers.

In this study, nitrogen leaching into groundwater is set to be 80kgN/ha/year, comparing the two estimations above.

The nitrogen load from precipitation was calculated using the average nitrogen concentration in precipitation and the infiltration rate, as calculated by the aforementioned method. The nitrogen load from water mains was calculated using the average nitrate and nitrite concentrations in tap water (2.3mgN/L) and the leakage rate of water mains, which were estimated for each ward by proportionately allocating the total amount of leakage according to the population. The leakage rate of Tokyo Metropolitan Water Bureau was 4.4% in 2004, which is equivalent to 86.1mm/year. This is comparable to the rainwater infiltration rate of 127mm/year.

7.4.6 Relationship Between Estimated and Monitored Nitrogen Concentrations

Estimated and monitored nitrogen concentrations

Estimated and measured nitrogen concentrations were compared to evaluate the methodology used for the estimation (Fig. 7.7).

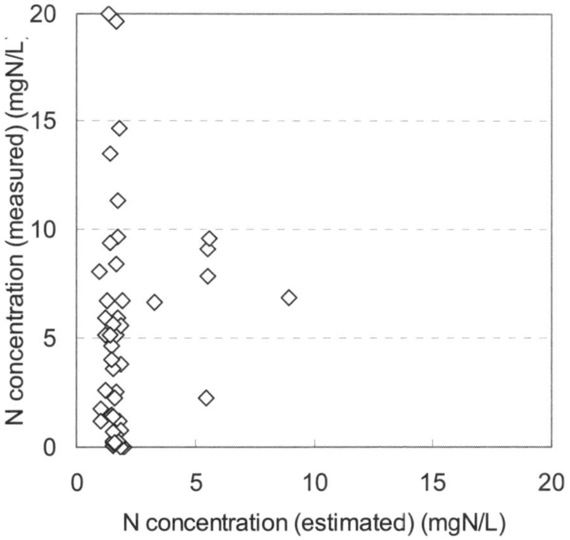

Fig. 7.7. Estimated and measured nitrogen concentrations

At boreholes that have no agricultural uses in their recharge area, the estimated nitrogen concentration ranged between 1 and 2mgN/L, with an average concentration of 1.6mgN/L. At these boreholes, the nitrogen concentration was calculated assuming that groundwater is a mixture of precipitation (0.72mgN/L) and water main leakage (2.3mgN/l). At boreholes that have agricultural fields in the recharge area, nitrogen concentration increased up to 8.9mgN/L. On the other hand, measured nitrogen concentration varied significantly between 0 and 20mgN/L. At 30 boreholes, i.e. 59% of the unconfined boreholes, the measured nitrogen concentration was higher than the estimated concentration.

At these boreholes, it was considered that there were other sources of nitrogen apart from precipitation, main leakage and fertilisers. On the other hand, at boreholes where the measured concentration was lower than the estimated concentration, denitrification could have lowered the nitrate concentration.

Source Estimation of Excessive Nitrogen Load and the Relationship with Land Use

At boreholes where measured nitrogen concentration exceeded estimated concentration, the excess nitrogen loads were compared to land use data in order to estimate the possible sources of excess nitrogen loads. The excessive nitrogen loads per area were calculated and then the correlation coefficient with land use in each recharge area was calculated (Table 7.3).

Table 7.3. Correlation coefficient between excessive nitrogen load and land use in each recharge area

Land use	Ratio to total area	Rank correlation coefficient, r
Forest and trees	2.5%	0.43**
Fields	2.3%	0.06
Developed land	0.0%	0.09
Open space	3.7%	-0.01
Industrial area	0.8%	-0.25
Normal low-rise residential area	23.7%	0.16
Dense low-rise residential area	5.6%	-0.01
Middle- and High-rise residential area	3.5%	0.02
Commercial area	15.5%	-0.25
Roads	15.0%	-0.19
Parks	7.6%	0.13
Other public area	19.5%	0.26
Schoolyard with pavement	-	-0.10
Schoolyard with soil	-	0.36*
Rivers and lakes	0.1%	-0.06
Others	0.0%	-0.15

**: $p<0.01$, *: $p<0.05$

The correlation coefficient with excessive nitrogen load was significantly positive in two cases: 1) forests and trees, and 2) schoolyards with soil. There was no significant correlation between excess nitrogen load and fields, industrial areas, residential areas, commercial areas or roads. The positive correlation between excessive nitrogen load and forested areas is quite interesting because nitrogen is often consumed rather than leached out to groundwater in temperate forest areas (Kunimatsu 1989).

The excess nitrogen load amounted to 6,204kgN/year. When divided by the 24 hectares of fields in the recharge area, excessive nitrogen load corresponds to 262kgN/ha/year. Using Eq. 7.6, nitrogen in fertilisers applied to fields would total 788kgN/ha/year, which is apparently too much. Therefore, there must be other sources of nitrogen than fertilisers. Pollution from point sources would not be probable, because no correlation was observed between excessive nitrogen and industrial areas. Thus, possible sources of this excessive nitrogen are leakage of domestic wastewater or nitrogen originating from the soil itself.

Table 7.4. Nitrogen loads per area

	N loads (KgN/year)	Loaded area (ha)	N load unit (Kg/y/ha)	Loaded area description
Fertiliser	1,897	24	80.0	Fields
Precipitation	715	283	2.5	Permeable areas (including pavement)
Mains leakage	1,747	581	3.0	Residential areas, roads, public areas, etc.
Unknown	6,204	581	10.7	Residential areas, roads, public areas, etc.
Total	10,563	783	13.5	All recharge areas

Nitrogen loads and the loads per area are shown in Table 7.4. If excessive nitrogen loads (unknown loads) are distributed in an urbanised area (residential areas, commercial areas, roads and public areas), the load per area is calculated as 10.7kgN/ha/year. Using the nitrogen discharge unit of 12gN per capita per day, a population density of 135 per square kilometre, and a ratio of residential area to total area of 56.6% in the central Tokyo area, nitrogen loads from residential areas are calculated as 1,045kgN/ha/year. The unknown nitrogen load of 10.7kgN/ha/year corresponds to 1% of the total nitrogen discharge from households. This result suggests that, at most, 1% of domestic wastewater is leaking into the groundwater.

7.5 Conclusion

Groundwater contamination poses serious problems for urban areas. Groundwater contaminants in urban areas include inorganic, organic and microbial contaminants. There are multiple sources and pathways of those contaminants. So far, various tools have been proposed to estimate pollution sources of urban groundwater. They include mass balance analysis, stable isotopes, marker species and statistical approaches. However, there is still considerable uncertainty regarding the estimation of sources. Therefore, these methods need to be further elaborated by systematic data collection and case studies.

A case study was carried out to elucidate nitrogen contamination and to estimate its possible sources in central Tokyo. The mass balances of water nitrogen revealed that the measured nitrogen concentration exceeded the estimated nitrogen concentration in 59% of sampling wells. If the excessive nitrogen load was distributed over the urbanised area, the nitrogen load was calculated to be 10.7kgN/ha/year, which corresponded to 1% of domestic wastewater if it is assumed to be the only source of nitrogen. For a more accurate estimation of nitrogen contamination sources, further investigation needs to be carried out using markers such as stable nitrogen isotopes.

Acknowledgements

The authors gratefully acknowledge the support of the ward offices of Tokyo and local caretakers of the wells for helping our groundwater quality monitoring.

References

Arai T (1987) Hydrological environments in cities (in Japanese). Kyoritsu Shuppan, Tokyo, p 37

Barrett MH, Hiscock KM, Pedley S, Lerner DN, Tellam JH, French MJ (1999) Marker species for identifying urban groundwater recharge sources: A review and case study in Nottingham, UK. Water Res 33(14):3083-3097

Bath A, Richards H, Metcalfe R, McCartney R, Degnan P, Littleboy A (2006) Geochemical indicators of deep groundwater movements at Sellafield, UK. J Geochemical Exploration 90:24-44

California Department of Health Services (2000) California Drinking Water: NDMA-Related Activities

(http://www.dhs.cahwnet.gov/ps/ddwem/chemicals/NDMAindex.htm, downloaded on January 14, 2000)
Corbitt RA (1998) Standard Handbook of Environmental Engineering, 2nd edn. McGraw-Hill, pp 7.21-7.22
Dagan G (1989) Flow and transport in porous formation. Springer, Berlin Heidelberg New York, p 465
Deutsch CV, Journal AG (1998) GSLIB: Geostatistical software library and user's guide. Oxford University Press, New York, p 369
de Marsily G (1986) Quantitative hydrogeology: groundwater hydrology for engineers. Academic, London, p 440
Ellis PA, Rivett MO (2007) Assessing the impact of VOC-contaminated groundwater on surface water at the city scale. J Contaminant Hydrol 91:107-127
Fenner RA (1990) Excluding groundwater into new sewers. J IWEM 4:544-551.
Fenz R, Blaschke AP, Clara M, Kroiss H, Mascher D, Zessner M (2005) Quantification of sewer exfiltration using the anti-epileptic drug carbamazepine as marker species for wastewater. Water Science and Technology 52:209-217
Ford M, Tellam JH (1994) Source, type and extent of inorganic contamination within the Birmingham urban aquifer system, UK. J Hydrol 156:101-135
Fujita S (1986) Experimental sewer system for reduction of urban storm runoff (in Japanese). Environ Eng Res 22:175-185
Fukada T, Hiscock KM, Dennis KF (2004) A dual-isotope approach to the nitrogen hydrochemistry of an urban aquifer. Applied Geochemistry 19:709-719
Goovaerts P (1997) Geostatistics for natural resource evaluation. Oxford University Press, New York, p 483
Heaton THE (1986) Isotopic studies of nitrogen pollution in the hydrosphere and atmosphere: a review. Chem Geol 59:87-102
Iqbal MZ, Krothe NC, Spalding RF (1997) Nitrogen isotope indicators of seasonal source variability to groundwater. Environ Geol 32:210-218
Jackson RE (1998) The migration, dissolution and fate of chlorinated solvents in the urbanized alluvial valleys of the southwestern USA. Hydrogeol J 6(1):144-155
Japanese Ministry of the Environment (2005) Annual results of the investigation of groundwater quality in 2005 (in Japanese) (http://www.env.go.jp/water/report/h18-08/index.html)
Kessler A, Rubin H (1987) Relationships between water infiltration and oil spill migration in sandy soils. J Hydrol 91:187-204
Kuroda K, Fukushi T, Takizawa S, Aichi M, Hayashi T, Tokunaga T (2007) Source estimation of nitrogen contamination in groundwaters in Tokyo Metropolitan Area (in Japanese). Environmental Engineering Research, 44:31-38
Kunimatsu (1989) Modelling analysis of pollutant loads in rivers (in Japanese). Gihodo Shuppan, Tokyo, pp 59-63
LaMotte AE, Greene EA (2006) Spatial analysis of land use and shallow groundwater vulnerability in the watershed adjacent to Assateague Island National Seashore, Maryland and Virginia, USA. Environmental Geology 52(7):1413-1421

Lee SM, Min KD, Woo NC, Kim YJ, Ahn CH (2003) Statistical models for the assessment of nitrate contamination in urban groundwater using GIS. Environmental Geology 44:210-221

Liu Y, Zhu WH (1991) Environmental characteristics of regional groundwater in relation to fluoride poisoning in North China. Environmental Geology 18:3-10

Luebker DJ, Case MT, York RG, Moore JA, Hansen KJ, Butenhoff JL (2005) Two-generation reproduction and cross-foster studies of perfluorooctanesulfonate (PFOS) in rats. Toxicology 215:126-148

Matsui Y, Takeda T, Takizawa S, Wongrueng A, Wattanachira S (2006) Evaluation of nanofiltration process for fluoride removal from groundwater in the Chiang Mai basin (in Japanese). J JSCE G 62:403-414

Meschke JS, Sobsey MD (2003) Comparative reduction of Norwalk virus, poliovirus type 1, F+ RNA coliphage MS2 and Escherichia coli in miniature soil columns. Water Science and Technology 47:85-90

Miyake N (1978) Runoff characteristics of a small watershed in a hill. In: Ichikawa M, Kayane I (eds) Water balance in Japan (in Japanese). Kokin Shoin, Tokyo, pp 77-88

Murakami M (2005) Adsorption and desorption of heavy metals in infiltration facilities receiving urban runoff, Ph D Dissertation (in Japanese), Graduate School of Engineering, the University of Tokyo.

National Research Council (1972) Accumulation of nitrate. National Academy of Sciences, Washington, DC

Nga TTV, Inoue M, Khatiwada NR, Takizawa S (2003) Heavy metal tracers for the analysis of groundwater contamination: case study in Hanoi city. Water Science and Technology 3:343-350

Ogawa Y (2000) Nitrate pollution in groundwater and change of farming methods (in Japanese). Noubunkyo, Tokyo, pp 68-71

Olea R (1991) Geostatistical glossary and multilingual dictionary. Oxford University Press, New York

Pankow JF, Cherry JA (1996) Dense Chlorinated solvents and other DNAPLs in groundwater. Waterloo Press, Portland Oregon

Powell KL, Taylor RG, Cronin AA, Barrett MH, Pedley S, Sellwood J, Trowsdale SA, Lerner DN (2003) Microbial contamination of two urban sandstone aquifers in the UK. Water Research 37:33-352

Reynolds JH (1994) Environmental protection-a pipe dream or reality? In: Proceedings of the institution of civil engineers-municipal engineer 103:121-128

Rivett MO, Feenstra S, Cherry JA (2001) A controlled field experiment on groundwater contamination by a multicomponent DNAPL: creation of the emplaced-source and overview of dissolved plume development. J Contam Hydrol 49:111-149

Schellekens GAP (1974) Olieverontreiniging pompstation Geulhen. H20 7:140-143

Schirmer M, Butler BJ (2004) Transport behaviour and natural attenuation of organic contaminants at spill sites. Toxicology 205:173-179

Schwille F (1975) Groundwater pollution by mineral oil products. IAHS-AISH, Publ No 103.

Smedley PL, Kinniburgh DG (2002) A review of the source behaviour and distribution of arsenic in natural waters. Applied Geochemistry 17:517-568

Tokyo Metropolitan Government (2006) Current status of groundwater abstraction in Tokyo Metropolitan (in Japanese)

Tsukada K (1978) Hydrological characteristics of the Shirasu plateau. In: Ichikawa M, Kayane I (eds) Water balance in Japan (in Japanese). Kokin Shoin, Tokyo, pp 88-103

UNEP/WHO/UNICEF (2002) Children in the new millennium. United Nations Environmental Programme (http://www.unep.org/ceh/main01.html)

Wakida FT, Lerner DN (2005) Non-agricultural sources of groundwater nitrate: a review and case study. Water Res 39:3-16

WHO (1985) Guidelines for the study of dietary intake of chemical contaminants, WHO Offset Publication No 87

WHO (1998) Nitrate and nitrite in drinking water http://www.who.int/water_sanitation_health/dwq/chemicals/Nitratenitrite.pdf

Wieringa K (1995) Environment in the European Union 1995. European Environment Agency, Luxembourg

Wolf L, Eiswirth M, Hotzl H (2003) Assessing sewer-groundwater interaction at the city scale based on individual sewer defects and marker species distributions, ICGGE Bled 2003

Yang Y, Lerner DN, Barrett MH, Tellam JH (1999) Quantification of groundwater recharge in the city of Nottingham, UK. Environmental Geology 38:183-198

Vanlooke R, DeBorger R, Voets JP, Verstraete W (1975) Soil and groundwater contamination by oil spills-problems and remedies. Int J Environ Stud 3:99-111

8. Detection of Microbial Contamination in Groundwater

Hiroyuki Katayama

Department of Urban Engineering, the University of Tokyo

8.1 Introduction

The water quality of groundwater is normally better and more stable than surface water, and it is sometimes used without treatment as a drinking water source. The water quality of groundwater is dependent on the local geological settings, land use and human activities. Pertaining to the anthropogenic contamination of groundwater, it is important to protect wellheads by regulating land use and limiting or eliminating specific activities that may contaminate groundwater.

However, due to high levels of fluctuation in microbial concentrations in groundwater, as well as due to the serious health impacts of outbreaks of infectious diseases, it is sometimes very difficult or dangerous to rely solely on protection of the source water. Beyond protection strategies, it is important sometimes to disinfect or treat groundwater. In Japan, chlorination is mandatory by law in all water supply systems. However, in European countries and in parts of the U.S., there is a tradition of drinking non-treated groundwater. Hence, due to different regulations and people's customs, there are different strategies for the management of microbial risks in groundwater in each country. In this chapter, the regulations of microbial contamination, cases of pathogenic outbreak, and the latest methods for microbial monitoring, as well as the sources, fate and transport of infectious microbes, are presented.

8.2 Pathogen control strategy

8.2.1 Regulations in US and Japan

In the U.S. and some European countries where chlorination is not mandatory in water supply systems, pathogenic bacteria are a great concern. Enteric viruses are also problematic in both countries. Due to the different barriers against microbial contamination in each water supply system, the microbial agents of concern are also different among countries.

In the United States, more than 50 percent of the people receive tap water from public water systems that use groundwater sources, and approximately 18 percent of the people drink groundwater without disinfection. The United States Environmental Protection Agency (EPA) published the Ground Water Rule in the Federal Register on November 8, 2006. The purpose of the rule is to provide increased protection against microbial pathogens in public water systems that use groundwater sources. The EPA is particularly concerned about groundwater systems that are susceptible to faecal contamination that may include disease-causing pathogens. The Ground Water Rule made it mandatory to carry out a sanitary survey for all groundwater supply systems, wherein source water is tested for indicators to assess the possibility of microbial contamination; if there is risk of infection, the public water authorities must install a disinfection or treatment system to decrease the contaminant level by 4 log.

In Japan, chlorination is mandated for all water supply systems. Most microbial infections via water supplies using groundwater have been caused by the deficiency of chlorination. In addition to the problem of how to improve chlorination practice, public health departments and water utilities are generally only concerned about *Cryptosporidium*, which has high tolerance against chlorination. After an outbreak of gastrointestinal disease due to *Cryptosporidium* in Japan in 1996, the Ministry of Health, Labour and Welfare set a tentative guideline against *Cryptosporidium* contamination in water. If any source water is found to be susceptible to contamination of *Cryptosporidium*, the guideline requires the water utility in charge of water treatment to remove turbidity to less than 0.1 mg/L. In 2007, UV irradiation was approved of as an effective treatment against *Cryptosporidium* in water supply systems using groundwater sources.

8.2.2 Microbial agents in Groundwater

Pathogenic microorganisms in groundwater are classified into two categories. One category includes bacteria, protozoa and viruses that are transmitted via the faecal-oral infectious route, and the other consists of bacteria and amoeba that grow in the environment outside of the host bodies.

For the pathogens of the faecal-oral route of transmission, the microorganisms replicate in a host, which then sheds the pathogens in faecal material that then enters a water body used for drinking water. Finally, another host ingests the water containing the pathogenic organisms and becomes infected. Pathogens may have other routes of infection such as food contamination. Waterborne pathogenic enteric viruses include polioviruses, Coxsackieviruses, enteroviruses, adenoviruses, rotavirus, hepatitis A virus, hepatitis E virus, astroviruses, noroviruses and sapoviruses. Certain species of *Shigella, Vibrio, Salmonella, Yersinia, Campylobacter, E. coli,* and *Arcobacter* bacteria also have been detected in faecal samples.

Legionella, *Mycobacterium* and some other bacteria may be detected in distribution systems and/or in groundwater. However, the infectivity of these bacteria occurring in natural sources is not easily determined. The extent and significance of these infectious microbes are still under investigation.

8.2.3 Faecal Contamination Indicators

Faecal contamination in groundwater can be estimated in two ways: 1) by monitoring pathogens directly, or 2) by monitoring non-pathogenic faecal indicators. The presence of indicator microbes implies faecal contamination of the water and therefore the potential presence of pathogens. However, pathogens are often difficult to culture from environmental samples, and because their concentrations in source waters are usually very low, filtering and analysis of a large volume of samples is required. On the contrary, a wide range of analytical methods for indicator organisms is available. These methods are more commonly used and are significantly less expensive than the methods for monitoring each of the pathogens.

As indicators for pathogenic bacteria of faecal-oral infectious route, *Escherichia coli*, faecal coliforms (95% of which are *E. coli*, incubated at 44.5±0.5°C) and faecal streptococci are recommended by the World Health Organization as principal indicators. However, there are many reports that these indicators cannot represent the presence/absence of enteric viruses. A variety of indicator microorganisms for enteric viruses have been proposed in scientific literature (IAWPRC Study Group, 1991). Bacterial and viral indicators are shown in Table 8.1.

Table 8.1. Microbial indicator organisms

Waterborne Indicator Bacteria	Waterborne Indicator Viruses
Heterotrophic Plate Count Bacteria (HPC)	Bacteriophage
	Bacteroides phage
Total Coliform, Faecal Coliform	Coliphage
Escherichia coli (*E. coli*)	FRNA phage
Faecal Streptococci	Somatic coliphage
Enterococci	
Anaerobic Spores	
Clostridium perfringens	

Among the viral indicators in Table 8.1, somatic coliphages are more likely to be present in a given sample, whereas the Bacteroides phages and FRNA phages tend to be less abundant. Bacteroides phages are recommended because the host bacteria are rather animal species specific. This enables differentiation of the faecal source. FRNA coliphages are similar to human enteric viruses in shape, size, morphology and composition. FRNA phages are expected to grow only in warm-blooded animal viscera because *E. coli* produces F pili only at the temperature range of 30-45°C, which is needed for FRNA phages to cause infection.

8.3 Source of Faecal Contamination of Groundwater

8.3.1 Generation and Sources of Faecal Contaminants

Faecal contamination of groundwater can be associated with human waste, animal waste and sources contaminated by such wastes (e.g., stormwater runoff). All types of sewage and solid waste may carry enteric pathogens. In general, both animal and human sources generate bacterial pathogens. Viral pathogens that infect humans, in contrast, are typically generated only by human sources. Viruses that infect the animal gut normally do not cause illness in humans. Because enteric pathogens are excreted by both humans and animals, the following discussions include both pathogen types associated with humans and animals. It is assumed that local faecal contamination sources may contain any or most of the faecal indicator organisms or waterborne pathogens. Potential sources of human faecal contamination that can infiltrate and contaminate groundwater include: 1) improperly located, designed, constructed, operated, and/or maintained septic systems, 2) unlined sewage ponds, 3) inadequately treated sewage treatment plant effluents used to irrigate crop land, 4) unlined or leaking sewage treatment plant lagoons 5) land application of improperly treated biosolids (sewage treatment plant residue), 6) ruptured, leaking, or overflowing sewer pipes, and 7) combined sewer overflow (CSO), that is, untreated sewage mixed with stormwater (EPA 2006).

8.3.2 On-Site Wastewater Systems (Septic Tanks and Drainfields)

The major sources of faecal contamination of groundwater are considered to be septic tanks or sewerage systems. The septic tank is designed to retain septage to remove solids either by settling or by liquefaction. This system is most common in developing countries, as well as in rural areas of developed countries. When properly located, designed, constructed, operated, and maintained, on-site wastewater treatment systems are effective in preventing faecal contamination of water supply wells because microbial contaminants become inactivated or are removed by the soil. Overflow or seepage from improperly located, designed, constructed, operated, or maintained septic tanks, drainfields, and cesspools are among the most frequently reported sources of groundwater contamination linked to disease outbreaks in the United States (Yates et al. 1985).

8.3.3 Wastewater Collection Systems

A wastewater collection system consists of a network of pipes that collect sewage and transfer it to a wastewater treatment plant. Sewer lines are constructed from pipe segments that may leak at connection points or may leak simply as a result of pipe deterioration with age. When leaky sewage pipes are in close proximity to the wells, source water contamination may result. Wastewater contains faecal bacteria and enteric viruses. Wastewater differs from septage because the much greater number of people served by the sewer system make it more likely that a wider variety and perhaps higher density of pathogens (and indicator organisms) will be present in the wastewater. In most of the studies, viral concentration in sewer mains enumerated by culturing methods ranged up to 400 plaque forming unit (PFU)/L (EPA 2006). Recently, more than 200 reverse-transcriptase (RT) polymerase chain reaction (PCR) unit/ml of norovirus genomes were found in a sewer system by using the PCR technique where levels of faecal coliforms were 8.4×10^4 colony forming unit (CFU)/ml (Katayama et al., in press).

8.3.4 Land Application of Treated Wastewater and Biosolids

Land application of secondary effluent and biosolids from sewage treatment plants and septic tank residual material (sludge) is a possible disposal/reuse method. If it is done improperly, however, these practices may cause microbial groundwater contamination. Most of the pathogens in raw sewage are adhered to activated sludge in biological treatment processes. Excess sludge is the residue generated by the treatment of domestic wastewater and is sometimes used as a soil conditioner and fertilizer. One case of disease outbreak at a resort in northern Arizona in the U.S. was linked to wastewater effluent or biosolids disposal that contaminated drinking water supply wells (Lawson et al. 1991).

8.3.5 Stormwater Infiltration

A study by Rose et al. (2000) evaluated waterborne disease outbreak data from 1971 through 1994 for groundwater and surface water in 2105 U.S. watersheds. Between 20 and 40 percent of the outbreaks were associated with extreme precipitation. This relationship was statistically significant for both surface and groundwater, though it was more apparent with surface water outbreaks. Therefore, stormwater run-off during extreme storm events may be directly linked to waterborne disease outbreaks.

In urban settings, stormwater may infiltrate into the subsurface from retention ponds. Leaking sewer mains are other possible sources of groundwater contamination by stormwater. Contaminated stormwater may also enter into aquifers via improperly constructed or maintained wells, as well as from abandoned wells. Municipal solid waste dumpsites, animal waste, effluents from slaughter houses, and graveyards are other potential sources of groundwater contamination (EPA 2006).

In some cities in developing countries, flood events recur every year due to poor drainage systems. The wells in many poor communities in Jakarta, Indonesia are often contaminated with viruses due to inundation with contaminated floodwater. In a study conducted by Phanuwan et al. (2006) in a slum area in Jakarta, it was revealed that all the groundwater sources found positive for viruses also contained *E. Coli*; hence, the presence of *E. Coli* is a conservative indicator of the occurrence of pathogenic viruses in groundwater contaminated with floodwater.

8.4 Fate and Transport of Pathogens

The fate and transport of bacteria and viruses in the subsurface environment determines the risk of human exposure to these microbial contaminants. The survival rates of microorganisms in the subsurface environment vary significantly among different microbial species, thus influencing the potential risk of human exposure. Transportation of microbes is affected by both microbial characteristics and hydrogeology of the contaminated sites. In principle, subsurface layers attenuate microbes by retaining them in the soil matrix, but the amounts of microbes retained in the soil matrix are dependent on microbe-soil interaction. In this section, factors affecting the fate and transport of viruses are described.

8.4.1 Factors Affecting the Fate and Transport of Viruses in the Subsurface

The fate and transport of viruses in subsurface layers is influenced by numerous factors including hydrogeological conditions, soil properties, pH, temperature, iron content, inorganic ions/salt species, and microbial activity (EPA, 2006).

Hydrogeological conditions

High precipitation may cause high water tables in an unconfined aquifer. As a result, the distance and time required to enter the shallow aquifer may be reduced for viruses. Other factors affecting virus/bacteria fate and transport are complex, interrelated, and poorly understood.

Soil properties

In the laboratory, soil columns are typically used to study and observe virus/bacteria transport under controlled conditions, though the applicability of soil column studies to natural hydrogeologic conditions is uncertain. Typically, soil column studies are conducted to understand transportation of the virus/bacteria over a short time period to minimize virus die-off during the experiment. Virus/bacteria survival assessment is typically conducted in the laboratory using natural groundwater to represent actual water chemistry conditions better. These survival experiments are designed to measure virus/bacteria survival in static groundwater and do not typically consider the effects of their interactions with solid material.

Water pH

Generally, viruses tend to adsorb on the soil at low pH and desorb at high pH since they tend to be slightly negatively charged under ambient conditions (Gerba 1984). Higher ionic strength also increases virus adsorption, whereas soluble organic matter competes with viruses for adsorption on the soil.

Temperature

Temperature is the most important factor on virus/bacteria survival. In general, viruses/bacteria survive longer at lower temperatures. Nasser and Oman (1999) studied temperature effects on virus inactivation in groundwater, wastewater, and buffered saline water. FRNA phages persisted in groundwater at 4 and 37°C, whereas *E. coli* was inactivated in the same water. Therefore, male-specific bacteriophages were found to be a better indicator of infectious viruses in groundwater than *E. coli*, especially at lower temperatures.

Inorganic ions and organic compounds

It has long been assumed that the electric charge on the viral outer coat governed the subsurface mobility of viruses, provided that all other factors were equal. The effect of various electrolytes on the adsorption of poliovirus was measured (Lance and Gerba 1984). Viruses suspended in deionised water moved much farther through the soil than those suspended in tap water. Salts contained in tap water and sewage water promoted virus adsorption in the soil matrix, but evidently the organic compounds in sewage lowered adsorption. Among the various species of cations, Al^{3+} was found to be most effective to retain viruses in the soil.

8.4.2 Survival of Bacteria in Subsurface

The transport and persistence of a bacterium in the subsurface is affected by most of the same parameters governing virus transport. In comparison to bacteria, viruses are longer lived in groundwater and saturated and unsaturated soils than bacteria.

Coliforms are measured in most of the monitoring practices, as an indicator of faecal contamination. Bacteriophages are more relevant as an indicator of enteric viruses. Chemical tracers have also been proposed as indicators, among which carbamazepine is promising since it is hardly removed by biological wastewater treatment nor adsorbed to soil (Fenz et al. 2005).

8.4.3 Well structure and conditions near the wellhead

Microbial contamination of groundwater can be caused by inappropriate location and/or poor construction of the wells. A water supply well should be located away from possible contaminant sources. In addition, an improperly constructed water supply well or broken well may bring about seepage of surface runoff into the wells. Shallow wells are more vulnerable to these types of contamination than deep wells.

Construction work and new groundwater source development sometimes alter the flow of groundwater, which results in the emergence of new contaminants in groundwater. The flow paths of contaminants in some hydrogeological settings, such as karst, are especially complicated, and thus the installation of wells in such geological settings requires special attentione (EPA, 2006).

8.5 Outbreaks of Waterborne Disease by Drinking Groundwater

It is very difficult to assess the microbial infectious risks via water from epidemiological surveillances. The number of identified and reported outbreaks in government statistics is the tip of the iceberg, representing only a small fraction of actual waterborne disease outbreaks, both for surface water and groundwater. In practice, most waterborne outbreaks in water supply systems are not recognised by officials until a sizable proportion of the population gets sick. Some of the reasons for the lack of recognition and reporting of outbreaks are: 1) poor or no surveillance systems, 2) difficulty in recognizing the outbreak, 3) difficulty in identifying the infectious route, 4) not all the sick seeking medical assistance, and 5) difficulty in diagnosis to identify the infectious agents.

For all of the reasons mentioned above, the aggregate outbreak data are probably not representative of the actual occurrence of microbiological contamination. Endemic waterborne diseases may not be counted in official reports of outbreaks. Therefore, in the following section, official and unofficial reports on waterborne diseases are summarised.

8.5.1 Outbreak studies in US

One useful data source is the waterborne-disease outbreak case study summarised by the United States Centers for Disease Control and Prevention (Liang et al. 2006). The outbreak data from groundwater systems collected between 1991 and 2000 indicate that the major problem in groundwater systems was source water contamination, which caused more than 79% of the outbreaks among people using groundwater systems. These data, however, do not include outbreaks caused by protozoa such as *Giardia* and *Cryptosporidium*. In most of the outbreaks caused by protozoa, the source waters are identified after the outbreaks as "groundwater under the direct influence of surface water." In the U.S., however, groundwater sources directly influenced by surface water are not officially classified as groundwater. Of the 68 outbreaks related with groundwater systems, 14 were associated with specific bacterial pathogens. Etiologic agents were not identified in 39 out of the 68 outbreaks. The diversity of pathogenic agents in groundwater was greater than that of treated surface water, which was dominated by protozoa and viruses. This result indicates that pathogenic bacteria can be more easily removed by surface water treatment than viruses and protozoa. Pathogenic bacteria, however, can exist in groundwater if the source protection is not adequate.

The U.S. EPA estimated that a majority of the outbreaks where the agent was not determined were virus-caused, given the fact that it is generally more difficult to analyze for viral pathogens than for bacterial pathogens. Recent advancements in analytical methods for viruses suggested that noroviruses are responsible for a large percentage of outbreaks, and the number of norovirus outbreaks is possibly increasing.

Among the waterborne disease outbreaks in 2003 and in 2004 associated with drinking water in the U.S., seven cases were reported in areas using groundwater sources and only one case was reported in an area served with treated surface water. Among the seven cases where groundwater was the source, three were from untreated groundwater and four were associated with treatment deficiencies (Liang et al., 2006). Hence, groundwater may carry a higher risk than treated surface water.

8.5.2 Outbreaks Related to Groundwater in Japan

In March 2003, there was an outbreak among the customers of a karaoke bar in Niigata Prefecture, which used a private groundwater well. In this outbreak, 151 cases were reported out of 227 total customers (Tamura et al. 2005). The well was a shallow well, and was only 12m away from the septic tank. The chlorination tank was empty at the time of inspection after the outbreak occurred; hence it was estimated that the outbreak took place because no chlorine was added into the groundwater.

Norovirus outbreak via a water supply system occurred in Akita Prefecture in March 2005 (Saito et al. 2005). There were 29 cases from 14 families in a small village, out of a total of 258 people in 94 families. 16 patients were tested for norovirus GII, and 11 turned out to be positive. An epidemiological survey pointed towards water as the source of the virus. The community water supply system was tested for norovirus and was found to be positive. The groundwater well was 6m deep and located 2m from a river, which was influenced by wastewater of the residents. The community water supply system was equipped with a chlorination unit but it was not properly operated during the outbreak. The influence of surface water was also suspected as a possible microbial contamination source.

The cause of these outbreaks can be attributed to the deficiency of chlorination. Considering that there had been no such outbreak cases when chlorination units were operated, chlorination seems to be effective at reducing the risk of norovirus infection.

8.6 Monitoring of pathogens and indicator microbes

8.6.1 Monitoring Strategy

The Ground Water Rule in the U.S. recommends measurement of indicator bacteria rather than detection of specific viruses for regular monitoring of groundwater. This is mainly due to fluctuation of virus concentration, but also is due to the lack of detection methods for a broad type of infectious viruses. Faecal indicator analysis is more reliable than pathogen analysis to predict a potential risk of infection.

8.6.2 Total Coliforms and E. coli

The m-ColiBlue24 (Millipore) test is a membrane filtration method that simultaneously determines the presence or absence of total coliforms and *E. coli*. TC colonies growing on the lactose-based medium are identified by a non-selective dye, 2,3,5-triphenoltetrazolium chloride (TCC), which produces red colonies. The selective identification of *E. coli* is based on the detection of the beta-glucuronidase enzyme. The test medium includes the chromogen 5-bromo-4-chloro-3-indoxyl-beta-D-glucuroindice (BCIG) which is hydrolysed by the enzyme, releasing an insoluble indoxyl salt that produces blue colonies (*E. coli*). The test involves filtering a 100mL water sample through a membrane filter, which is associated with an absorbent pad saturated with m-ColiBlue24 broth. After incubation at 35° C for 22±2 hours, the membrane is examined for colony growth.

Chromocult agar (Merck) is also a good medium to determine both total coliforms and *E. coli* from the environment. Since this method is not linked with filtration, its application to groundwater samples may be limited due to sample volume limitation.

8.6.3 Coliphages

To determine FRNA phages and somatic coliphages in water, the single agar layer procedure is widely used. This method is a quantitative procedure that uses a single agar layer technique for determining coliphage density in groundwater samples. It requires at least a 100mL sample to which magnesium chloride, the host bacteria, and an equal volume of double-strength molten tryptic soy agar medium are added. The host bacteria are log-phase WG49 for FRNA phages and WG5 (*E coli* C) for somatic coliphages. *Salmonella* Typhimurium (WG49) is not susceptible to somatic coliphages that tend to interfere with detection of FRNA phages when *E. coli* hosts are used. The total volume of the mixture is poured onto 5 to 10 plates and incubated overnight. All plates from a single sample are examined for plaque formation (circular lysis zones of bacterial host lawn clearing where each plaque represents one bacterium initially infected by a phage). The quantity of coliphages in a sample is expressed as plaque forming units (PFU/100mL).

8.6.4 Enteric viruses

Concentration method

Concentrating viruses in water by adsorption to and subsequent elution from a positively charged membrane (Sobsey and Jones 1979) is currently considered to be one of the most useful methods for contaminant analysis (American Public Health Association 1995). This method has been applied to river water, lake water, secondarily treated sewage, tap water and groundwater (Abbaszadegan et al. 1993).

In conventional concentrating methods, including the above method, beef extract was often used as an eluate from various adsorbents. However, contents of beef extract are suspected to have some inhibitory effects on PCR detection for viruses, especially after re-concentration (Abbaszadegan et al. 1993; Schwab et al. 1996).

There have been many studies on virus concentration and purification methods from water samples. In the 1990s, when molecular techniques were developed and applied to the determination of viruses, many studies focused on purification aimed at removing PCR inhibitors from eluates obtained in a virus concentration method, such as the gel filtration technique (Abbaszadegan et al. 1993; Schwab et al. 1995), immuno-capture method (Schwab et al. 1996), protein coagulation technique (Jaykus et al. 1996), and guanidine purification (Shieh 1995).

Recently, Katayama et al. (2002) developed a virus concentration method by introducing an acid rinse step between adsorption to and elution from a negatively charged membrane. The advantage of this method over the conventional methods is that the virus concentrate can be subjected to RT-PCR without any additional purification, even after re-concentration using a centrifugal ultrafiltration unit Centriprep YM50 (Millipore, Tokyo).

The virus concentration with acid rinse method was modified for a large volume of freshwater by replacing the $MgCl_2$ addition with an $AlCl_3$ addition prior to filtration (Haramoto et al. 2004). As a result, noroviruses were successfully detected in a large volume (*ca.* 300 litres) of properly treated tap water. The modified method is promising for concentrating viruses from groundwater samples.

Detection method

There are two kinds of detection methods for viruses in water. One is to cultivate a possible virus in a mammalian cell to see if there is cytopathic effect. Another method is to detect genomic DNA or RNA of viruses using molecular biological methods such as PCR.

Cell culture methods are more likely to have correlation with infectious risk because this method only can detect infectious viruses. On the other hand, not all the viruses are culturable so this method may underestimate the risk of infection. Culturing methods require a long time before the result is obtained. Moreover, these methods cannot identify the virus type.

In most of the outbreak studies, PCR methods are used to detect the disease causing agents. PCR is more easily applied to the samples after virus concentration from groundwater. However, PCR cannot distinguish infectious viruses from non-infectious ones. This limitation is normally accepted in the outbreak studies where high concentrations of viruses are detected, but arguments remain regarding the use of PCR as part of regular monitoring of groundwater quality.

8.7 Treatment technology

8.7.1 Chlorination

Chlorination is widely used in many countries to disinfect groundwater. Normally, groundwater quality is better than that of surface water so that chlorination is good enough as a treatment method. Chlorination is very effective on enteric bacteria such as *Vibrio cholerae* or coliforms, but not so much on *Cryptosporidium*. In Japan, chlorination is mandatory by the water supply law, but additional treatment or disinfection methods must be applied to remove or inactivate *Cryptosporidium*.

8.7.2 Membrane Filtration

Membrane filtration is now the most popular water purification method. According to the pore size, a membrane is categorized as microfilter (MF), ultrafilter (UF), nanofilter (NF), or a reverse osmosis membrane (RO). MF and UF are commonly used for water filtration; NF is used for the treatment of brackish water and salts in groundwater; RO is used for seawater desalination. Most suspended solids in the source water are retained by MF or UF membranes. *Cryptosporidium*, as well as bacteria, is completely retained by these membranes. Thus, membrane filtration processes are considered to be the best for the treatment of source waters that are subject to microbial contamination including protozoa. Viruses, however, can be completely retained only by UF because the pore size of MF is greater than some species of viruses. However, it is known that once a MF membrane is clogged, it shows a better retention rate of viruses due to the cake layer and/or pore clogging on the membrane surface (Otaki et al. 1998).

8.7.3 Ultraviolet Irradiation

UV light irradiation is an effective disinfection method for bacteria, viruses, and protozoa in water and wastewater. UV disinfection systems are easy to maintain, do not require additional chemical inputs, and produce no hazardous by-products. Therefore, water treatment plants that utilize UV disinfection processes have increased in number in recent decades.

In the 1990s, *Cryptosporidium* was the most important pathogens targeted by the water treatment industry. The viability of *Cryptosporidium* oocyst was examined by the excystation method, but in the late 1990s it was examined by the infection method to the sensitive mouse or cell culture with a specific cell line. The new methods revealed that UV is very effective to inactivate *Cryptosporidium* (Abbaszadegan et al. 1997).

Inactivation of microorganisms by UV irradiation is effected through the formation of lesions in the genomic DNA of the organisms. However, UV irradiation is not always perfect because some organisms are known to possess the ability to repair their DNA by mechanisms such as photoreactivation and dark repair. *Cryptosporidium* were tested for the ability of photoreactivation and turned out to have no such abilities, however (Oguma et al. 2001).

In 2007, the Japanese Ministry of Health and Welfare approved using UV irradiation as a barrier against *Cryptosporidium* for water supply systems using groundwater.

8.8 Case Study of Tokyo

8.8.1 Introduction

Occurrence of viruses in groundwater in Tokyo was studied using the samples collected in the research described in Chapter 7.

For the detection of viruses in groundwater in Tokyo, adenoviruses were selected as the target virus because of their abundance among the viruses throughout the year (Katayama et al. in press). Noroviruses are also abundant in the winter, but much less so in summer.

8.8.2 Materials and Methods

Sample collection

Sampling and analysis of microbial contaminants were carried out between November 2005 and January 2006. In total, 46 groundwater samples were collected from both confined and unconfined aquifers located in the eastern lowland area of Tokyo. Groundwater samples were stored in disposable plastic bags sterilised by gamma-irradiation, and tested for adenovirus, total coliforms and *E. coli*.

Virus concentration

The method used for the concentration of wastewater samples was described previously (Katayama et al. 2002). Briefly, a 2000 mL sample was added with 2.5 M $MgCl_2$ to obtain a final concentration of 25 mM, then passed through an HA filter (0.45 μm pore size and 90 mm diameter; Millipore, Tokyo) attached to a glass filter holder (Advantec, Tokyo). The filter was rinsed with 200 mL of 0.5 mM H2SO4 (pH 3.0), followed by elution of viruses with 10 mL of 1 mM NaOH (pH 10.8). The filtrate was recovered in a tube containing 50 μL of 100 mM H2SO4 (pH 1.0) and 100μL of 100× Tris-EDTA buffer (pH 8.0) for neutralization, followed by centrifugation using a Centriprep YM-50 (Millipore) at 2500 rpm for 10 minutes. The remaining portion (approximately 2 mL) was further applied to ultrafiltration at 2500 rpm for 5 minutes to obtain a final volume of 700μL.

Virus detection

For the detection of adenoviruses, 200 μL of the final concentrated sample was applied to DNA extraction using a QIAamp DNA mini kit (QIAGEN, Tokyo) to obtain 200 μL of the resulting DNA. The TaqMan reagents and procedures were used for amplification of the viral genomes. All the aliquots (approximately 80 tubes) of 5 μL of each DNA sample were mixed with 45 μL reaction buffer containing 25 μL of 2× TaqMan universal PCR master mix (Applied Biosystems), 400 nM of each primer, and 300 nM TaqMan probe. The primer pairs and the TaqMan probes used for the detection of AdVs (Heim et al. 2003) are designed to amplify a variety of strains of adenovirus.

Total coliforms and Escherichia coli

Total coliforms and *E. coli* in 1 mL of the samples were quantified by a single-agar-layer method using chromocult-coliform agar (Merck, Tokyo). After incubation at 37°C for 24 hours, blue colonies were counted as *E. coli*, while the sum of blue and red colonies were counted as total coliforms.

8.8.3 Results and Discussion

Among 46 total samples, only four samples were positive for adenoviruses. Among the positive samples, only one sample gave two positive results out of 80 tubes tested, whereas only one positive from 80 tubes in other three positive samples. The concentration of adenovirus was discovered at 1 or 2 PCR unit/2000ml at the highest. *E. coli* was detected only from two samples, in which no virus was detected. Occurrence of adenovirus and total coliforms (TC) were summarised in Table 8.2. The probability of occurrence of adenovirus was the same between TC positive and negative groups. Hence, the indicator bacteria could not predict the presence of adenoviruses. Based on the results of both ammonia and nitrate analyses presented in Chapter 7, it seems that the groundwater in Tokyo is contaminated with wastewater. However, the detection rates of both adenovirus and TC were very low, which indicated that the decay and retention of these microbes are important factors that reduce the detection rates of these microbes.

Table 8.2 Occurrence of adenovirus and total coliforms in groundwater in Tokyo

	TC positive	TC negative
Adenovirus positive	1	3
Adenovirus negative	8	34

8.9 Conclusion

Groundwater is a preferred source of drinking water because of easy access and its generally good quality. In the U.S. and many other countries, people drink groundwater without any treatment or only with simple disinfection such as chlorination. The etiologic agents of outbreaks related with groundwater are more diverse than those of surface water. In Japan, microbial safety of drinking water has been basically guaranteed by chlorination. Some outbreak incidences in Japan, however, reveal that poor maintenance of the disinfection process can easily lead to outbreaks of infectious microbes. The emergence of chlorine-tolerant protozoa is another threat to the safely of water systems using groundwater. Membrane filtration processes and UV irradiation have proven highly effective against protozoan contamination, and thus can protect the quality of water systems using groundwater. Genetic methods for the detection of viruses are advancing and improvement of the pre-treatment process can make it possible to concentrate viruses in a large volume of water, which enables detection of viruses at lower concentration levels. The fate and transport of microbes are dependent on many factors including the survival rates of microbes and hydrogeological conditions of the subsurface layer. There is no good indicator so far, however, regarding the fate and transport of microbes, which leaves room for future research efforts.

References

Abbaszadegan M, Hasan MN, Gerba CP, Roessler PF, Wilson BR, Kuennen R, Dellen EV (1997) The disinfection efficacy of a point-of-use water treatment system against bacterial, viral and protozoan waterborne pathogens. Wat Res 31:574-582

Abbaszadegan M, Huber MS, Gerba CP, Pepper IL (1993) Detection of enteroviruses in groundwater with the PCR. Appl Environ Microbiol 59:1318-1324.

American Public Health Association (1995) Standard methods for the examination of water and wastewater, 19th edn. American Public Health Association, Washington DC

Fenz R, Blaschke AP, Clara M, Kroiss H, Mascher D, Zessner M (2005) Quantification of sewer exfiltration using the anti-epileptic drug carbamazepine as marker species for wastewater. Wat Sci Tech 52:209-217

Gerba CP (1984) Applied and theoretical aspects of virus adsorption to surfaces. Advances in Applied Microbiology 30:133-168

Haramoto E, Katayama H, Ohgaki S (2004) Detection of Noroviruses in tap water in Japan by means of a new method for concentrating enteric viruses in large volumes of freshwater. Appl Environ Microbiol 70:2154-2160.

Heim A, Ebnet C, Harste G, Pring-Akerblom P (2003) Rapid and quantitative detection of human adenovirus DNA by real-time PCR. J Med Virol 70:228-239

IAWPRC Study Group on Health Related Water Microbiology (1991) Bacteriophages as model viruses in water quality control. Wat Res 25:529-545

Jaykus LA, De Leon R, Sobsey MD (1996) A virion concentration method for detection of human enteric viruses in oysters by PCR and oligoprobe hybridization. Appl Environ Microbiol 62:2074-2080

Katayama H, Shimasaki A, Ohgaki S (2002) Development of a virus concentration method and its application to detection of Enterovirus and Norwalk virus from coastal seawater. Appl Environ Microbiol 68:1033-1039

Katayama H (2007) One-year monthly quantitative survey of noroviruses, enteroviruses, and adenoviruses in wastewater collected from six plants in Japan. Wat Res, doi:10.1016/j.watres.2007.10.029

Lance JC, Gerba CP (1984) Effect of ionic composition of suspending solution on virus adsorption by a soil column. Appl Environ Microbiol 47:484-488

Lawson HW, Braun MM, Glass RIM, Stine S, Monroe S, Atrash HK, Lee LE, Engelender SJ (1991) Waterborne outbreak of Norwalk virus gastroenteritis at a southwest U.S. resort: role of geological formations in contamination of well water. Lancet 337:1200-1204

Liang JL (2006) Surveillance for waterborne disease and out-breaks associated with drinking water and water not intended for drinking - United States, 2003-2004, CDC MMWR, December 22, 2006/55(SS12), pp 31-58

Nasser AM, Oman S (1999) Quantitative assessment of the inactivation of pathogenic and indicator viruses in natural water sources. Wat Res 33:1748-1752

Oguma K, Katayama H, Mitani H, Morita S, Hirata T, Ohgaki S (2001) Determination of pyrimidine dimers in Escherichia coli and *Cryptosporidium parvum* during ultraviolet light inactivation, photoreactivation and dark repair. Appl Environ Microbiol 67:4630-4637

Otaki M, Yano, Ohgaki S (1998) Virus removal in membrane separation process. Wat Sci Tech 37(10):107-116

Phanuwan C, Takizawa S, Oguma K, Katayama H, Yunika A, Ohgaki S (2006) Monitoring of human enteric viruses and coliform bacteria in waters after urban flood in Jakarta, Indonesia. Wat Sci Tech 54(3):203-210

Rose JB, Daeschner S, Easterling DR, Curriero FC, Lele S, Patz JA (2000) Climate and waterborne disease outbreaks. Journal AWWA 92(9):77-87

Saito et al. (2005) Norovirus outbreak via community water supply system in Akita prefecture (in Japanese). Infectious agents surveillance peport 26:150-151

Schwab KJ, De Leon R, Sobsey MD (1995) Concentration and purification of beef extract mock eluates from water samples for detection of enteroviruses, hepatitis A viruses, and Norwalk viruses by reverse transcription-PCR. Appl Environ Microbiol 61:531-537

Schwab KJ, De Leon R, Sobsey MD (1996) Immunoaffinity concentration and purification of waterborne enteric viruses for detection by reverse transcriptase PCR. Appl Environ Microbiol 62:2086-2094.

Shieh YSC, Wait D, Tai L, Sobsey MD (1995). Method to remove inhibitors in sewage and other fecal wastes for enterovirus detection by the PCR. J Virol Methods 54:51-66.

Sobsey MD, Jones BL (1979) Concentration of poliovirus from tap water using positively charged microporous filters. Appl Environ Microbiol 37:588-595

Tamura et al. (2005) Norovirus outbreak via drinking water in Niigata prefecture (in Japanese). Infectious agents surveillance report 26:330-331

US EPA (2006) Occurrence and monitoring document for the final ground water rule http://www.epa.gov/safewater/disinfection/gwr/pdfs, as of Oct 1, 2007

Yates MV, Gerba CP, Kelley LM (1985) Virus persistence in ground water. Appl Environ Microbiol 49:778-781

Part III Efficient Use of Groundwater Resources

9. Environmental Systems Analysis of Urban Water Systems

Satoshi Ishii

Department of Urban Engineering, the University of Tokyo

9.1 Introduction

Conventionally, urban water systems have been selected to meet water quality standards and to keep the costs of construction low. It has recently become more important, however, to take environmental impacts into consideration when water systems are designed and chosen for a city. In this chapter, water supply systems using groundwater is compared with the existing water supply system in terms of their respective carbon emissions, which are calculated based on four different scenarios of water use.

9.2 Water Supply Systems in Cities

9.2.1 Cities and Their Water Use

Table 9.1 lists various types of water uses, the amount of water consumed and the quality of the water needed. It is well known that urban water uses become more diversified during the course of a city's development. A water supply system for a mega-city collects water from multiple watersheds to meet the large water demand. It was estimated that only 17.7% of the annual water supply of 1,006 mm/year comes from water resources within its own watersheds and the rest is transported from others (MLIT 2002a). At present, twelve dams and reservoirs constructed within four watersheds supply water to the Tokyo metropolitan region (Bureau of Waterworks, TMG 2007).

Table 9.1. Type of Water Use in a City

Usage	Required Quality	Required Quantity	Demand variation	Standards[a)]	Parameters for the standards[a)]
Drinking water	- Safe for human consumption	- Depends on the number of people and their life-style	- Constant	- Drinking water quality standard	- 50 legal parameters, - 27 recommended parameters - 40 candidate parameters
Industrial processes	- Less stringent than drinking water (depends on the purpose of water use)	- Depends on the type of industry and its process	- Constant	- Industrial water quality standard	- Chloride, Iron, pH, Temperature, Total Alkalinity, Total Hardness, Total Solids, Turbidity
Toilet flushing	- Less stringent than drinking water - Recycled water from wasatewater treatment plants	- Depends on the number of people and type of system installed.	- Constant	- Reclaimed water standard for toilet flushing	- Colour, E. coli, Odour, pH, Turbidity
Environmental uses (e.g. creeks and ponds)	- Less stringent than drinking water - Some need bathing water quality	- Depends on the type of landscape	- Constant	- Reclaimed water standard for water recreation / landscaping - Water quality standard for swimming pools	- Colour, E. coli, Odour, pH, Turbidity - pH, Turbidity, Chlorine, Oxidisability, Total Coliform bacteria
Gardening / Urban green spaces / Street cleaning	- Less stringent than drinking water, toilet flushing and industrial processes	- Depends on local climatic conditions and the season	- Constant / Occasional (depends on the season)	- Reclaimed water standard for spraying to the street	- Colour, E. coli. Odour, pH, Turbidity
Emergency stock / Fire fighting	- Generally, low quality required	- Depends on the number of people	- Rare	Regulation by local municipalities	- Colour, Odour, pH, Turbidity[a)]

[a)]: Standards and candidates for the standards in Japan
[b)]: Required parameters to register an emergency well (City of Yokohama, 2007)

Table 9.2. Water Treatment Technologies

	Technology	Description	General Characteristics
Sedimentation	Flocculation/ Sedimentation	- Turn small particles into large flocs by adding coagulants and mixing so as to remove them by sedimentation	- Conventionally the first step of treatment, followed by the filtration process
Filtration	Slow filtration	- Water is filtered through a sand filter where bio-films are formed and remove contaminants	- Low cost, and simple - Can remove dissolved contaminants at low concentration - Requires large footprint - Cannot be applied to contaminated raw water
	Rapid filtration	- Similar to the slow filtration but adds coagulating agents and removes contaminants more efficiently.	- Smaller footprint and higher treatment capacity than slow filtration - Needs careful operation and maintenance - Cannot remove dissolved contaminants (e.g. fertiliser)
	Membrane filtration (e.g. Micro-filter, Ultra-fine filter, Reverse Osmosis)	- Physically removes contaminants by using various pore sizes and materials of membranes	- Low labour intensity - Simple system, small plant size and footprint - Needs chemical cleaning of membranes - Energy intensive process (depends on pore size and filtration pressure)
Disinfection	Chemical disinfection	- Adding chemicals (e.g. Chlorine, Chlorine dioxide, Chloramines) for disinfection.	- Low cost and simple system - Formation of by-products (e.g. Chlorite, Trihalomethanes)
	UV disinfection	UV irradiation	- Low cost and simple system - No by-product formation
Advanced treatment	Ozone and granular activated carbon (GAC) treatment	- Ozone injection followed by activated carbon filtration	- Removes colour and odour - Reduce chlorination by-producets but produces other by-products (e.g. aldehyde and bromate) - Energy intensive process
	Biological treatment	- Biological degration of contaminants by microbes attached on media	- Removes ammonium nitrogen, manganese, etc. - Low reaction rates at lower temperature in winter
	Charcoal filtration	- Absorption of substances by charcoal filter	- Lower cost and simpler than ozone+GAC - Lower treatment efficiency than ozone+GAC

Source: compiled from Kawakita (2005), JWWA (2002)

9.2.2 Water Treatment Technologies

The water supply system is a necessary urban infrastructure and consists of a variety of technologies. Since more technologies have become commercially available in recent years, water supply companies confront the difficult task of choosing which treatment system (or systems) to adopt. Currently available technologies and their characteristics are summarised in Table 9.2.

For a city to be both water efficient and environmentally friendly, the choice of technologies should be given careful attention. Water supply systems potentially pose adverse environmental consequences through water resource development (e.g. dams and reservoirs), energy inputs for operations (e.g. pumps and processes) and water extraction from regional ecosystems (e.g. water transfer between watersheds). A city's water supply system should pursue not only system robustness, but also cost effectiveness and should avoid adverse environmental impacts. Therefore, urban water systems should be analysed in terms of available water sources, water demands, technologies and system design and performance.

9.3 Assessment of Water Supply Systems and Their Environmental Implications

9.3.1 Life Cycle Approach

One approach to address the assessment of system design is the Life Cycle Approach (LCA). The International Organization for Standardization (ISO) defines LCA as the "compilation and evaluation of the inputs, outputs and potential environmental impacts of a product system throughout its life cycle." Often this cradle-to-grave approach can be used to compare the accumulated environmental impacts among different products (Guineè 2002). An advantage of LCA lies in its capacity to analyse the "total" environmental impacts in a relatively simple manner. It should be noted, however, that there are some practical concerns with LCA. Firstly, since the target of LCA varies from a single product to a system that consists of several processes, the final analysis depends largely on what have been counted in the analysis and what have been left out. In general, LCA requires clarifying the boundaries of the analysis in order to avoid any misunderstandings. This also means that analytical flexibility can be ensured by adopting various boundaries for comparison, and thus LCA can be applicable even to a complex system. Secondly, since the original scope of LCA looks at holistic environmental impacts, it requires the consideration of many aspects (i.e. parameters) when it is used as a decision making tool. For instance, those environmental impacts include emissions of pollutants to various environmental media, energy consumption and ecological impacts. When LCA is carried out for product and/or system comparison, these aspects are supposed to be analysed in an integrated manner (e.g. monetary value). This may cause an over-simplification of some aspects.

One of the advantages of LCA is that LCA makes it possible to analyse different entities by a common yardstick so that the total energy and material inputs in construction, operation and demolition phases of the systems can be analysed. It is often the case that an entity consists of multiple products and components and thus it is not possible to breakdown and add all the parameters in each component of the entity. Therefore, in some cases, the effects of substituting one process by another are evaluated; for example, a relative energy saving of the current water supply system substituted by new water supply systems is calculated. Since it is a useful concept, the idea of "life cycle" has been adopted to analyse a variety of entities. Various methodologies have been developed to supplement LCA, namely input-output analysis, material flow analysis (MFA) and eco-efficiency. Such life cycle thinking also can be adopted to analyse cumulative impacts of particular aspects. Where economic aspects or CO_2 emission is the focus of an analysis, life cycle cost (LCC) or life cycle CO_2 ($LCCO_2$) approaches can be carried out for evaluation (Guineè 2002).

9.3.2 Life Cycle Analysis of Water Supply Systems

The water supply system in a city has environmental and socio-economic implications, and the degree of these impacts are of great concern. For instance, water treatment and distribution systems require different amounts of energy and costs. The withdrawal of a large volume of water in the upper basin of a watershed might cause an environmental disturbance by affecting the natural water cycle in the lower basin. Since more and more treatment technologies have become available in recent years, supply systems should be selected and assessed in view of the total system performance. A new and upgraded system should not increase the adverse environmental and socio-economic impacts as a whole.

Fig. 9.1 illustrates a typical urban water supply system that consists of several water resources and types of infrastructure. There are two analytical boundaries. The "Boundary 1" box may indicate the Tokyo case, in which water is supplied mainly through surface water resources. The water comes from reservoirs and is transferred from other watersheds into local water treatment works before being supplied to the city. In contrast, the "Boundary 2" box consists of pumps and onsite treatment facilities, and uses groundwater as its primary water source. It is clear that the systems based on "Boundary 1" and "Boundary 2" have different economic and environmental impacts.

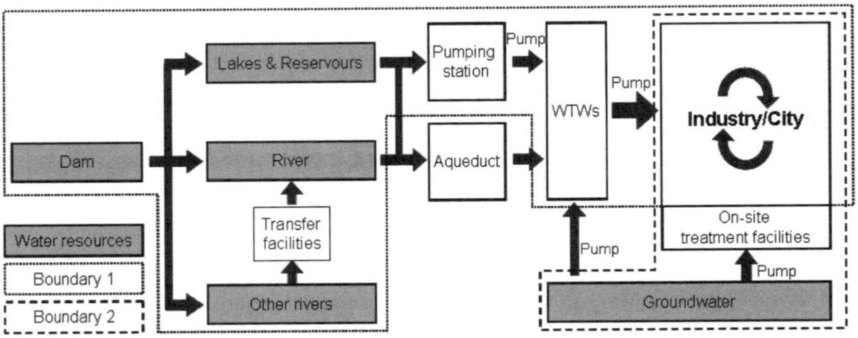

Fig.9.1. Typical urban water supply systems and analytical boundaries

By setting the analytical boundaries, the LCA visualises the issue in a holistic manner. If a water system requires large amounts of infrastructure or consumes a lot of energy for water transportation and operation, the cost and environmental impacts of supplying $1m^3$ of fresh water may be significant. This might be the case when large dams and/or a long aqueduct are necessary to bring the water to city, such as an aqueduct in Arizona. For instance, Suez Group reported that energy intensity of their water supply in 2006 was $0.430kWh/m^3$ (Suez 2007) while the Bureau of Waterworks in Tokyo reported a Fig. of $0.505kWh/m^3$ in 2005 (Bureau of Waterworks, TMG 2006).

As far as groundwater utilisation is concerned, the amount of energy used for water transmission might be minimal, as is the case with Boundary 2 in Fig. 9.1. Although the development of membrane technology is a key to decentralised water supply systems using groundwater, treating low quality water requires additional treatment processes with higher energy inputs, e.g. iron and manganese removal, utilisation of reverse osmosis membrane. The result is an increase in unit cost ($\$/m^3$) and carbon intensity ($t\text{-}CO_2/m^3$) of the water supply system. In this context, groundwater utilisation should be considered carefully and should be optimised to yield a better outcome for the city's water management. By understanding the pros and cons of water supply systems using groundwater, the adverse environmental impacts of the system can be minimised.

9.4 Environmentally Sound Use of Groundwater

As a source of water supply to a city, groundwater has a number of advantages over surface water resources, though it also has limitations and its use results in other kinds of environmental impacts. Table 9.3 summarises the advantages and disadvantes of using unconfined and confined groundwater in a city.

Table 9.3. Summary of advantages and disadvantages of utilising unconfined/confined groundwater in a city

Advantages	Disadvantages
Unconfined groundwater	
- Quality of water is generally good compared to surface water resources. - Close distance to the points of use allows de-centralised supply - Constant water temperature.	- Available amount may depend on local geological and climatic conditions. - Suitable only for a small water demand. - Water quality may vary. - Vulnerable to contamination (e.g. heavy metals,, N-products from fertiliser).
Confined groundwater	
- Distant from contamination sources and water quality can be even better than groundwater from an unconfined aquifer. - Close distance to the points of use allows de-centralised supply - Constant water temperature. - Utilising the appropriate amount could avoid adverse impacts on underground infrastructure.	- Over-abstraction may lead to various environmental impacts including land subsidence and saltwater intrusion. - Not suitable for a very large water demand. - Large influence of local geology on water quality (e.g. F, Mg, Fe etc.). - Vulnerable to land use change in groundwater recharge areas.

In general, the environmentally friendly use of groundwater calls for more information compared to surface water. For instance, more information on local hydrogeology, water quality of aquifer(s), potential sources of contamination (e.g. sewerage pipes, agricultural activities, industrial activities) and the extractable amount must be obtained. Nevertheless, groundwater is used for the following purposes.

Groundwater as a Main Source of Drinking Water

The presence of groundwater allows for a decentralised drinking water supply system. Although there is a constraint on its extractable volume, the quality of groundwater, especially from a confined aquifer, is better than that of rivers and lakes. This suggests that a relatively simple treatment system can be adequate. From a practical point of view, there is also the advantage that the temperature of groundwater is relatively constant. Since water supply by groundwater requires little energy input and infrastructure for water distribution, the costs for installation and maintenance, as well as CO_2 emissions, are low. It should be noted that groundwater is vulnerable to contamination, so there must be a careful risk assessment of the source before it can be used for human consumption.

Groundwater for Environmental Restoration

By taking advantage of its ubiquitous nature, groundwater can be used effectively for urban landscaping, parks, recreation ponds, small creeks and so forth. It does not require transmission mains or large treatment plants; hence the investment cost can be minimal. Where surface water has been contaminated, discharging 'cleaner' groundwater can be an effective countermeasure to pollution, as has been done in the rehabilitations of Shinobazu-no-ike Pond and the Tachiai River. Contaminants in the groundwater should be monitored carefully, and water quality should be ensured according to its usages. Since groundwater used for environmental restoration purposes usually requires only a small investment, this is often a preferred use of groundwater.

Groundwater as a Heat Source

The constant temperature of groundwater is a significant advantage for its use as a heat source or sink for air conditioning (see Chapter 10). Compared to the air temperature, groundwater is cooler in the summer and warmer in the winter. Such a difference in temperature can be used in heat exchange devices and utilised for various purposes. This is recognised as a renewable energy source, which might be able to be substituted for energy sources with higher carbon intensities. However, it should be noted that continuous heat discharge into the underground environment should be carefully assessed in order to avoid adverse impacts on ecosystems.

9.5 Case study: Use of Shallow Groundwater in a Decentralised Water System in Tokyo

This section analyses the environmental and cost implications of utilising shallow groundwater in Tokyo based on various scenarios and technology options. The study took two steps; firstly, the spatial distribution of demands for drinking and toilet-flushing water and the quality of shallow groundwater were described. Then appropriate treatment technologies were selected in order to attain the required water quality.

9.5.1. Material and methods

Tokyo metropolitan area and availability of shallow groundwater

Tokyo is the capital of Japan and the central Tokyo area consists of 23 wards with an area of 620 km^2 and a population of 8.6 million. The present analysis focused on only shallow groundwater as a water source. The extractable amount of the water in each ward was considered to be balanced by the groundwater input so that there were few adverse impacts on the natural ecosystem or on the regional water cycle. The water balance in central Tokyo indicates that the total water input to the shallow groundwater is 246 mm/year. Inputs include precipitation (359 mm/year) and leakage from water supply pipes (90 mm/year). The outputs are extraction for drinking (84 mm/year) and industrial (53 mm/year) purposes, and emission through sewerage pipes (66 mm/year) (MLIT, 2002a). These parameters were used for calculating the available groundwater for each ward.

Scenarios for the analysis

Four scenarios were constructed based on two conditions: the first is water use, i.e. drinking water and/or toilet-flushing, and the second is the ownership of the buildings utilising groundwater, i.e. public or private (Table 9.4). The scenarios were set to address the issue of 1) ownership of groundwater and 2) quality and quantity requirements among different water uses. Although the analysis only covered drinking and toilet-flushing water demand in the city, it should be noted that the quality required for toilet flushing does not differ significantly from other aesthetic uses, gardening and street cooling under Japanese regulations. Scenario 1 assumed that the groundwater is used only for drinking in both private and public buildings, while Scenario 2 considered only the water used for flushing toilets in both private and public buildings. Scenario 3 considered the water use for both drinking and toilet flushing but but only in public buildings, while Scenario 4 considered drinking and toilet flushing uses in both public and private buildings. Minimising cost and energy consumption are the targets of this analysis.

Table 9.4. Scenarios for Analysis

	Technological Options	Water uses	Building Ownership[a)]	Parameters
Scenario 1	5 levels	Drinking	Public and Private[b)]	Total CO_2 reduction
Scenario 2	5 levels	Toilet flushing	Public and Private[b)]	Total CO_2 reduction
Scenario 3	5 levels	Drinking Toilet flushing	Public	Min. Cost Max. CO_2 reduction
Scenario 4	5 levels	Drinking Toilet flushing	Public and Private[b)]	Min. Cost Max. CO_2 reduction

[a)]: 'Public' includes public agencies and educational buildings and parks, whereas 'Private' includes offices, shops, cultural buildings, bars & restaurants and multi-story housing. All educational buildings are considered public even though there are many privately owned schools and universities.
[b)]: When 'public and private' buildings are targeted, water allocation priority is given to public buildings

Spatial Distribution of Shallow Groundwater Quality

A horizontal distribution of shallow groundwater quality in Tokyo was presented in Kuroda et al. (2006), who sampled 45 shallow wells in central Tokyo between 2005 and 2006 (see Section 7.4.2, Fig. 7.1). The sampling protocol and analysis methodologies are illustrated in Section 7.4.3. In this chapter, the groundwater quality data obtained by Kuroda et al. (2006) were used to select the water treatment processes for each type of usage.

In this scenario analysis, only five water quality parameters, namely, turbidity (NTU), chloride (Cl), total manganese (Mg), total iron (Fe) and total nitrogen (NO_3-N), were taken into consideration. Since the quality of groundwater varies significantly, the quality of groundwater was classified into the categories with scale -1 to +3: not detected (= -1), detected but under the standard value for drinking water (= 0), above the standard value but less than twice as high as the standard value of drinking water (= 1), between two to ten times as high as the standard value (= 2), and over ten times the standard value (= 3) (Table 9.5). In order to assume the spatial distribution, a Voronoi diagram was drawn to clarify boundaries of the areas of different groundwater quality.

Table 9.5. Classification of groundwater quality

Category	-1	0[a)]	1	2	3
Total iron (Fe)	Not detected	< 0.3 mg/L	0.3 ~ 0.6	0.61 ~ 3.0	3.01 ~
Total manganese (Mg)	Not detected	< 0.05 mg/L	0.05 ~ 0.10	0.11 ~ 0.50	0.51 ~
Total Nitrogen (NO_3-N)[b)]	Not detected	< 10 mg/L	10 ~ 20	20.1 ~ 100	100.1 ~
Chloride (Cl^-)	Not detected	< 200 mg/L	200 ~ 400	401 ~ 2000	2001~
Turbidity (NTU)	Not detected	< 1.4 NTU	1.4 ~ 2.8	2.81 ~ 14.0	14.01 ~

[a)]: Standard for drinking water
[b)]: Nitrogen substances includes nitrate (NO_2^-) and nitrate ions (NO_3^-)

Spatial Distribution of Buildings and Their Water Demands

The 2001 building floor use survey in Tokyo (TMG, 2002a) and other national statistics were used to estimate the water demand of each building. In the survey, floor use within buildings was classified into 22 categories, namely, floor category i = 22, and water use in the following categories was estimated: public agencies, offices, commercial uses, hospitals, educational buildings, cultural buildings, public baths, bars & restaurants and multi-story residential buildings. Drinking water demand was calculated using the following equation:

$$\text{Water demand of building (m}^3\text{/day)} = \Sigma w_i \cdot c_i \cdot a_i \text{ or } \Sigma w_j \cdot a_j \quad (1)$$

where w_i: daily water consumption rate per capita (L/capita/day), c_i: the number of people occupying the unit floor area of floor category i (number of people/m^2), a_i: the floor area of floor category i (m^2), w_j: daily water consumption per unit of floor area (L/m^2/day), a_j: actual floor area of the building (m^2). Different drinking water consumption rates were given for each floor use category. For estimates of toilet-flushing water, a per capita estimation was used. Where available, gender was taken into consideration in the analysis; p_1 and p_2 stand for male and female, respectively. When the gender ratio was not available, the average values obtained from Tokyo household water use survey in 2002 (TMG, 2002b) were used. According to the actual installation of membrane technologies in Japan, the minimum treatment capacity of the plant is set as 5 m^3/day (JWRC, 2006), thus those buildings with water demand less than 5 m^3/day were excluded from the analysis.

Technological options

Table 9.6. Technology options for groundwater treatment[a)]

Level of treatment	Name	Composition	Treatment cost (Yen/m^3)	Energy input (kWh/m^3)
Level 1	Chlorine treatment	Strainer + Chlorine	8.0	0.10
Level 2	Normal membrane	Strainer + Membrane + Chlorine	16.3	0.17
Level 3	Advanced membrane	Fe/Mg removal + Membrane + (activated carbon) + Chlorine	72.5	0.45
Level 4	Normal RO	Strainer + Membrane + RO + Chlorine	172.3	3.5
Level 5	Advanced RO	Fe/Mg removal + Membrane + RO + Chlorine	228.6	4.0

a): Cost and energy values were obtained from Jinbo (2000), Nakao & Watanabe (2004).

Five levels of technological options were considered in the analysis (Table 9.6). By taking advantage of its suitability for on-site systems, most of the water treatment systems were used membrane filtration combined with other technologies. The cost and energy demands of the system were obtained from reviews of relevant literature and interviews to private companies. Considering the categories of water quality in Table 9.5, the procedure of selecting these technological options is shown in Fig. 9.2.

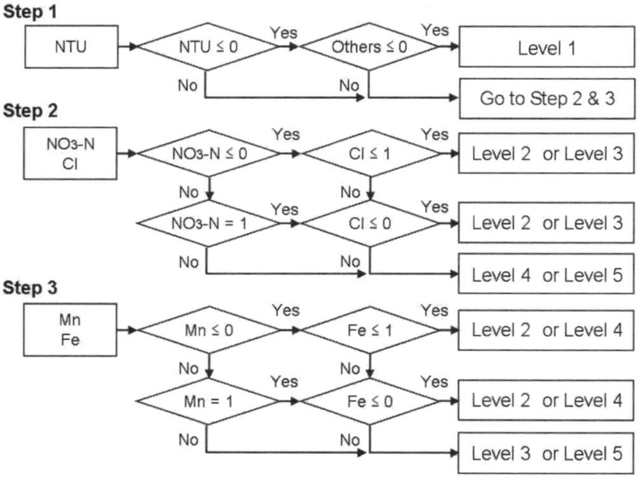

Fig. 9.2. Procedure of selecting treatment technologies

In order to calculate the CO_2 reduction potential of each scenario, the water supplied by the proposed system was assumed to take place of the present centralised water supply system in central Tokyo. The current CO_2 emission intensity of unit electricity generation and energy consumption per water supply in Tokyo were calculated to be 0.34 kg-CO_2/kWh (TEPCO, 2007) and 0.51 kWh/m^3 (Bureau of Waterworks, TMG, 2006), respectively.

9.5.2. Results and Discussion

Spatial Distribution of Water Demands and Groundwater Quality in Tokyo

Fig. 9.3 shows the spatial distribution of shallow groundwater quality in Tokyo, as expressed by a Voronoi diagram. It should be noted that each water quality score for the five parameters was added up to obtain the total water quality score. Dark areas of a high score indicated that the area had poor water quality, whereas the white areas had good water quality.

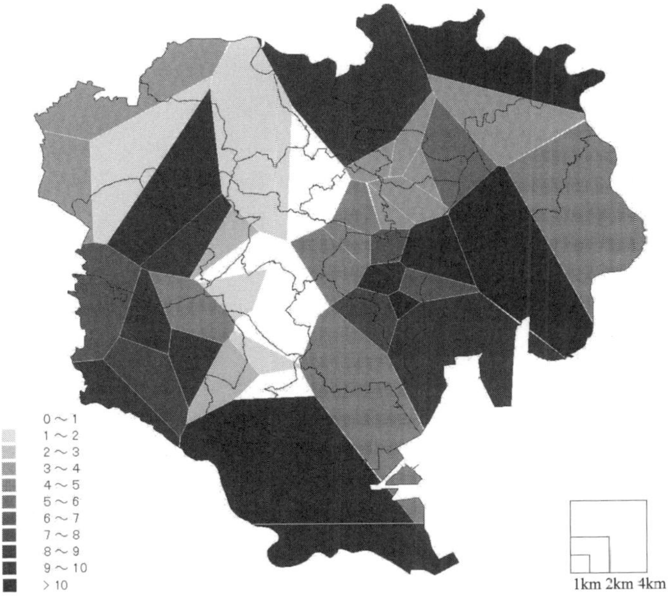

Fig. 9.3. Quality of shallow groundwater in central Tokyo (total water quality scores are the sum of each water quality score)

It was found that the quality of groundwater was low in the eastern and southern parts of Tokyo, whereas groundwater was of good quality in the central to northwest region of the city, namely, Shinjuku, Toshima, Itabashi, Meguro and Minato wards. It was noticed that the limited number of groundwater quality data constrained the resolution of analysis. Increasing the number of sampling points and providing detailed analysis on groundwater flow might overcome some of these constraints.

Table 9.7. Water availability and water demand of the buildings in 23 wards, Tokyo

Wards	Water availability (m^3/day)	Average water demand[a)] in public buildings (m^3/day)		Average water demand[a)] in private buildings (m^3/day)	
		Drinking	Toilet Flushing	Drinking	Toilet Flushing
Adachi	35855.3	14.1 (202)[b)]	11.8 (166)[b)]	22.8 (5367)	18.1 (1180)
Arakawa	6874.5	14.4 (23)	13.2 (5)	22.6 (1513)	16.3 (405)
Bunkyo	7622.6	16.9 (248)	15.5 (213)	39.5 (2181)	39.1 (455)
Chiyoda	7845	29.9 (268)	16.8 (207)	44.1 (2590)	31.6 (798)
Chuo	6840.8	18.2 (58)	15.4 (33)	34.1 (3460)	23.0 (1066)
Edogawa	33604.3	17.2 (216)	15.6 (172)	25.8 (4526)	19.1 (1179)
Itabashi	21681.7	14.4 (204)	13.3 (165)	24.6 (4437)	19.5 (1022)
Katsushika	23481.2	14.6 (138)	12.5 (113)	20.3 (3111)	16.0 (656)
Kita	13877.1	12.7 (177)	9.6 (126)	22.5 (2635)	18.3 (578)
Koto	26615.2	14.8 (148)	10.1 (115)	35.9 (3541)	24.0 (1145)
Meguro	9907.4	13.2 (130)	12.3 (101)	23.2 (2466)	21.4 (470)
Minato	13708.6	15.1 (227)	15.0 (162)	44.4 (5301)	28.0 (1839)
Nakano	10507.2	14.3 (121)	12.1 (96)	18.7 (2341)	16.7 (413)
Nerima	32458.5	11.8 (211)	10.9 (175)	18.1 (4799)	14.3 (897)
Ohta	40074.4	13.9 (239)	12.9 (172)	24.1 (5615)	18.4 (1343)
Setagaya	39144.3	13.2 (503)	12.3 (425)	17.8 (7142)	15.3 (1213)
Shibuya	10183.7	17.5 (246)	16.4 (209)	29.0 (4523)	21.5 (1165)
Shinagawa	15312.7	13.7 (174)	12.6 (125)	30.2 (3400)	21.9 (946)
Shinjuku	12286.5	25.9 (373)	18.9 (301)	48.4 (5435)	31.4 (1845)
Suginami	22928.5	14.0 (201)	12.3 (164)	15.7 (3997)	12.7 (667)
Sumida	9267.1	14.6 (102)	12.1 (81)	34.7 (2450)	30.1 (572)
Taito	6793.6	16.1 (114)	12.3 (81)	23.4 (3319)	18.3 (747)
Toshima	8768.4	17.4 (232)	15.4 (190)	29.0 (2908)	23.4 (694)

[a)]: Average water demands of the buildings whose water demands exceed 5 m^3/day
[b)]: Numbers in parenthesis indicate the number of buildings in each category.

Table 9.7 summarises groundwater availability, the number of public and private buildings and their average water demand for drinking and toilet flushing in each ward. The table also shows the characteristics of building floor distribution in Tokyo. There were 4,555 public buildings and 87,057 private buildings which had a drinking water demand of over 5 m^3/day. For toilet-flushing water, 3,597 and 21,295 buildings were selected for public and private buildings, respectively. Since some of these wards had a large potential for water demand, e.g. private buildings in Shinjuku and Minato wards, groundwater could be used more efficiently in these wards.

The table also indicates the variation of water consumption per building in the 23 wards. There were many public buildings with large water demands in Chiyoda and Shinjuku wards, probably due to a large number of existing national and metropolitan governmental buildings. It was noteworthy that the same trend was found for private buildings, indicating that these wards also had many office buildings. Except for these two wards, the average water demands in public buildings were in a similar range. Private buildings had varied totals of water consumption in each ward. There were many buildings in Ohta and Setagaya wards because the areas of these wards are larger than other wards.

Scenario analysis

The groundwater utilisation rate, defined as the percentage of the groundwater utilised out of available yield of groundwater, the annual cost of water supply and the annual CO_2 reduction in comparison to the current water distribution system are summarised in Table 9.8. The compositions of selected technological options are illustrated in Fig. 9.4.

The difference between Scenario 1 and Scenario 2 was whether the groundwater was utilised to supply only drinking or toilet-flushing, respectively. Because the wards located in the central to east side, i.e. Chuo, Taito, Sumida, Koto and Edogawa, required more expensive and energy intensive technologies (i.e. Level 5) to treat the groundwater, water treatment cost and CO_2 emission were substantially higher in these areas. It was found that the available groundwater was fully consumed by drinking water demands in all wards (416,000 m^3/year). For Scenario 1, the total annual water supply total cost was estimated to be approximately 12.4 billion yen and an additional 22,000 t-CO_2/year would be emitted to the atmosphere annually.

In contrast, the total annual cost for Scenario 2 was expected to be about 5.6 billion yen and CO_2 reduction from the current state would be by 8,200

Table 9.8. Water utilisation rate, Cost and CO_2 reduction in each scenario

	Scenario 1			Scenario 2			Scenario 3			Scenario 4		
Wards	GW use ratio (%)[a]	Cost (million Yen/yr.)	CO_2 reduction (t-CO_2/yr.)	GW use ratio (%)[a]	Cost (million Yen/yr.)	CO_2 reduction (t-CO_2/yr.)	GW use ratio (%)[a]	Cost (million Yen/yr.)	CO_2 reduction (t-CO_2/yr.)	GW use ratio (%)[a]	Cost (million Yen/yr.)	CO_2 reduction (t-CO_2/yr.)
Adachi	100%	948.8	284.4	65%	457.8	487.6	7%	53.4	-2.2	100%	788.6	589.4
Arakawa	100%	181.8	54.5	97%	144.6	113.3	3%	5.1	-2.1	100%	150	115.4
Bunkyo	100%	173.2	114.7	100%	47.3	351.9	47%	25.3	131.8	100%	44.7	358.1
Chiyoda	100%	210.6	35.9	100%	150.6	169.7	59%	108.4	-23.7	100%	101.1	263.3
Chuo	100%	356.6	-1467.8	100%	181	54.3	12%	26	-62.4	100%	181	55.1
Edogawa	100%	1890.6	-8415	75%	440.7	629.2	9%	59.9	-79.2	100%	661.9	697
Itabashi	100%	190.9	906.5	100%	187.9	961.5	12%	20	97.3	100%	73.8	1188.6
Katsushika	100%	637.7	44.2	51%	115.9	468.9	7%	21.7	34.9	100%	422.7	562.4
Kita	100%	332.6	180.1	85%	283.5	153.2	13%	37.5	3.4	100%	276	294.4
Koto	100%	2107.9	-11957.1	100%	704.3	211.6	6%	57.3	-212.7	100%	704.2	212.1
Meguro	100%	156.6	-404.1	100%	70.3	465.8	16%	12	48.5	100%	28.9	558.5
Minato	100%	357.3	103.6	100%	326.5	180.2	23%	74.5	-14.8	100%	40	772.7
Nakano	100%	201	236.6	77%	132.9	220.4	13%	22.6	21.1	100%	52.4	522.7
Nerima	100%	538.2	862	45%	252.3	378.3	7%	42.4	19.1	100%	353.9	1214.8
Ohta	100%	979	485.3	67%	608.9	427.3	7%	62.3	3.2	100%	754.9	949.2
Setagaya	100%	1116.4	-885.6	61%	429.8	573.9	16%	110	52	100%	787	792.7
Shibuya	100%	33.5	562.6	100%	44.7	543	38%	12.1	198.6	100%	29.7	575
Shinagawa	100%	332	272.1	100%	354.1	226.4	14%	50	-7.7	100%	200.7	542.1
Shinjuku	100%	88.7	576.6	100%	43.5	672.9	54%	34.6	296.9	100%	35.9	693.3
Suginami	100%	534.4	318.8	46%	209.8	211.4	11%	49.2	9.6	100%	363	644.1
Sumida	100%	763.7	-4422.1	100%	245.2	73.5	12%	36.9	-96.9	100%	245.2	74
Taito	100%	194.9	-76.9	100%	125.5	156.2	20%	34.4	-21.1	100%	90.2	223.1
Toshima	100%	44.9	434.1	100%	27	491.2	39%	11.7	169.9	100%	25.6	494.8
Total		12371.2	-22156.8		5584.1	8221.8		967.6	563.6		6411.4	12392.7

[a]: percentage of the groundwater utilised out of the total available groundwater in selected buildings, where water demand > $5 m^3$/day.

t-CO_2/year. It was noted, however, that the analytical framework in the present study did not include reverse osmosis technology (i.e. Levels 4 and 5) for supplying toilet-flushing. In addition, since the demand for toilet-flushing was not as much as the demand for drinking water, the available groundwater could not be fully utilised in some wards. As a result, the water utilisation rates in ten wards were estimated to be below 100%. In Nerima ward, only 45% of available groundwater would be utilised, although the total annual groundwater utilisation potential is 320,000 m^3/year. It was also noteworthy that the groundwater quality in Shibuya ward was so good that the technology option in Level 1 was sufficient for most of the toilet-flushing demands.

In Scenario 3, groundwater was only used in public buildings. Under this scenario, the water utilisation rates were all under 100%, suggesting that the total drinking water and toilet-flushing demands from public buildings were smaller than the available groundwater. In Arakawa ward, for instance, only 3% of available groundwater could be utilised. The highest water utilisation rates were recorded in Chiyoda and Shinjuku wards, where many public buildings are located. In terms of the water supply cost and CO_2 reduction, Fig. 9.4 shows the result of the most energy efficient choices of water use in target buildings. It is important to realise that the cost and energy use of each technological option showed the same trend, i.e. Level 1 < Level 2 < Level 3 < Level 4 < Level 5. Moreover, there was no merit of scale by plant size; that is, larger plants have no better efficiency in cost and energy use per amount of treated water than smaller ones. Hence, the most cost efficient selection of water use was theoretically equal to the most energy efficient one. If different levels of technologies were selected for drinking and for toilet frushing, the model would choose the latter because it would be more cost and energy efficient. Under this selection procedure for water uses, the model outputs for Scinarios 3 was that 61,000 m^3/year of groundwater was utilised and that the cost and CO_2 reduction would be approximately 970 million yen and 600 t-CO_2/year, respectively.

Scenario 4 considered the policy option that groundwater could be used in both public and private buildings. In this scenario, the water utilisation rates were all 100%, indicating that available groundwater in each ward was fully utilised to supply either drinking or toilet-flushing in all wards. In the analysis, the model calculated the water supply cost and its energy use for different water use in each building and gave priorities to public buildings with the most cost and energy efficient use of water. Thus public buildings with larger water demand were given priorities for receiving the water. The model also give preference to extracting groundwater from the area with better groundwater quality where different groundwater qualities were found within the same ward. The technology options in Levels 4 and 5 were, therefore, likely to be avoided, or the model chose to use the water for toilet flushing in order to avoid those technological options with higher cost and energy use (Fig. 9.4).

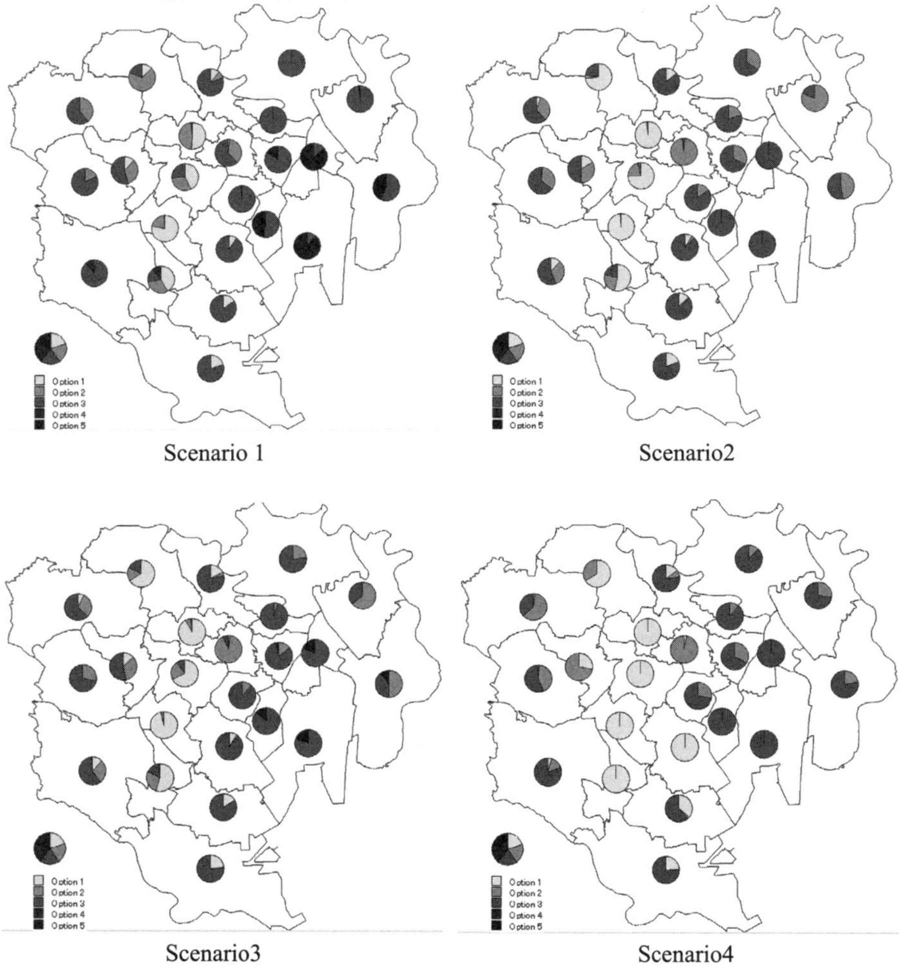

Fig. 9.4. Technology options selected under Scenario 1-4.

The result of Scenario 4 showed that 416,000 m^3/day of groundwater was supplied at a cost of 6.4 billion yen/year and the CO_2 reduction by 12,000 t-CO_2/year. Interestingly, Scenario 4 used approximately the same amount of groundwater as Scenario 1, though its composition of technological options was similar to that of Scenario 2 (Fig. 9.4). The cost of Scenario 4 was found to be approximately half of Scenario 1 and it was significantly preferable in terms of CO_2 reduction. A comparison between

Scenarios 3 and 4 indicated that the more buildings there were, the higher the cost and energy use efficiency in each ward. The present analysis shows clearly that systems analysis is very important in terms of designing more environmentally sustainable water supply systems in cities.

9.6. Conclusion

For the comparison of different alternatives for urban water systems, it is important to pay attention to the environmental impacts of those alternatives. The Life Cycle Approach is one tool for such an evaluation. In the present chapter, the case study addressed a comparison between the current water supply system using surface water in Tokyo and hypothetical decentralised water systems using groundwater, both in terms of cost and carbon emission. Since groundwater quality in the eastern region of Tokyo is of low grade, it requires a higher degree of treatment. Hence, the drinking water supply scenario gives negative effects of increased carbon emission. It was revealed that water uses that require lower degrees of treatment, such as toilet flushing, could maximise the advantages of decentralised water systems using groundwater. Although people in Japan have the rights to abstract groundwater, its usage in Tokyo is extensively regulated to avoid land subsidence. In this context, publicly controlled use, such as providing groundwater only to public buildings, was also examined in this chapter. Since the number of such public buildings is smaller than the privately owned buildings, the effects of groundwater use are quite small except in some wards, i.e. Shinjuku and Shibuya wards. Because this study was limited to the selection and environmental impact evaluation of water treatment processes, a further study incorporating supply networks would make it possible to conduct a more detailed comparison of different water supply systems.

References

Bureau of Waterworks, Tokyo Metropolitan Government (TMG) (2007) Environment plan 2007-2009

Bureau of Waterworks, Tokyo Metropolitan Government (TMG) (2006) Environmental report 2006

City of Yokohama （2007） (http://www.city.yokohama.jp/me/kenkou/ public_hygi/public_health/saigai/saigai.html)

Guinée JB (2002) Handbook on life cycle assessment. Kluwer Academic Publishers, the Netherlands, pp 5-13

Japan Water Research Center (2006) A list of water treatment plant with membrane technology. (http://www.jwrc-net.or.jp/tec/)

Japan Water Works Association (2002) Design criteria for water supply facilities 2000. JWWA, Tokyo

Jinbo Y (2000) Treatment technology of Fe & Mn containing water by organic membrane (in Japanese). In: Yoshida T (ed) New water treatment with membrane technology. Nikkei Print, Tokyo, pp 129-158

Kawakita K (2005) Drinking water engineering (Jyosuido kougaku) (in Japanese). Morikita Publishing Co Ltd, Tokyo, pp 89-139

Kuroda K, Fukushi T, Takizawa S, Aichi M, Hayashi T, Tokunaga T (2006) Groundwater contamination and its sources in the Tokyo metropolitan area. J Water and Waste 48(9):37-45

Ministry of Land Infrastructure and Transport (2002a) Outline of Tokyo water cycle master plan (in Japanese). (Report by MLIT)

Ministry of Land Infrastructure and Transport (2002b) Manual for standardization of treated wastewater utilisation (in Japanese). (http://www.mlit.go.jp/kisha/kisha05/04/040422_.html)

Nakao S, Watanabe Y (2004) New developments of water treatment technology with membrane (in Japanese). CMC, Tokyo

Suez Group (2007) 2006 activities and sustainable development report (http://www.suez.com/en/activites/environment/businesses/businesses/)

Tokyo Electric Power Company (2007) TEPCO environment highlights 2007 (http://www.tepco.co.jp/en/challenge/environ /highli-e.html)

Tokyo Metropolitan Government (2006a) Land subsidence and current condition of groundwater in Tokyo. TMG Bureau of Environment, Tokyo

Tokyo Metropolitan Government (2002b) Survey on household water consumption pattern in 2002. TMG Bureau of Waterworks, Tokyo

10. Groundwater Use for Thermal Energy

Ryozo Ooka and Yujin Nam

Institute of Industrial Science, the University of Tokyo

10.1 Introduction

The temperatures of the ground and groundwater a few metres beneath the earth's surface remain relatively constant throughout the year, even though the outdoor air temperature may fluctuate greatly every season. The ground temperature is almost the same as the annual average of the air temperature in each city. At a depth of approximately two meters, for example, the temperature of the soil in most of the world's regions is stable between 10°C and 20°C. This is why groundwater tastes cool even on a hot summer day. This temperature difference between groundwater and ambient air is often utilized as "temperature difference energy."

In winter, it is much easier to capture heat from the soil, which is at a moderate 12°C, than from the atmosphere when the air temperature is below zero. This is also why ground or groundwater heating systems encounter no difficulties sending comfortably warm air through a home's ventilation system, even when the outdoor air temperature is extremely cold. Conversely, in summer, the relatively cool ground absorbs waste heat more readily than the warm outdoor air. In this chapter, examples and methods of thermal energy use are elucidated.

10.2 Groundwater Heat Pump Systems

10.2.1 Outline of Groundwater Heat Pump Systems

Groundwater is often used as thermal energy for heating and cooling buildings. Heating and cooling are usually carried out using heat pump systems, which bring heat from a low temperature to a high temperature, against the natural heat flow. Generally, the heat source of this system is the atmosphere. Rooms are cooled or heated when the outdoor temperature is high or low, respectively. In such cases, the performance of the heat pump deteriorates if air is used as a heat source because the temperature difference is very large. If the ground or groundwater is used as the heat source, it is expected that the performance of the heat pump improves because the temperature of the ground or groundwater is higher or lower than that of the outdoor air in winter and summer, respectively. Thus, this system, called the Ground Source Heat Pump (GSHP) system, is expected to become an important energy saving system. When groundwater is used as the heat source, this system is called a Groundwater Heat Pump (GWHP) system. A large GWHP system that utilizes an aquifer and seasonal heat storage is called an Aquifer Thermal Energy Storage (ATES) system. GWHP and ATES systems for heating and cooling are popular in European and North American countries. GWHP systems are among the oldest GSHP systems.

GWHP systems are classified into two types according to their well systems: open loop systems and standing column systems. The schematics of these two systems are shown in Fig. 10.1.

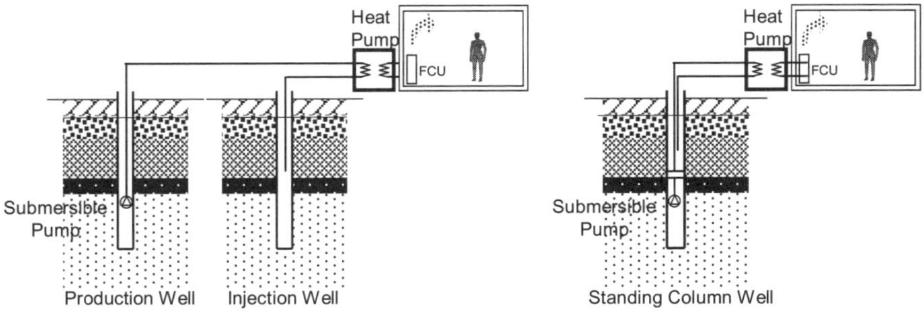

(1) Open Loop Well (2) Standing Column Well

Fig. 10.1. Groundwater heat pump (GWHP) systems

Open Loop System

This type of loop configuration is not commonly used, but may be cost-effective if groundwater is plentiful. Open loop systems, in fact, are the simplest to install and have been used successfully for decades in areas where local regulations permit. In this type of system, groundwater is piped directly from a well to the building, wherein the heat of the groundwater is transferred to a heat pump. After the groundwater leaves the building, it is pumped back into the same aquifer via a second well, i.e. a discharge well, located at a suitable distance from the first well. Because the open loop GWHP may affect the local environment, it is important to comply with regulations and to consult with the environmental agency in the region before installing an open loop system. Fig. 10.2 shows an example of an open loop system in Japan.

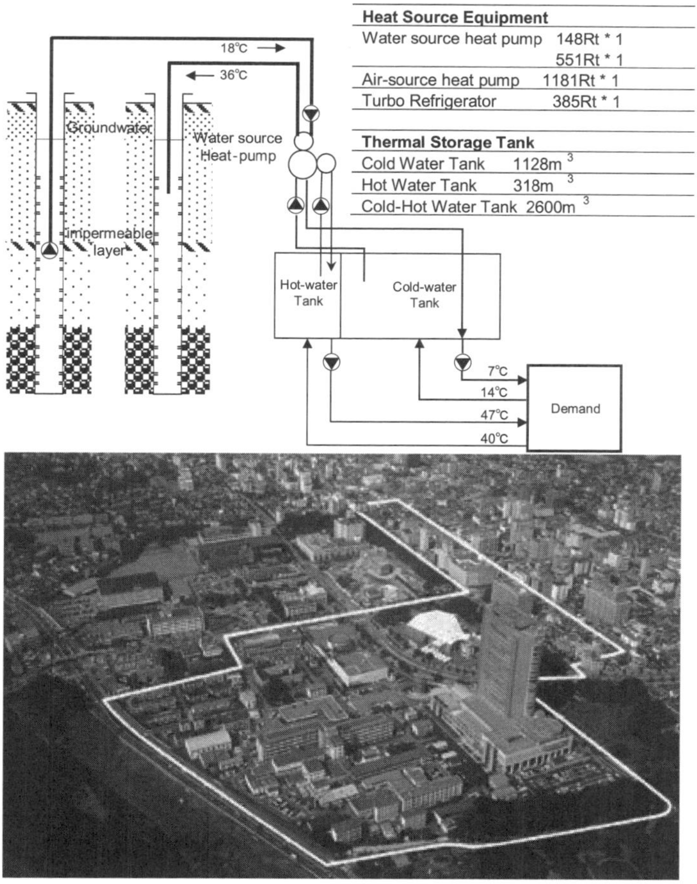

Fig. 10.2. Open loop GWHP system in Takasaki City, Japan

Standing Column Well System

Standing column wells, also called turbulent wells or Energy Wells™, have become an established technology. Standing wells are typically six inches in diameter and may be as deep as 500 metres. Groundwater from the bottom of the well is withdrawn, circulated through the heat pump's heat exchanger, and returned to the top of the water column in the same well. Usually, the well also provides potable water. However, groundwater must be plentiful for a standing well system to be operated effectively. If the standing well were installed where the water table is too deep, pumping groundwater would be prohibitively costly. The well water temperature can be brought back to the surrounding temperature by "bleeding" off some of the system flow and discharging the same amount of heated water to some other wells or watercourse. Whether high or low temperature water can be returned to the aquifer without any serious environmental impact depends on local environmental conditions, but usually such an impact is small because the water volume is often small and it contains no chemical substances.

ATES

An ATES uses aquifers as seasonal heat storages. This system is similar to the open loop system, which uses two wells connected with hydraulic coupling. One well is for the warm water and the other one is for the cold water. In winter, warm groundwater is pumped, transfers its heat and is then recharged into the cold well. In summer, the process is reversed and cold water is used for cooling. The advantage of this system is that it is environmentally sound because the water circulating from underground to the heat exchangers and then back into another well, will not be contaminated as it is closed in the system. Moreover, there is no net loss of water from underground. The only problem is that this system can only be used in areas that lie above aquifers suitable for ATES.

10.2.2 Field Test of Performance of GWHP

GWHP Experimental Facilities

An experimental facility was built at the Chiba experimental field of the University of Tokyo. Chiba is located to the east of Tokyo, and the annual average air temperature there is about 15.4°C, with a monthly average air temperature of 26.4°C in August and 5.4°C in January. The Plan of the experimental facilities and the system configuration are shown in Figs. 10.3 and 10.4, respectively. There are five wells (100mm in diameter, 20m in depth), of which two are utilised for pumping and injection. The wells consist of unplasticised polyvinyl chloride (PVC-U) pipes and a screen manufactured with a slit (10mm in diameter) 9.2m below ground level. In the production well, a submersible pump (23m lift, 47 L/min, 600W) and a U-type strainer to remove sand from the groundwater were installed. Those wells are connected to two rooms in which a fan coil unit and a radiation air conditioner are installed. Thermostat and electrical valves control the amount of water supplied to the fan coil unit. The system employed in this experimental equipment consists of a water-to-water heat pump with a reciprocating compressor (4.6kW in cooling, 5.7kW in heating). Cold and hot water circulates through a fan coil unit and a radiation panel, respectively, in these two examination rooms. During heating or cooling operations, groundwater is pumped from the production well and injected into the return well after heat exchange by the heat pump.

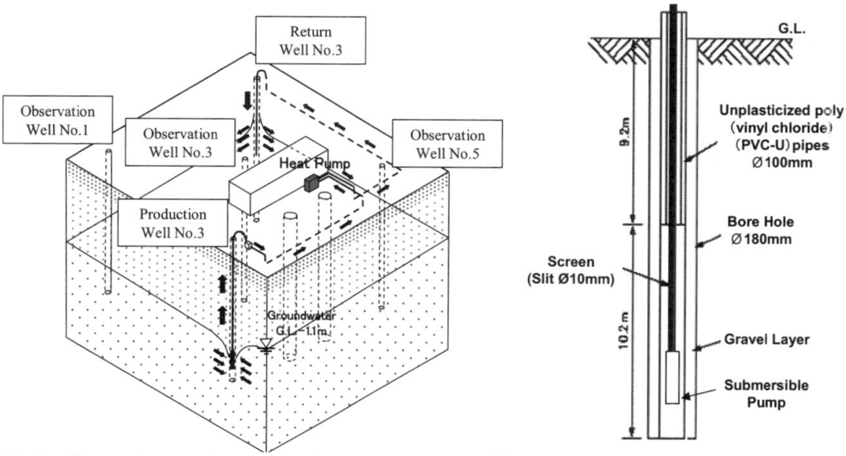

(a) Outline of experiment equipment (b) Section of production well

Fig. 10.3. Outline of the heat exchange system

Table 10.1. Monitoring parameters of GWHP system

Parameters	Measuring equipment (permissible range)	Measurement point
Groundwater temperature	T-type thermo couples (±1°C)	Wells No. 2, 4, & 5
Groundwater level	Micro-pulse water gage (±3 mm)	Wells No. 1 & 3
Heat source/sink Water temperature	Platinum measurement resistor (±0.5°C)	In the pipe
Cold and hot water temperature	Platinum measurement resistor (±0.5°C)	In the pipe
Water flow	Flow meter (±2%)	In the pipe
Electrical power used	Electric power meter	Power panel
Outside temperature, relative humidity, wind velocity, wind direction, quantity of solar radiation, rainfall		

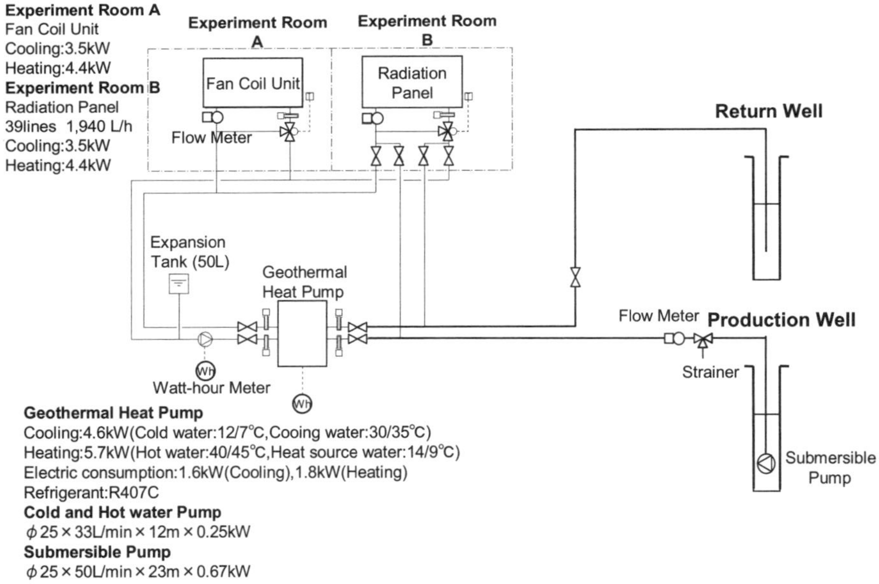

Fig. 10.4. Schematic of the experimental GWHP system.

Methods of Experiment

Wells number 2 and 4 were installed 32 m apart from each other, and were used as the production and return wells, respectively. The heat pump in this system was operated from 9:00 to 18:00, Monday to Friday, as in typical office buildings. During the cooling experiment (from 9th Oct. to 22nd Oct.), heat was discharged into the groundwater. Because the cooling load in October was insufficient compared with that in summer, two electric heating pads (750W × 2) and a room air-conditioner (2.8kW) were used to produce artificial heat loads. Conversely, during the heating period (from 4th Dec. to 12th Feb), heat was extracted (sourced) from the groundwater. During this period, the variation of system performance by changing the pumping rate was considered. Table 10.1 shows the parameters monitored in this experiment.

Experimental Results

The temperature of the groundwater from the production well remained at approximately 18°C for the entire experimental period. Fig. 10.5 shows the results of the cooling experiment using an artificial heat load. The heat exchange rate was calculated from the pumping rate of the groundwater and the temperature difference between the inlet and outlet at the heat pump. In the cooling experiment, the average heat exchange rate was 5.5kW with an average pumping rate of 47 L/min (28.2 m^3/day). However, the heat exchange rate peaked at the start of operation, and then decreased gradually as the room temperature cooled down. Conversely, it achieved greater than 10kW during the heating experiment when the heating load was raised by opening a window. The average heat exchange rate while heating is shown in Fig. 10.6.

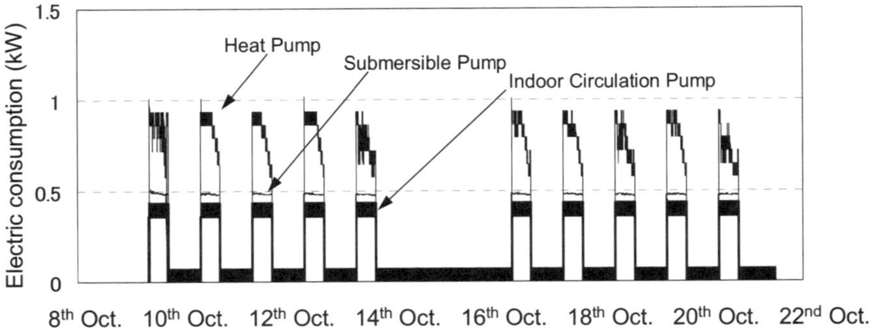

(a) Electric consumption

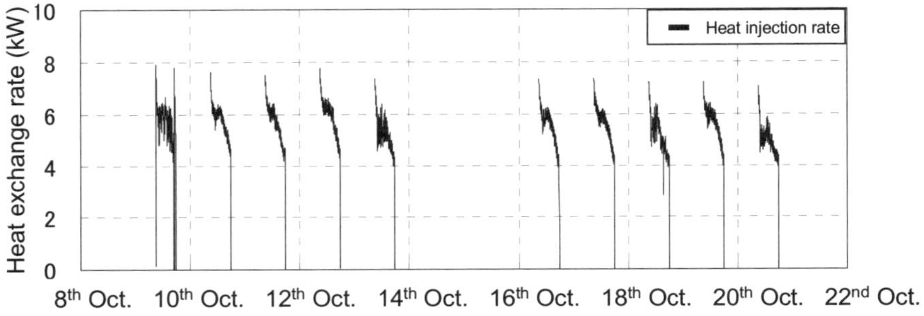

(b) Heat exchange rate

Fig.10.5. Results of the cooling experiment

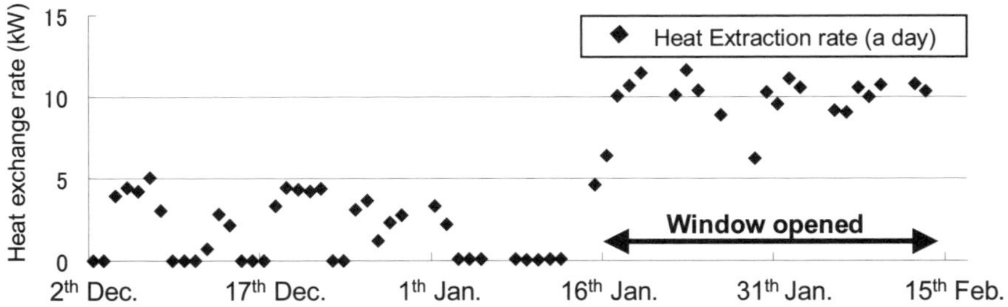

Fig.10. 6. Heating experiment results

Table 10.2 shows the results of the heating and cooling experiments. The value of cooling COP (Coefficient of Performance), the ratio of cooling ability to the electricity consumption of the heat pump, reaches 6.6 in this experiment. This value is much higher than that of an ordinary air-source heat pump. This means the GWHP system has a high potential for energy saving compared with an ordinary air conditioning system. However, the SCOP (System COP), which includes electric consumption from pumping the groundwater, was 3.2. This means that the pumping energy accounts for a fairly large portion of the total energy consumption of this system. Therefore, it is very important to develop an energy-saving method for pumping groundwater. The value of heating COP changes from 4.2 to 5.6 with the change of the heat exchange rate. The heat exchange rate corresponds to the loading rate of a heat pump. With the maximum heat load, the COP shows a high value, while the COP value falls with a partial load on the heat pump. Thus, the optimum operation method of the GWHP system should also be developed for saving energy.

Table 10.2. Experiment Results

Period	Operation Mode	Pumping rate (L/min)	Heat exchange rate (kW)	Electrical power (kW)		Coefficient of Performance (COP)	
				Pump	HP	HP	S.COP
9th Oct 2006 ~ 22nd Oct	Cooling	47	5.5	0.49	0.83	6.6	3.2
4th Dec ~ 10th Dec	Heating	46	4.2	0.48	1.01	4.2	2.3
11th Dec ~ 17th Dec	Heating	22	2.8	0.40	Technical Failure (Overheated)		
18th Dec ~ 24th Dec	Heating	9	5.3	0.34	1.10	4.8	3.0
18th Dec~ 3rd Feb 2007	Heating	46	6.2	0.48	1.10	5.6	3.2

10.3 Practical Application of the Thermal Energy of Groundwater

10.3.1 Snow Melting System

Groundwater is also used for snow melting systems in regions with heavy snowfall. In cold regions, heavy snow and road freezing causes various problems. To melt such snow or road ice, direct heating systems, such as boilers or electrical heating, are often used. However, these systems consume much energy. Because groundwater has very large amount of heat storage, its heat can be used for snow melting. There are two methods of melting snow with groundwater. First is the direct use of groundwater, i.e. sprinkling of groundwater. Fig. 10.7 shows the sprinkling of groundwater for snow melting. The temperature of groundwater is high enough to melt snow even in very cold regions. One of the major drawbacks of this system is that it requires a lot of groundwater to melt the snow. In many cities in Niigata Prefecture, Japan, over abstraction of groundwater has led to a rapid drawdown of the groundwater table. In these cities, the amount of groundwater abstraction must be regulated to prevent the adverse affects of groundwater over-abstraction. In addition, a groundwater-based snow melting system requires extensive maintenance of pipe networks to spray groundwater.

The second system utilizes a GWHP system. Figure 10.8 shows an example of the snow melting system using a GWHP. Obviously, the advantage of a GWHP system over groundwater spraying system is that while it does extract groundwater, it then recharges groundwater back into the aquifer so that there is no net amount of groundwater taken out of the aquifer. Hence, a GWHP system does not bring about a drawdown of the groundwater table. The downside of this system is that it cannot be easily applied to cover wide areas or network systems such as roads.

Fig. 10.7. Groundwater sprinkling for melting snow

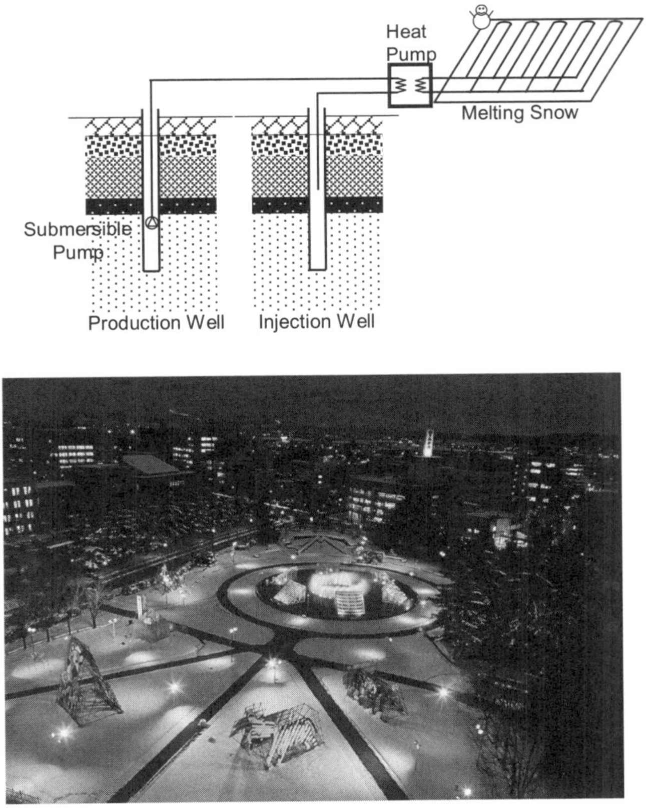

Fig. 10.8. Snow melting system using a GWHP

10.3.2 Hot Spring

A hot spring is produced by the emergence of geothermally heated groundwater from the earth's crust, and is the most elementary application of geothermal energy for space heating and agricultural and industrial uses. In general, the temperature of rock under the ground increases with depth according to the geothermal gradient. The formation of thermal groundwaters requires, that precipitation infiltrates into great depths, that there is a sufficiently long time of contact with the rock, and that there is a possibility for the water to ascend to the earth's surface again. Thermal groundwater sources become available for use when an uplift force moves the groundwater to the earth's surface. In some places where hydrogeological settings give pressure to the groundwater, it flows freely from the earth's surface. Depending on the properties of rock and the duration of contact between groundwater and rock minerals, thermal groundwaters have different chemical compositions and temperatures.

There are many hot springs all over the world, on every continent and even under the sea. Hot springs have been utilized in many countries, including the U.S., Iceland, Indonesia and Japan. There are over 115 major geothermal spas in the U.S., and thousands of hot springs in total (NOAA, 1980). The majority of these are located in the volcanic regions of the western states, but several famous ones exist in the east. The major spas are estimated to have an annual energy use of 1.53 x 10^{12}kJ, or an equivalent of 340,000 barrels of oil (Lund, 1996). Japan is a volcanic country, being located in the "Pacific Ring of Fire."There are over 10,000 thermal mineral springs, more than in any other country in the world, and over 3,000 hot springs (or *onsen*).Fig. 10.9 shows two examples of famous hot springs in Japan. *Onsen* resorts are a central feature of the Japanese tourism. Some amount of thermal energy from hot springs is also used in agriculture and for space heating.

Fig. 10.9. Hot Springs in Japan (*Noboribetsu Onsen*, left and *Fukuchi Onsen*, right).

10.4 Conclusion

Growing concerns of energy efficiency and the need to reduce carbon emissions have made us look into the ways to find more energy-efficient ways of living. The use of thermal energy from groundwater is one such method for energy efficiency. While the COP of GWHP systems is high, the power needed for pumping groundwater partly offsets the COP for GWHP systems, bringing down the GWHP. Hence, it is also important to find energy-efficient groundwater pumping systems.

Groundwater's heat capacity is also used for other purposes such as melting snow. It should be noted that excess abstraction of groundwater brings about the drawdown of the water table; methods for more efficient use of groundwater's heat capacity should be explored.

References

Grenville Y, Vladimir M (1995) Analysis of the effect of induced groundwater flow on heat transfer from vertical open-hole concentric-tube thermal well. ASHRAE Transactions 101(1):173-185

Hatten M (1992) Groundwater heat pumping lessons learned in 45 years at one building. ASHRAE Transactions 98(1):1031-1037

Kavanaugh SP, Rafferty K (1997) Ground-source heat pumps, design of geothermal systems for commercial and institutional buildings. ASHRAE Inc., pp 72-113

Knoblich K (2000) Thermal springs in Germany and Middle Europe. Geo-Heat Center Quarterly Bulletin 23(1):7-9

Lund J (2000) Balneological use of thermal water in the USA. Geo-Heat Center Quarterly Bulletin 23(1):31-34

Nakamura M, Yamaoka M, Higuchi M, Hokoi S (2002) Study on thermal energy storage system utilizing finite aquifer (Part 2) (in Japanese). Japan Arch Plan. Environ Eng 554:41-46

Ochifuji K, Nakamura M, Nagano K, Ikenaga Y, Kobata T (1992) Investigation of long-term heat storage in the aquifer (in Japanese). Transaction of SHASE of Japan 50:53-61

Rafferty K (2004) Water chemistry issues in geothermal heat pump systems. ASHRAE Transactions 110(1):550-555

Rafferty K (2001) Dual set point control of open-loop heat pump systems. ASHRAE Transactions 107(1):600-604

Rees SJ, Spitler JD, Deng Z, Orio CD, Johnson CN (2004) A study of geothermal heat pump and standing column well performance. ASHRAE Transactions 110(1):3-13

Sekine K, Ooka R, Yokoi M, Shiba Y, Hwang S (2007) Development of a ground source heat pump system with ground heat exchanger utilizing the cast-in-place concrete pile foundations of buildings. ASHRAE Transactions 113(1):1-9

Theodore B, Leila H (2000) Spas in Japan. Geo-Heat Center Quarterly Bulletin 23(1):29-30

Umemiya H, Aoyagi T (1991) Basic study of aquifer thermal energy storage (in Japanese). Journal of Japan Society of Mechanical Engineers 91:3543-3350

11. Bioremediation of Groundwater and Soil in Urban Areas

Futoshi Kurisu

Research Centre for Water Environment Technology, the University of Tokyo

11.1 Introduction

Microbial activity definitely plays an important role in material cycles on the earth, especially in water and subsurface environments. Faeces and carcasses of animals and plants can be easily degraded by microorganisms in the natural environment. The natural environment of the earth is protected by microbial activity.

Since the industrial revolution, human beings have clearly come to exist beyond being a mere member of the ecosystem and have been producing enormous quantities of waste. Consequently, the environment can become polluted, unless the waste is properly disposed of or treated. Some pollutants can be cleaned up by microorganisms, which maintain a sustainable earth. However, other pollutants which cannot be degraded by microorganisms have been accumulating.

Bioremediation, in the broadest sense of the word, is a term to describe the method to cleanup polluted wastewater or the environment by using or enhancing the ability of microorganisms to degrade or convert substances. The activated sludge process to degrade wastewater is one of the biggest advance in bioremediation, which was developed in the early 20th century. It was empirically found that wastewater could be cleaned by aeration, followed by removal of suspended matter through sedimentation. Later, it was found that in this process the major part of the organic matter removal was done by aerobic microorganisms. Now, various kinds of biological processes have been developed and applied to various kinds of wastewater.

Similarly, biological remediation techniques to cleanup groundwater and soil have been developed recently. In general, the term "bioremediation" specifically refers the groundwater or soil remediation. This chapter describes the bioremediation of groundwater and soil. The characteristics of bioremediation and comparisons with other cleanup technologies will be presented. The treatment of volatile organic compounds is described in detail as an example of bioremediation.

11.2 Bioremediation of Groundwater and Soil

11.2.1 Choice of Bioremediation of Groundwater and Soil Treatment

There are many technological options to treat groundwater and soil contamination. They include construction methods, without actual treatment, to minimize the effect of the contamination, as well as cleanup technologies. Both methods are done either in situ, without soil excavation or groundwater pumping, or ex situ. Cleanup technologies are based on either physical, chemical and biological methods or a combination of the three. These options are summarized in Table 11.1. An example of the physical methods is adsorption by activated carbon; chemical treatments include chemical reduction of chlorinated compounds by zero-valent iron; biological treatments employ the biodegradation or bioaccumulation of the contaminants by microorganisms or plants. The treatment technologies that were employed for contaminated sites in Japan in 2004 are shown in Figure 11.1 (Ministry of Environment Japan 2006). When compared to heavy metal contamination, volatile organic compound (VOC) contaminations are more commonly treated in situ.

Recently, bioremediation technologies have become more popular. Figure 11.2 shows the details of the treatment methods for VOC, which are indicated as "in situ treatment" in Figure 11.1 (Ministry of Environment Japan 2006). The number of sites treated by bioremediation has been increasing lately. This is because bioremediation has advantages over other physicochemical methods in terms of cost and environmental impact. It is suitable for relatively low concentrations of contamination, as microorganisms cannot degrade high concentrations of the compounds. In addition, it takes longer time to complete the cleanup than other methods, as it proceeds at modest rates. Thus, bioremediation is suitable for the treatment of a widely spread contamination of low concentration, which may be found after the treatment of the core contamination by another method.

In general, contaminated sites can be treated by a combination of treatment methods. Figure 11.3 shows a schematic diagram of a contaminated site and choice of treatment processes. For example, for a highly contaminated area, soil is excavated and clean soil is back-filled. The groundwater is pumped up, treated and injected back into the aquifer. Around the highly contaminated area, there is an area of lower contaminant concentration. This area is treated by injecting substances into the groundwater or soil. The treatment method can be either chemical or biological, but biological method can be the best choice for an area with lower contaminant concentration.

Table 11.1. Groundwater and soil treatment methods

Cleanup/ Containment	In situ/ Ex situ	Process	Explanation
Clean-up Technologies	In situ	Pump and Aeration	Removal of volatile compounds by aerating extracted groundwater followed by adsorption treatment.
		Soil Vapor Extraction	Extraction of the contaminated gas of volatile compounds.
		Oxidation/ Reduction	Injecting an oxidizing or reducing agent into the ground, or oxidizing groundwater with ultraviolet light.
		Bioremediation	Injecting nutrient compounds for microorganisms into the ground, or injecting microorganisms to degrade contaminants.
	Ex situ	Soil Washing	Washing contaminated soil with water or solvent. Normally used together with size classification to concentrate the contaminants.
		Biopile	Bioremediation method with an enhanced supply of air and nutrients by piling up contaminated soil.
		Heat Treatment	Recovering low boiling-point compounds by heating. Incineration is a similar method.
Containment/ Removal	In situ	Pavement/Embankment	This method is suitable where only direct contact with the contamination must be avoided.
		Sealing	Enclosing contaminants by installing sheet pile or concrete.
		Insolubilization	Injecting chemicals to physically or chemically prevent dissolution of contaminants.
	Ex situ	Soil replacement	Substitution of contaminated soil with clean soil.
		Backfilling insolubilized soil	Excavating contaminated soil, and returned it after insolubilization treatment.

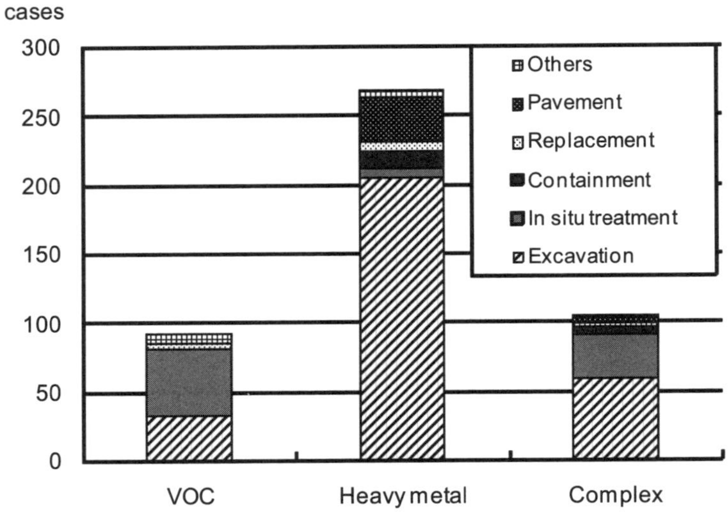

Fig. 11.1. Treatment methods used for VOCs, heavy metals and complex (sites with both VOCs and heavy metal contamination) in Japan, in 2004 (Ministry of Environment Japan 2006)

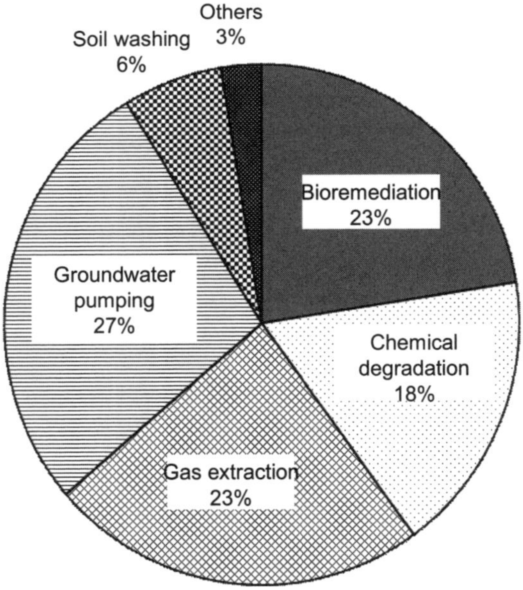

Fig. 11.2. Details of in situ treatments for VOCs in Figure 11.1 (Ministry of Environment Japan 2006).

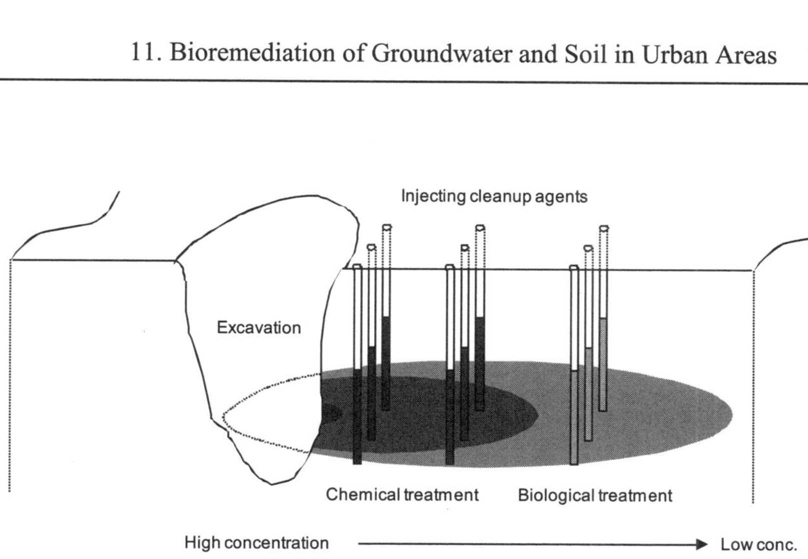

Fig. 11.3. Schematic diagram of soil/groundwater clean-up strategy

11.2.2 Bioremediation Concepts

Bioremediation can be achieved through one of three different concepts, depending on the existence and capacity of the degrading microorganisms. The required remediation time and available time for the treatment must also be taken into account to decide the method.

Biostimulation is a method to "stimulate" and activate microorganisms to degrade contaminants by adding some compounds to contaminated sites. Biostimulation is suitable if the microorganisms to degrade the target compounds exist in the site but the degradation rate is not sufficient. The compound used for the stimulation must be something that can speed up the degradation. They can be inorganic nutrients like nitrogen and phosphorus, substrates to the microorganisms like methane and low molecule organic compounds, and/or oxygen to assure an aerobic condition. This type of biostimulation is the most popular method. There are a variety of construction methods to add the compounds.

If there are no microorganisms to degrade the target compounds, it is necessary to add the microorganisms from outside. This addition of the microbes is called bioaugmentation. The microorganisms to be added to the site can be pure cultures, which are isolated and well characterized to degrade the compounds, or mixed cultures, such as enriched cultures known to have a good ability to degrade the compounds. However, it is necessary to evaluate the safety of the exogenous microorganisms. The microorganisms must not be pathogenic. It is also required to evaluate the effect of adding the microbes on the site's microbial ecology. Public acceptance by residents around remediation sites is also a big issue for bioaugmentation; hence it is more difficult to implement than other bioremediation methods.

If sufficient microorganisms to degrade the compounds are found in the site and it is estimated that the compounds can be cleaned by the required time, it is not necessary to conduct active remediation. In such a case, it is enough to monitor the contamination periodically and to confirm that the concentration reduction is occurring as projected. This method is called monitored natural attenuation (MNA). If the concentration reduction is not enough and it is expected that the cleanup will not be finished within the period, active remediation can be introduced afterwards.

11.2.3 Engineering / Construction Methods for Bioremediation

Many technical methods for bioremediation have been developed. They can be classified into solid phase bioremediation, slurry phase bioremediation and in situ bioremediation. Although solid and slurry phase bioremediation are used for soil treatment, it can also affect the groundwater quality indirectly by preventing or reducing the elution from contaminated soil. In situ bioremediation treats soil and groundwater onsite without excavation. These treatment methods are mainly developed for biostimulation, but most of them can also be applied to bioaugmentation methods. Schematic diagrams of some bioremediation methods are illustrated in Figure 11.4.

Solid Phase Bioremediation

In solid phase bioremediation, an approximately one-meter-high pile of contaminated soil (biopile) or excavated soil onsite (land farming/land treatment) is aerated, mixed and supplied with water and nutrients such as nitrogen and phosphorus to activate aerobic microorganisms. The effectiveness of the treatment is affected by water permeability, water content, type of soil and contaminants, and climate. This method is suitable for the treatment of petroleum contaminations.

Slurry Phase Bioremediation

Water is added to the contaminated soil to make it a slurry. Then the slurry is transferred to a reaction tank supplemented with nutrients and microorganisms, if necessary. This technique is suitable for the treatment of highly concentrated recalcitrant compounds, such as 2,4-dichlorophenoxyacetic acid (2,4-D) and pentachlorophenol.

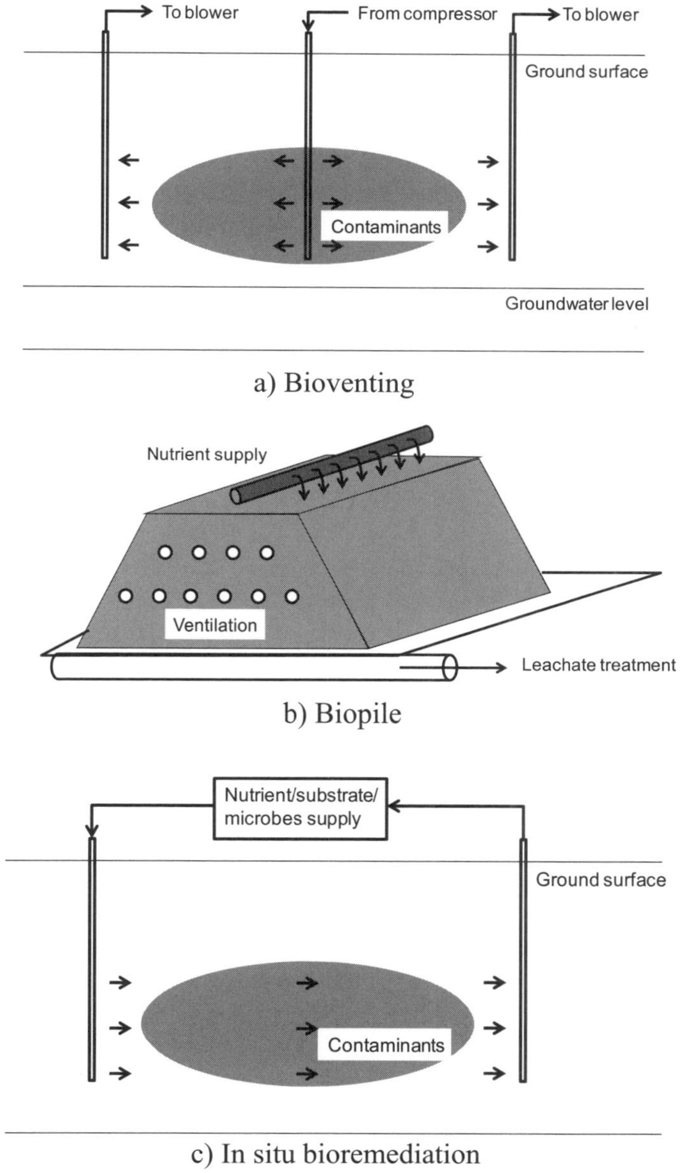

Fig. 11.4. Schematic diagrams of bioremediation methods: a) Bioventing, b) Biopile, and c) In situ bioremediation

Bioventing

This technique is designed to activate microorganisms in contaminated sites by aerating the vadose zone, i.e. unsaturated soil. For aeration, air injection and vacuum extraction is done simultaneously. It becomes more effective if nutrients like nitrogen and phosphorus are added. This method is not suitable for less permeable soil such as silt.

Biosparging

In contrast to bioventing, biosparging aerates the saturated zone to activate microorganisms. It is effective for the treatment of petroleum products like diesel fuel and gasoline. It is often applied in conjunction with vacuum extraction and bioventing.

Direct Injection Method

Nutrients, oxygen or oxygen supplying compounds like hydrogen peroxide, and organic matter like methane, toluene and molasses, are injected to the subsurface through vertical or horizontal wells to activate microorganisms.

Groundwater Circulation

Groundwater is circulated by extraction and injection wells for this method. The groundwater is extracted from downstream of the contaminated zone, and then injected back into the upstream wells with nutrient supplements.

Permeable Reactive Barriers

A vertical, permeable wall of nutrients and oxidative or reductive compounds is constructed downstream of the contaminated zone. The groundwater is treated when it goes through the wall. The chemical compounds used for the wall are selected to enhance chemical reaction and/or biological reaction.

Phytoremediation

Phytoremediation is a method that uses plants to extract contaminated compounds from the soil through the roots and activates rhizospheric microorganisms. Some of the extracted compounds can be degraded in the plant body, and some of them can be removed from soil by harvesting highly accumulated plant body. The extraction rate is generally slower compared to microbial bioremediation.

11.3 Bioremediation of Chlorinated Ethylene Contaminations

11.3.1 Chlorinated Ethylene Contaminations and their Behaviour in the Environment

Volatile organochloride compounds such as tetrachloroethylene (PCE) and trichloroethylene (TCE) have been used widely in industries such as dry cleaning and semiconductor manufacturing as solvents because of their flame resistance and degreasing power. They are toxic to human beings and other living things and thus, environmental standards for soil and groundwater have been established for PCE, TCE, cis-1,2-dichloroethylene (cis-DCE) and 1,1-dichlroethylene (1,1-DCE) in Japan. Inappropriate handling during their use and storage, as well as illegal dumping, have caused soil and groundwater contamination. These compounds are quite persistent in subsurface environments because they are heavier than water and intrude into the vadose zone, and are hardly degradable in the environment. They are difficult to cleanup as the contamination is usually widespread due to their high mobility.

11.3.2 Aerobic Degradation

Aerobic degradation of chlorinated ethylenes (CEs) can be achieved through the co-metabolism of some toluene degrading bacteria, phenol degrading bacteria and methane oxidizing bacteria. One example is a methane oxidizing bacterium *Methylocystis* sp. strain M, which has been isolated as a trichloroethylene degrading bacteria (Uchiyama et al. 1989).

The known pathway of TCE degradation by strain M is shown in Figure 11.5. The degradation is initiated by the oxidation of TCE by methane monooxygenase. The methane monooxygenase has relatively low substrate specificity, and thus it can also oxidize TCE. Strain M has a strong ability to degrade TCE to carbon dioxide with trace amount of residual byproducts. Like strain M, there are other aerobic CE-degrading bacteria that can also degrade CEs by co-metabolism of oxygenases that have wide substrate specificity, such as phenol monooxygenase, toluene monooxygenase, and toluene dioxygenase.

The aerobic treatment can be employed for all the CEs except tetrachloroethylene (PCE). Up to the present, there have been no reports that demonstrate PCE reduction under aerobic conditions. It is most probably due to PCE being too oxidised for the oxygenases to oxidise. Moreover, because CE-degrading bacteria cannot use CEs as their carbon source for growth, it is necessary to feed on other growth substrates. Thus, treatment of CEs must be coupled with the supply of their substrates, or be done with the bacteria grown separately using the substrates prior to the treatment.

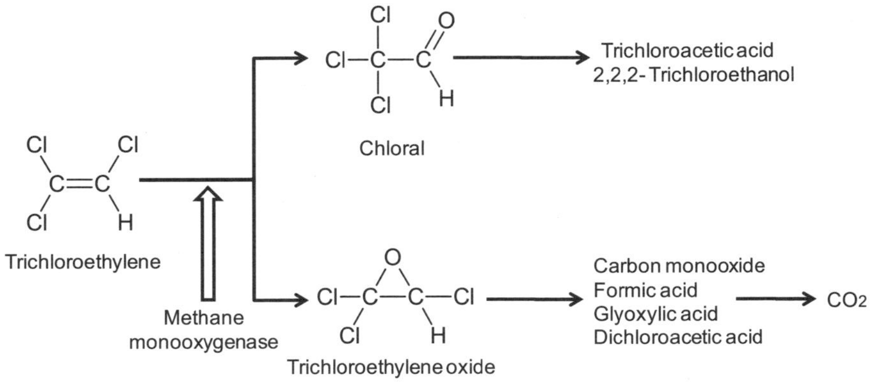

Fig. 11.5. Trichloroethylene degradation pathway by *Methylocystis* sp. M (Nakajima et al. 1992)

11.3.3 Anaerobic Condition

PCE and TCE can be degraded rather easily to cis-DCE under anaerobic conditions. This process is called anaerobic dechlorination. Several diverse kinds of bacteria and methanogens have been reported to perform the reaction. However, it is often found that the anaerobic dechlorination ends up with the accumulation of cis-DCE in the environment. Recently, Maymo-Gatell et al. (1997) isolated *Dehaloccoides ethenogenes*, which can dechlorinate CEs to vinyl chloride (VC) and to ethylene. Up until now, *Dehalococcoides* sp. was believed to be the only kind of bacteria which could dechlorinate cis-DCE to further VC and to ethylene. The anaerobic dechlorination pathway is shown in Figure 11.6.

Dehalococcoides uses hydrogen as a direct electron donor to dechlorinate the CEs as electron acceptors. As CEs are used for terminal electron acceptors, the reaction is also called dehalorespiration. Hydrogen, which is necessary for the reaction, can be supplied as a fermentation product of organic matters under anaerobic condition. Although *Dehalococcoides* is found widely in many places in the world and both its growth and the dechlorination can be stimulated by adding organic compounds in soil, the factors affecting its growth in soil have not yet been clarified well.

The remediation of CEs can be achieved by adding electron donors to accelerate the dechlorination by *Dehaloccoides*. As hydrogen gas itself is expensive, explosive, and difficult to supply in situ, organic compounds that can produce hydrogen as a fermentation product are normally used as the electron donors. Hendrickson et al (2002) and Lu et al (2006) demonstrated the strong relationship between the presence of *Dehalococcoides* and the dehalogenation in the environment. Thus it is informative to monitor the number of *Dehalococcoides* before and during the remediation. If *Dehalococcoides* is not found in the contaminated site, it is possible to add it to the soil and treat it by a bioaugmentation method.

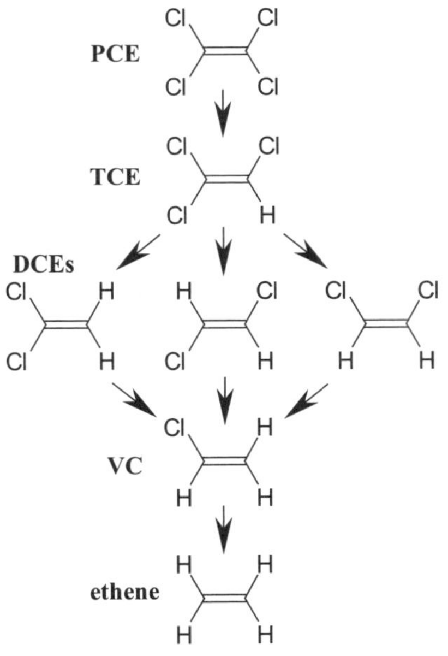

Fig. 11.6. Anaerobic dechlorination of chlorinated ethylenes

11.4 Bioremediation of Aromatic Hydrocarbons

11.4.1 Aromatic Hydrocarbon Contaminations and their Behaviour in the Environment

Petroleum product contamination is often found at places where it is used, for example, at a gas station due to the leakage from underground storage tanks. Oil is a mixture of hydrocarbons, both aliphatic and aromatic. Among them, benzene is one of the most important compounds in terms of being a human health risk. Benzene is carcinogenic, mutagenic and chronically toxic. The Japanese Environmental Standard regulates it in groundwater to be lower than 0.01 mg/L. Benzene is the only compound in petroleum products regulated by environmental standards in Japan. Benzene is rather soluble in water (1.8g/L, 25°C) and thus the contamination can spread widely in groundwater. Benzene and its derivatives, toluene and xylene, are called BTX, which are toxic and should be cleaned up. Furthermore, attention is also paid to the polynucleic aromatic hydrocarbons (PAHs) in oil.

11.4.2 Aerobic Degradation

In practice, aeration of contaminated soil/groundwater is normally enough to clean up the benzene contamination. Aliphatic hydrocarbons and BTX can be easily degraded under aerobic conditions and many reports have been done on their degradation. The biodegradability of PAHs is dependent on their molecular sizes; higher molecular sized compounds are more difficult to be degraded.

The degradation of aliphatic hydrocarbons generally occurs through beta-oxidation. Aromatic hydrocarbons are degraded from either mono-oxidation or di-oxidation of aromatic compounds. Although the initial process of oxidation may differ for different aromatic compounds, most of the degradation pathways fall in the catechol-like compounds, followed either by ortho- or meta-cleavage of the aromatic ring (Maier 2000). Those pathways are illustrated in Figure 11.7.

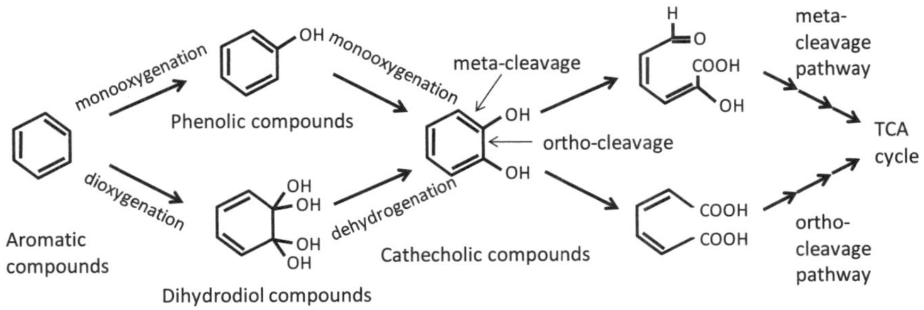

Fig. 11.7. Aerobic degradation pathways of aromatic compounds

11.4.3 Anaerobic Degradation

Anaerobic degradation of the aromatic compounds, especially benzene, is still a big challenge. There are only a few studies reporting the anaerobic degradation of benzene. The degradation has been observed under different oxidation-reduction conditions, such as nitrate reduction, ferric iron reduction, sulphate reduction, and methanogenesis. Among them, two articles reported isolated bacteria able to degrade benzene under nitrate reducing condition. Coates et al. (2001) reported *Azoarcus* sp. strain DN11 and strain AN9 and Kasai et al. (2006) did *Dechloromonas* sp. strain RCB and strain JJ. All strains can use nitrate as a terminal electron acceptor and benzene as an electron donor. However, there have been no reports that isolate benzene-degrading bacteria under other conditions.

Sakai et al. (2006) established the enriched culture that can degrade benzene under complete anaerobic conditions. They analysed the microbial consortia and found that some novel bacteria in delta-proteobacteria may be involved in the benzene degradation under methanogenic conditions using a combination of a stable isotope and molecular microbiology (Sakai et al. 2006). The combination of anaerobic CEs degradation and benzene degradation could become a very promising technology if we can achieve it by understanding the mechanisms of anaerobic degradation of both CEs and benzene.

11.5 Conclusion

At present, bioremediation processes have already been established for some compounds that are rather easily degraded by microorganisms. As most of organic compounds can finally be mineralised by aerobic microorganisms, aerobic processes can be generally applied to many kinds of contamination. On the other hand, anaerobic processes are more attractive as they require much less energy. Although anaerobic detoxification of chlorinated ethylenes is a good example of anaerobic bioremediation, it is still challenging to develop remediation processes for aromatic hydrocarbons.

Development of new technologies to degrade more recalcitrant compounds is also in progress. The goals for research include finding a condition to degrade hardly degradable organic compounds and to retrieve microbes to degrade them. In addition, some innovative ideas have already been reported to remediate metal contamination by biological means. They mostly utilize the change of chemical characteristics through biological transformation. Arsenic and mercury contamination can be remediated by converting the compounds to volatile ones (Okino et al. 2000; Islam et al. 2007). Some heavy metal contamination can be stabilized by oxidizing or reducing the compounds.

For further progress of bioremediation technology, it is necessary to develop methods to estimate the degradation. Efficient and reliable remediation will be possible if we can assess the degradation rate before initiating the remediation. One approach is to find the relationship between degradation performance and the microbial community before the remediation. DNA microarray technology can be a promising tool for the comprehensive analysis (Iwai et al. 2007). Ultimate goal of the research works on bioremediation is to completely understand the mechanisms of a microbial community in bioremediation. As the subsurface environment is very heterogeneous and microbial communities in the environment are very complicated, overall mechanisms are still far from being understood. However, fragmentary but useful knowledge is accumulating and is awaiting future integration.

References

Coates JD, Chakraborty R, Lack JG, O'Connor SM, Cole KA, Bender KS, Achenbach LA (2001) Anaerobic benzene oxidation coupled to nitrate reduction in pure culture by two strains of Dechloromonas. Nature 411:1039-1043

Hendrickson ER, Payne JA, Young RM, Starr MG, Perry MP(2002) Fahnestock S, Ellis DE, Ebersole RC: Molecular analysis of *Dehalococcoides* 16S ribosomal

DNA from chloroethene-contaminated sites throughout North America and Europe. Appl Environ Microbiol 68:485-495

Islam SM, Fukushi K, Yamamoto K, Saha GC (2007) Estimation of biologic gasification potential of arsenic from contaminated natural soil by enumeration of arsenic methylating bacteria. Arch Environ Contam Toxicol 52:332-338

Iwai S, Kurisu F, Urakawa H, Yagi O, Furumai H (2007) Development of a 60-mer oligonucleotide microarray on the basis of benzene monooxygenase gene diversity. Appl Micorbiol Biotech 75: 929-939

Kasai Y, Takahata Y, Manefield M, Watanabe K (2006) RNA-based stable isotope probing and isolation of anaerobic benzene-degrading bacteria from gasoline-contaminated groundwater. Appl Environ Microbiol 72:3586-3592

Lu X, Wilson JT, Kampbell DH (2006) Relationship between *Dehalococcoides* DNA in ground water and rates of reductive dechlorination at field scale. Water Res 40:3131-3140

Maier RM (2000) Microorganisms and organic pollutants. In: Maier RM, Pepper IL, Gerba CP (eds) Environmental Microbiology. Academic press, pp 363-402

Maymo-Gatell X, Chien Y, Gossett JM, Zinder SH (1997) Isolation of a bacterium that reductively dechlorinates tetrachloroethene to ethene. Science 276:1568-1571

Ministry of Environment, Japan (2006) Survey on execution status of the soil contamination countermeasures law, and contamination survey and countermeasure in FY2004 (in Japanese) (http://www.env.go.jp/water/report/h18-06/index.html)

Nakajima T, Uchiyama H, Yagi O, Nakahara T (1992) Novel metabolite of trichloroetylene in a methanotrophic bacterium, *Methylocystis* sp. M, and hypothetical degradation pathway. Biosci Biotech Biochem 56: 486-489

Okino S, Iwasaki K, Yagi O, Tanaka H (2000) Development of a biological mercury removal recovery system. Biotechnol Lett 22: 783-788

Sakai N, Kurisu F, Yagi O, Yamamoto K (2006) Selective cultivation of benzene degrading microbial consortia under anaerobic condition and their analysis (in Japanese). Environ Eng Res 43: 161-168

Uchiyama H, Nakajima T, Yagi O, Tabuchi T (1989) Aerobic degradation of trichloroethylene by a new methane utilizing bacterium strain M. Agric Biol Chem 53: 2903-2907

Part IV Groundwater Management in Asian Cities

12. Groundwater Management Policies in Asian Mega-Cities

Yatsuka Kataoka and Tetsuo Kuyama

Freshwater Project, Institute for Global Environmental Strategies (IGES)

12.1 Introduction

Groundwater is an easily accessible, cheap and high quality water resource. Because of its suitability for various purposes, groundwater is used in the course of socio-economic development of modern cities, and is especially used in the early stages of development. However, groundwater exploitation is not well managed or controlled in many cases, and groundwater sources are exploited intensively without consideration of their recharge rates. As a result, such intensive use causes drawdowns of water tables that are associated with other environmental problems such as land subsidence and saltwater intrusion. Such problems caused by intensive groundwater use are often irreversible or take a long time to remedy. Therefore, immediate actions for mitigation or prevention should be taken. However, it is often difficult for national and/or local governments to regulate groundwater abstraction for various reasons, including limited water resources available in a city, increasing demand and failure to resolve competition among different beneficial uses. This chapter describes the current status of groundwater management in Asian mega-cities through case studies of three cities, namely Bandung (Indonesia), Bangkok (Thailand) and Osaka (Japan), which face problems related to groundwater use. These three cities are at different stages of socio-economic development and have different political backgrounds. On the other hand, the cities have common features such as: 1) having a main river running through the city, 2) being under the influence of a monsoon climate, and 3) having domestic and industrial sectors as the major users of groundwater. A comparison of groundwater management policies in these cities reflects the different stages of socio-economic development in these cities.

12.2 Groundwater Management in Bandung: Struggling for Effective Implementation

12.2.1 Groundwater Use and Problems

Geography of the Bandung Metropolitan Area

The Bandung Metropolitan area (hereafter referred to as Bandung) is located in the western part of Java Island. The total area of Bandung is 2,340.88 km^{2}, and the total population is about 6 million. Bandung is divided into four administrative areas, namely Bandung Regency, Sumedang Regency, Bandung City, and Cimahi City. The potential of water resources, especially surface water, is relatively high in Bandung, though there is seasonal fluctuation. About 45 percent of the study area is hilly land that encircles the flat land in the city centre. The main river is the Citarum River, which also has ten tributaries. It is generally recognised that surface water resources are abundant in the area, but it is estimated that only 53 percent of available surface water is utilized in the city. Public water supply coverage differs in each local administration, ranging from about 40 percent in Bandung City to only 8.2 percent in Bandung Regency and Cimahi City (Wangsaatmaja et al. 2007).

Groundwater Use

Groundwater in Bandung is abstracted mainly by the industrial sector, which uses about 80 percent of the total abstraction volume. The dependency rate of the industrial sector on groundwater reaches 60 percent. 20 percent of the total groundwater is used for domestic consumption, which includes individual household wells, as well as the public water supply managed by the Water Supply Enterprise (WSE). Groundwater abstraction rapidly increased starting in the 1970s along with the city's economic development and reached its peak in 1996, as shown in Fig. 12.1. The sudden drop of abstraction volume in 1997 was due to downturn of the economic activities caused by the Asian Economic Crisis. There is a strong correlation between regional GDP (RGDP) and groundwater abstraction volume, which demonstrates the close relationship between industrial activities and groundwater consumption.

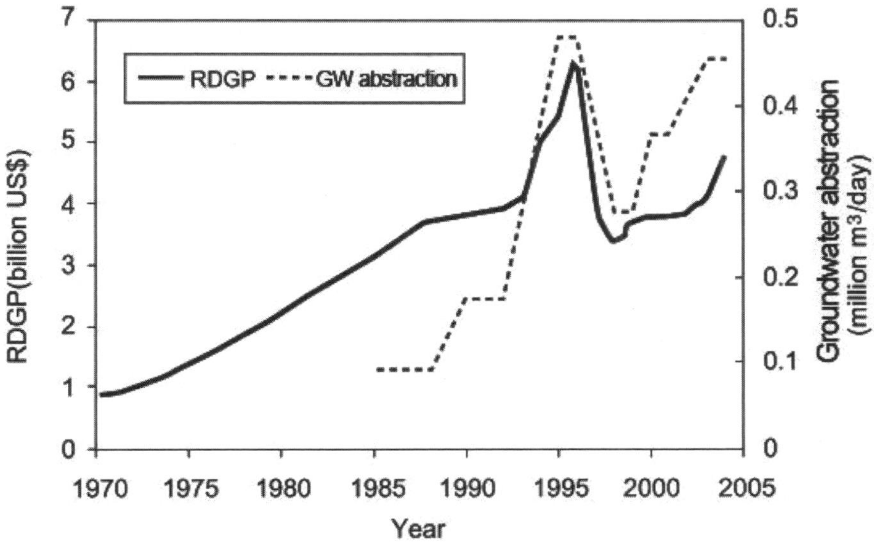

Fig. 12.1. Time course of groundwater abstraction and regional GDP in Bandung. (IGES 2006)

Groundwater Problems

The rapid increase of groundwater abstraction without adequate management resulted in drawdown of the groundwater table. According to the monitoring data of the Directorate of Environment and Geology in Cimanggung, the static groundwater level decreased by more than 60m in the decade from 1994 to 2004. The monitoring data obtained by a Global Positioning System (GPS) showed that the annual rate of land subsidence was about 24cm, but this had not yet been recognised as a region-wide problem.

12.2.2 Groundwater Management

Groundwater Rights

If the rights of the landowners to abstract groundwater underneath their lands are granted, it is difficult for the policy makers to regulate groundwater abstraction. In Indonesia, the Fundamental Act of 1945 stipulates that the national resources including water resources are considered as public commodities that must be fairly managed and used for the benefits of the Indonesian people. "Act Number 11, Year 1974, on Watering" also defined groundwater as a public commodity that shall be optimally used for the wellbeing of the people.

Organisational Arrangements

At the national level, the Ministry of Energy and Mineral Resources is responsible for groundwater management, while the Ministry of Public Works is responsible for the management of other water resources and coordination of water related policies. Presidential Decree Number 64/1972 stipulated that the provincial governments were primarily responsible for the implementation of groundwater laws; in the case of Bandung, the West Java Provincial Government is responsible. Later in 1999, (National) Act Number 22 regarding decentralization was introduced, which gave local governments the responsibility of groundwater management, including the right to impose a groundwater tax. However, the responsibilities which were delegated to the city governments were not accompanied by appropriate funding, infrastructure, and human resource arrangements, so the city governments faced difficulties in accomplishing their responsibilities (Wangsaatmaja et al. 2007). Furthermore, the law related to the groundwater tax (Act 34/2000) designated that the tax should be a part of the revenue of the provincial governments, which made the taxation system of groundwater inefficient. Such confusion continued until 2004 when groundwater management responsibility was returned back to the provincial government by National Act 32 on the revision of decentralization.

Legislation for Groundwater Control

Since the 1990s, when groundwater depletion was identified, various measures to control groundwater use have been introduced in Bandung. In 1990, the Plan of Determination of Zoning by the Geology Environment Office under the Ministry of Mining and Energy Resources was announced at the national level. Following this announcement, the zoning plan in Bandung was issued by a Minister of Mining and Energy Resources Decree in 1994. A permission and reporting system of groundwater abstraction for industrial use was introduced in 1995 by West Java Provincial Rule No. 9 on Groundwater and Surface Water Monitoring, which included a ban on the construction of new wells in critical areas where groundwater depletion was identified. In 2001, a licensing system to control domestic groundwater use was introduced by Provincial Regulation No. 16/2001 on Groundwater Management. According to the regulation, users who abstract more than 100 m^3/month of groundwater from wells between 40 and 60m in depth need a license for abstraction, and they must report the amount of groundwater abstraction. However, it is estimated that there are many non-registered groundwater abstraction wells, especially for domestic use.

Charging System to Groundwater Use

The groundwater abstraction charge was introduced as the groundwater tax stipulated by Act Number18/1997. However, the groundwater tax is imposed only on the industrial sector, while other uses such as agricultural and domestic uses were exempt from the tax. The tax price of groundwater is calculated based on a price index that includes three components: a natural resource component that is determined by the zone of abstraction point, the class of water quality, alternative resource availability, and the depth of the aquifer; a recovery and compensation component, which is determined based on abstraction volume and usage; and a raw water price component, which is fixed prices by the cubic metre for deep or shallow groundwater, respectively. The method to calculate the groundwater tax was designed to reflect the economic value of groundwater as a natural resource based on social, economic and environmental values. On the other hand, the effectiveness of the groundwater tax as a means to discourage groundwater abstraction by the industrial sector is weak because the calculated groundwater tax is always lower than the tariff of public water supply (Wangsaatmaja et al. 2007).

Development of Other Water Resources

To reduce groundwater abstraction and mitigate drawdown of groundwater tables, new water sources need to be found to satisfy the growing water demand. Bandung has great potential for surface water development and construction of small dams in the upper basins, and such planning is now underway. However, the major beneficiary of newly developed surface water would be the agricultural sector, and the industrial sector may not be able to get water from the new water development projects. If water from other sources cannot be provided to industries at a reasonable price, they cannot help but to continue abstraction of groundwater or to buy water from private water vendors, who sell water at higher prices than the public water supply charges.

12.2.3 Groundwater Management Challenges

Although a groundwater management scheme has existed in Bandung since the 1990s, the volume of abstracted groundwater is increasing again following the economic recovery, as shown in Fig. 12.1. The biggest reason for this is that industries, the major groundwater user, do not have alternative water sources that would allow for a reduction of groundwater use. The public water supply does not have enough capacity to supply water to the industries. New surface water development is underway, but the industries do not always receive the benefit of the new developments. Hence, groundwater continues to be the only reliable water source for the industrial sector.

To cope with the situation, the West Java Provincial Government has been promoting water recycling and improving water efficiency in the industrial sector. A water-recycling programme was initiated by the provincial government, but responses of industries have not always been positive. According to a questionnaire survey conducted by the West Java Environmental Protection Agency in January 2007, many of the 33 industries that took part in the survey answered that they don't see the economic benefit of water recycling. Some industries reported a lack of incentive mechanisms by the government to motivate industries to promote water recycling. In general, water recycling is a promising option to reduce water use in the industrial sector. However, it may not be the case in Bandung because about 50 percent of the industries in Bandung are textile companies, in which water-recycling potential is low compared with other industries such as steel and chemical companies. The strategy for water recycling needs to be elaborated considering the structure of industries.

The questionnaire survey also revealed that some industries are facing serious water shortages, especially in the dry season, and that they purchased/planed to purchase water from private vendors at the price of 20,000 to 25,000 Indonesian Rupiah (IDR)/m^3 (2 - 2.5 USD/m^3). The price is much higher than the water from the public water supply system, which ranges between 4,725 and 9,600 IDR/m^3 (0.5 - 1 USD/m^3) (Wangsaatmaja et al. 2007). If industries need to rely on water from such private vendors, high water prices may become a burden for the industrial development of Bandung.

12.3 Groundwater Management in Bangkok: Regulating Groundwater Abstraction by Economic Instruments

12.3.1 Groundwater Use and Problems

Geography and Water Supply in the Bangkok Region

The Bangkok Region includes Bangkok and six other provinces in its vicinity, namely Nonthaburi, Samut Prakarn, Pathumthani, Samt Sakhon, Nakhon Pathom and Ayuttaya Provinces. It has long been the centre of the socio-economic development of Thailand. The total area of the Bangkok Region is 10,315 km^2, and the population is about 11 million people. The seven administrative areas share the same groundwater basin. The Bangkok Region is located in the lower Chao Phraya Basin in central Thailand, and comprises about 30 percent of the basin's total area. The main water source in the Bangkok Region is surface water from Chao Phraya River and its tributaries. There are two main water supply authorities in the Region: the Metropolitan Waterworks Authority (MWA) and the Provincial Waterworks Authority (PWA). The MWA is responsible for water supply to Bangkok, Nonthaburi and Samut Prakan (hereafter called "Bangkok Metropolitan Area") and the PWA is responsible for the rest of the region. In 2003, the MWA provided water to approximately 89 percent of the total population in its service area, while service coverage of the PWA was low. For example, the population served by the water supply in Samut Sakhon was just 6 percent of the total population (Babel et al. 2007).

Groundwater Use

There are eight confined aquifers between 50m and 550m in depth in the region, and groundwater is mainly abstracted from the second aquifer (Phara Pradaeng Aquifer, 100m in depth), the third aquifer (Nakhon Luang Aquifer, 150m in depth) and the fourth aquifer (Nonthaburi Aquifer, 200m in deoth). In recent years, abstraction from the deeper aquifers has increased, especially for industrial use. Groundwater use in the Bangkok Region began to rapidly increase in the mid-1950s following economic development and the advancement of pumping technologies. Groundwater use in Bangkok, Nonthaburi and Samut Prakarn increased from 8,300m^3/day in 1954 to 937,000m^3/day in 1976. In the 1950s to the early 1960s, groundwater was mostly used to supplement the public water supply. Private uses, especially industrial uses, have increased since the late 1960s. Currently, the industrial sector accounts for more than 60 percent of the total groundwater abstraction in the Bangkok Region (IGES 2006). Fig. 12.2 shows the correlation between the groundwater abstraction and RGDP in the Bangkok Metropolitan Area. Although the RGDP dropped suddenly in 1997 due to the Asian Currency Crisis, the groundwater abstraction volume did not decrease that much after 1997. This is somewhat different from what was observed in Bandung. As mentioned in the previous section, groundwater use is closely correlated with the fluctuation of RGDP before and after 1997 in Bandung. In Bangkok, however, the groundwater abstraction volume decreased only after the year 2000, when an increase in the rate of the groundwater tax was announced.

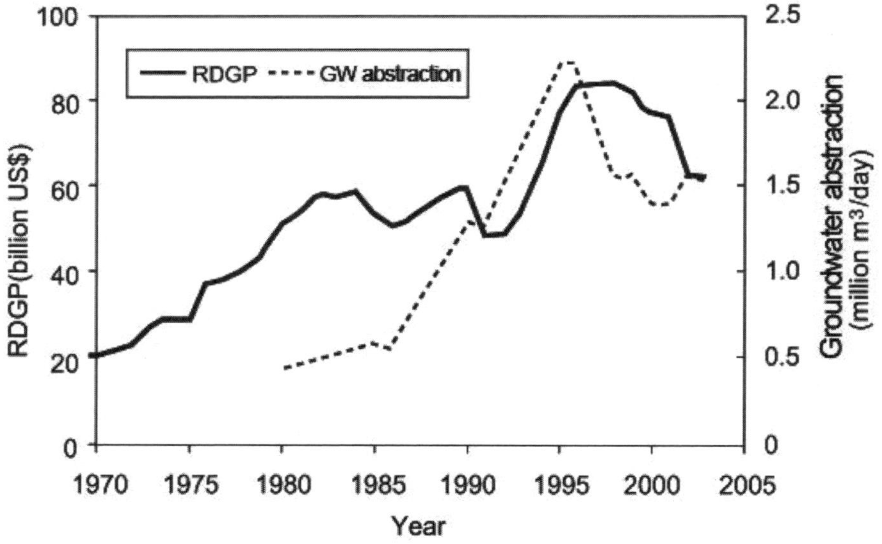

Fig. 12.2. Time course of groundwater abstraction volume and regional GDP in Bangkok. (IGES 2006)

Groundwater Problems

As a result of the intensive abstraction of groundwater in the Bangkok Region, groundwater depletion and land subsidence have become social, as well as economic, problems since the late 1960s. The groundwater level has declined more than 40m since the late 1960s in the Bangkok Aquifer (Babel et al. 2007). According to a survey by the National Environmental Board of Thailand conducted between 1978 and 1981, the most severe subsidence was observed in the eastern part of Bangkok City, which saw an annual subsidence rate of 10cm. Cumulative subsidence from 1978 to 1999 ranged from 50cm to 1m in Bangkok City. The drawdown of the water table and land subsidence was observed in the central part of Bangkok and also was spread throughout the entire region. Chloride concentration increased in Phra Pradeng, Nakhon Luan, and Nontaburi Provinces, where groundwater is most extensively used according to the Department of Groundwater Resource (DGR). However, the reduction of groundwater abstraction volume following the hikes in groundwater charges since the late 1990s has lessened the impacts of groundwater over-abstraction in the Bangkok Region.

Fig. 12.3. Evidence of land subsidence in Bangkok

12.3.2 Groundwater Management

Groundwater Rights

Water rights for groundwater are not clearly defined in statutory form in Thailand. However, it is generally recognized that water is a public commodity that provides the benefits for all.

Organisational Arrangements

The Groundwater Committee was set up in 1977 by the Groundwater Act as a coordinating body of national groundwater management policy. The Committee consists of representatives from different government agencies related to groundwater. The Department of Groundwater Resources (DGR) takes primary responsibility for conceiving and implementing groundwater management plans, including coordinating groundwater policy under the Groundwater Committee. DGR belonged to the Ministry of Industry when it was established, but is now under the Ministry of Natural Resources and Environment.

Legislation for Groundwater Control

Following rapid depletion of groundwater associated with severe land subsidence in the Bangkok Region, the Thai Government took initiative in 1977 through the promulgation of the Groundwater Act to control groundwater abstraction and mitigate problems caused by intensive groundwater use. The Groundwater Act provided the basis of groundwater management in Thailand, including the organisational arrangement for groundwater management as mentioned above. The Groundwater Act also introduced a licensing system of groundwater drilling and abstraction, but it did not have any working measures to control groundwater abstraction at that time. As a result, groundwater abstraction increased and land subsidence became more serious even after the promulgation of the law.

To cope with intensified groundwater problems, the cabinet in 1983 passed a resolution entitled "Mitigation of Groundwater Crisis and Land Subsidence in Bangkok Metropolis," which was followed by a Ministerial Decree in 1985. Following these governmental decisions, the DGR has strengthened groundwater control since 1983 by implementing the following policies:

- designation of critical areas, with tighter control of groundwater abstraction in those areas. A critical area is defined by the Groundwater

Act as an area where more groundwater is abstracted than the natural recharge, which causes a drawdown of groundwater tables, land subsidence, salt-water intrusion, and/or other adverse effects on the environment

- reduction of groundwater use as a source of water supply by the MWA
- introduction of a groundwater abstraction charge

The Groundwater Act was amended in 1992 and in 2003 to strengthen control of groundwater abstraction. The 2003 amendment introduced a new and additional charge entitled the "groundwater preservation charge."

Reduction of Groundwater Use for Public Water Supply

Until the mid-1960s, more than 90 percent of groundwater abstraction in the Bangkok Metropolitan Area was carried out by the MWA for the publci water supply, as shown in Fig. 12.4. The peak of groundwater abstraction volume by the MWA was 510,137 m^3/day in 1981, which was approximately 30 percent of the total groundwater abstraction in the Bangkok Metropolitan Area. Since 1983, groundwater abstraction by the MWA has decreased in response to the governmental policy to mitigate land subsidence. However, groundwater abstraction by the MWA increased again in the early 1990s because of a serious water shortage. In 2000, the abstraction volume again went down, and the present pumping rate is only 6,500m^3/day.

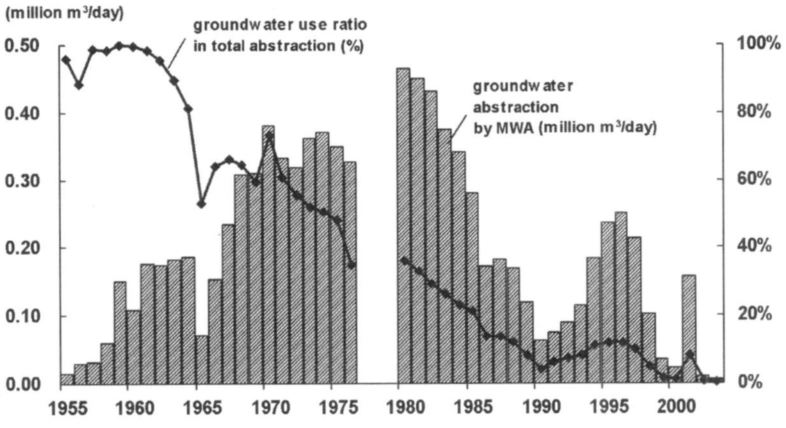

Fig. 12.4. Groundwater abstraction by the MWA and its share of total abstraction in Bangkok Metropolitan Area

Charging System

The groundwater charge was introduced in 1985 in the Bangkok Region, except for Nakhon Pathom and a part of Samut Sakhon, as a measure to discourage groundwater use. However, the original charge was only 1 Baht/m^3 (0.03 USD/m^3). Because this rate was still cheaper than the tariff on the piped water supply by the MWA, it had little effect on the reduction of groundwater abstraction. The charge was increased to 3.5 Baht/m^3 (0.1 USD/m^3) in 1994, but amount of groundwater abstraction was still not reduced. The charge was raised again incrementally up to 8.50 Baht/m^3 (0.25 USD/m^3) in the Bangkok Region between 2000 and 2003. Table 12.1 shows the current groundwater charge in the Bangkok Region. As shown in the table, the groundwater charge is imposed only where the public water supply is available to use. The effectiveness of the charging system largely depends on the expansion of public water supply systems.

Table 12.1. Current groundwater charge scheme in the Bangkok Region

User type	Area with public water supply service	Area without public water supply service
1. Domestic consumption	8.50 Baht/m^3	Exempt
2. Businesses		
Businesses not using agricultural products (proclaimed by minister) as raw material	8.50 Baht/m^3	Discount Charge 75% of groundwater volume (8.50 x 75% = 6.375 Baht/m^3)
Businesses using agricultural products (proclaimed by minister) as raw material	8.50 Baht/m^3	Discount Charge 30% of groundwater volume (8.50 x 30% = 2.55 Baht/m^3)
3. Agriculture		
Crop cultivation	8.50 Baht/m^3	Exempt
Animal farms with groundwater use license not more than 50 m^3/d	8.50 Baht/m^3	Exempt
Animal farms with groundwater use license greater than 50m^3/d	8.50 Baht/m^3	Discount Charge 30% of groundwater volume (8.50 x 30% = 2.25 Baht/m^3)

Furthermore, in 2004, the DGR introduced a new and additional charge entitled "groundwater preservation charge" to the Bangkok Region, which is designated as a critical area according to the amendment of Groundwater Act in 2003. The preservation charge gradually increased over three years from 1 Baht/m^3 to 8.5 Baht/m^3. Separate from the groundwater charge, the revenue from the groundwater preservation charge is designated to be utilised only for conservation of groundwater through the Groundwater Development Fund, which was also established by the amendment of the Groundwater Act in 2003. The Fund is supposed to be used only for groundwater conservation purposes, such as hydrogeological studies, investigation of groundwater-related issues, creation of master plans and provision of grants for activities related to groundwater conservation.

In 2006, the sum of groundwater charge and preservation charge became more than the water tariff of the public water supply. The higher price of groundwater has given groundwater users good motivation to reduce their groundwater abstraction and use more water from the public water supply systems. Fig. 12.5 shows that the total groundwater use in the Bangkok Region has declined since the groundwater charge was raised in 2000.

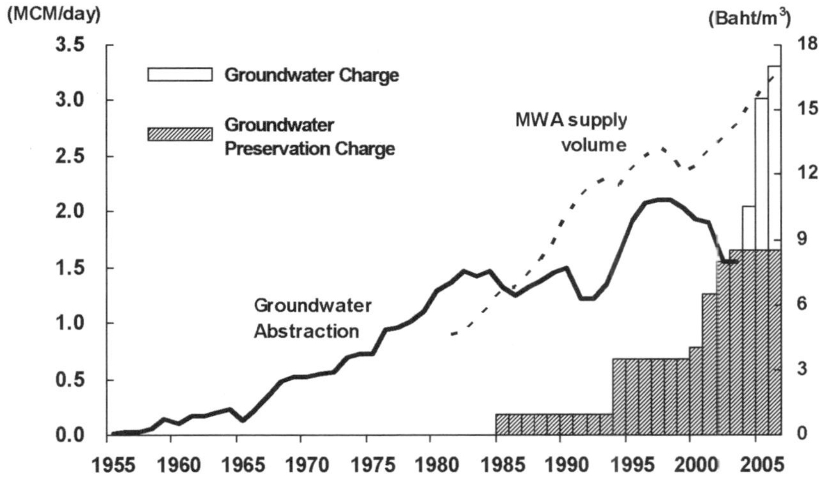

Fig. 12.5. Change in groundwater abstraction, MWA supply volume and the Groundwater Charge (Babel 2006)

12.3.3 Future Challenges of Groundwater Management

The increase in capacity of the public water supply scheme in conjunction with the increase in groundwater charges has recently been effective in controlling groundwater use. However, industries in Bangkok have raised a concern regarding the reliability of the piped water supply by the MWA in terms of quality and quantity.

A survey of 85 factories in Samut Prakan that use more than 300 m^3/day of groundwater was conducted in 2006 by the Asian Institute of Technology to record their opinion on groundwater management policies. 55 percent of factories use only groundwater or use it in combination with the piped water supply. One of the major reasons that they use groundwater was its quality and the stability of supply in comparison with the piped water supply. 33 percent of the factories using water from the MWA expressed doubts on the reliability of quantity and 37 percent had concerns about quality. Chlorine used for water treatment is the main concern in terms of quality. 75 percent of those surveyed answered that they would purchase more water from the MWA if the reliability of water supply were improved in terms of both quantity and quality. Hence, it was revealed that improving the reliability of the public water supply systems is the key to the transfer of industrial groundwater uses to the public water supply systems.

Due to strengthened groundwater management policy, the groundwater level has recovered and the rate of land subsidence has been eased in recent years in the Bangkok Region. Corresponding to the recent alleviation of the impacts of groundwater abstraction, a slight change of groundwater management policy is now being explored. The new direction may seek sustainable use of groundwater rather than strict reduction of groundwater abstraction. For example, the exemption of groundwater preservation charge for domestic purpose and the permission of groundwater use for specific industries, such as the food industry, that require more stable and higher quality water sources for production could be introduced. The future challenge in Bangkok may be how to use groundwater in more efficient and sustainable ways under a comprehensive policy of water management.

12.4. Groundwater Management in Osaka City, Japan: Control of Groundwater Use by Providing Alternative Water

12.4.1 Groundwater Use and Problems

Water Supply in Osaka City

Osaka City is located in the western part of Honshu Island in Japan. The administrative area is 221.96 km^2 and the population was 2.6 million in 2002. The Yodo River runs through the centre of Osaka City and is the main source of drinking water in the public water supply system. Public waterworks were first constructed in 1885 and supply coverage reached 100 percent in 1970 (Osaka City Waterworks Bureau 2003).

Groundwater Use

While domestic use mostly depends on surface water sources, industries have long used groundwater as a cheaper and reliable water source in the course of industrial development. In 1953, estimated total groundwater abstraction volume was 21 million m^3/year, which increased up to 123 million m^3/year. About 80 percent of the total groundwater abstraction in Osaka City was made by industries in 1962. The purpose of groundwater use by industries was mainly cooling (43 percent), followed by processing (29 percent) and washing (20 percent).

Non-potable groundwater use in buildings has also increased since the 1950s. Groundwater is used for cooling and toilet flushing in large buildings such as office and commercial buildings and apartments (Japan Society of Civil Engineers Kansai Chapter 2002).

Groundwater Problems

The intensive use of groundwater during the economic growth period in the late-1950s to the 1960s resulted in the rapid drawdown of the groundwater table and was associated with severe land subsidence. Cumulative land subsidence between 1935 and 1970 in the coastal area of Osaka City reached 280cm. Because of land subsidence, the height of dykes became lower and thus their ability to protect the city from high tide and flooding diminished. Damage to public infrastructure such as bridges and railway stations by land subsidence was also identified at this time. As countermeasures against land subsidence, Osaka City spent about 2.5 million USD between 1955 and 1969 on projects to reinforce the dykes and raise bridges (Committee on Comprehensive Countermeasure against Land Subsidence in Osaka 1972).

12.4.2 Groundwater Management

Groundwater Rights

The Japanese Civil Code stipulates that the rights of landowners reach above and below the land. The groundwater rights of landowner are based on this code. Hence, the right to abstract groundwater may not be constrained by local authorities such as the Osaka City Government unless there is a law to override the Civil Code (Committee on Comprehensive Countermeasure against Land Subsidence in Osaka 1972).

Organisational Arrangement

In principle, local governments are responsible for groundwater management in Japan. Hence, Osaka City is in charge of groundwater resources within its administrative boundary. In most municipalities in Japan, the environmental departments are the responsible sections for groundwater management both in terms of quantity and quality. Where groundwater abstraction is restricted, such as Osaka City, the local municipalities control new well drilling and compile reports on the amounts of groundwater abstraction by the licensed users.

Legislation to Control Groundwater Use

As mentioned, groundwater is regulated by the local governments in Japan. However, when Osaka City took action to regulate groundwater use in the 1950s for the first time, there was no existing law that gave the local governments such authority to regulate groundwater. Therefore, the city government worked together with other local governments that suffered from land subsidence such as the Tokyo Metropolitan Government to request the national government to formulate a national law to control groundwater use (Committee on Comprehensive Countermeasure against Land Subsidence in Osaka 1972).

As a result, two national laws to regulate groundwater use were promulgated. The first was the Industrial Water Law in 1958 and the second one was the Law Concerning the Regulation of Pumping-up of Groundwater Use in Buildings (hereafter called the "Building Water Law") in 1962 (See also Chapter 2).

Industrial Water Law: The Industrial Water Law of 1958 was the first law related to controlling groundwater use. One of the important features of the law was that it had the dual purpose of promoting sound industrial development and mitigation of land subsidence in the designated areas. To achieve these objectives, the establishment of a new and additional water supply scheme called the Industrial Water Supply Works (IWSW) was stipulated as a prerequisite for the enforcement of regulation on groundwater abstraction in the designated areas. In many cases, it took some time to construct infrastructure for the IWSW. Therefore, the regulations on groundwater use could not be implemented immediately after the promulgation of the Industrial Water Law.

As a way to conserve groundwater, the law stipulated a permission system of groundwater abstraction in designated areas and gave local governments the authority to implement the law. However, since the law regulated only newly constructed wells, groundwater abstraction from existing wells continued. As a result, the law was not truly effective until its amendment in 1962, when abstraction from wells with a borehole size more than $6cm^2$ and a depth less than 500–600m was prohibited. This meant that almost all the existing wells had to be closed and no new industrial wells could be constructed (Committee on Comprehensive Countermeasure against Land Subsidence in Osaka 1972).

Building Water Law: The Building Water Law, which regulated groundwater abstraction for building and commercial use in designated areas, was promulgated in 1962. The law banned both the use of existing wells and the construction of new wells in the areas designated by the law. Because there were technical options available to reduce water used for cooling and flushing, groundwater abstraction was quickly phased out and replaced by water from the existing public water supply.

Industrial Water Supply Works (IWSW)

As mentioned above, the Industrial Water Law stipulated the establishment of the IWSW, which provided industries with water at reasonable price. Because water provided by IWSW was and is not intended for drinking, the water treatment cost is less than the treatment cost for drinking water (Aya and Matsumoto 2003). In addition, subsidies by the central and local governments made the water price of IWSW much lower than the water tariff of the municipal drinking water supply.

In 1954, before the Industrial Water Law was promulgated, the Osaka City Government had already constructed IWSW as a measure to minimise groundwater use by the industrial sector, though the initial coverage area was very limited. After the Industrial Water Law was promulgated in 1958, the city, with financial support from the national government, expanded the coverage area by the construction of new infrastructure. The tariff on water supplied by the IWSW was also reduced from 6.8 yen/m^3 (about 0.02 USD/m^3 at the currency rate in 1962) to about 4 yen/m^3 (about 0.01 USD/m^3 at the currency rate in 1962) in 1962 due to a subsidy provided by the national government. The estimated cost of groundwater abstraction at that time was about 3–4 yen/m^3 (Committee on Comprehensive Countermeasure against Land Subsidence in Osaka 1972), so the reduction of the IWSW water price motivated industries to change their water sources from groundwater to that supplied by the IWSW. With strict enforcement of the ban on groundwater use by the amendment of the Industrial Water Law in 1962, water provided by the IWSW replaced groundwater in a short time. As a result, land subsidence was mitigated, as shown in Fig. 12.6. However, the water supply volume by the IWSW has decreased since the mid-1970s following the reduction of water demand by the industrial sector. One of the reasons for the decrease of IWSW demand was that the industrial sector promoted water recycling and reuse to save on costs for water and wastewater (Osaka City Waterworks Bureau 2005, Aya and Matsumoto 2003).

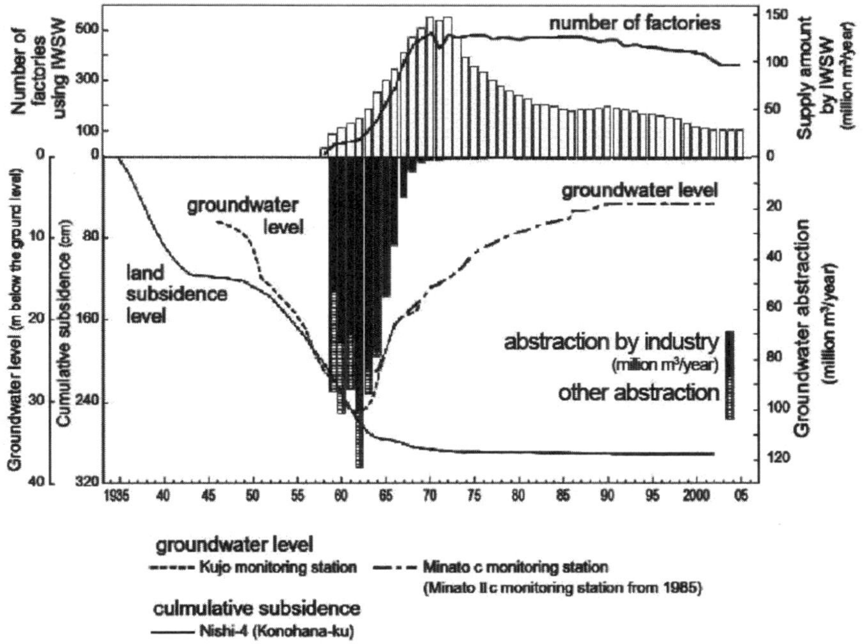

Fig. 12.6. Reduction of groundwater abstraction through the provision of alternative water resources in Osaka City (IGES 2006)

12.4.3 Challenges to Groundwater Management

Osaka City is an example of successful groundwater management to mitigate groundwater over-abstraction between the 1950s and the 1970s. Although Osaka City, as well as other cities in Japan, faced difficulty in controlling the amount of groundwater abstraction at first, public support for Osaka City's policy on groundwater protection has made it possible to implement strict regulations on groundwater abstraction by industries. A low priced water supply by the Industrial Water Works has also given industries time to adjust to the new water supply scheme, and also to improve the recycling rate of industrial water. As a result of three decades of groundwater abstraction regulation, the groundwater table has recovered. However, the recovery of the groundwater table has brought about a new problem; underground infrastructures that have been built in the last three decades are now receiving bouyancy force due to rising groundwater table. How to cope with this new issue has become a new challenge in groundwater management.

12.5 Conclusion

Groundwater has played a major role in the development of some Asian cities. However, the resource is often overexploited without due consideration of the sustainability of the resource. To cope with the groundwater-related problems in the case study cities, measures to control groundwater use have been introduced. A permission or licensing system is the basis of good management practice and is useful for control of groundwater exploitation as well as obtaining information of abstraction volume. One of the challenges is how to minimise the number of unregistered groundwater users.

Bangkok's experience shows the effectiveness of economic instruments to reduce groundwater use, but the effectiveness of the system there depends on whether the minimum volume of existing water demand can be satisfied by other water resources. The introduction of an economic instrument to charge for abstraction is useful to make groundwater users realise the value of groundwater, which had been neglected for a long time. On the other hand, the current groundwater price is not determined based on the true value of groundwater and externalities. It is set in comparison with other water prices such as the tariff of public water supply. How to reflect the true value of groundwater to the groundwater charge should be considered in the future.

As shown in the three cases, groundwater management cannot be effective if it is planned and implemented in consideration of only groundwater. It should be planned as part of a comprehensive picture of water management including surface water development and a water supply scheme. In many cases, the industrial sector is a major user of groundwater in megacities; hence, industrial structure has a great impact on groundwater management. Promoting industrial water reuse and recycling is a very important way to minimise groundwater abstraction.

Acknowledgements

This paper is based on the findings of the research project entitled "Sustainable Water Management in Asia (SWMP)" conducted by the Freshwater Resources Management Project of the Institute for Global Environmental Strategies (IGES), Japan. The authors wish to thank to the West Java Environmental Protection Agency (Indonesia) and Asian Institute of Technology (Thailand) for their contribution to the case studies in Bandung and Bangkok, respectively.

References

Aya H, Matsumoto K (2003) Water for sustainable development of industry – pole and contributions of industrial water supply works in Japan (in Japanese). Kogyo Yosui 537:2-12

Babel MS, Gupta AD, Sto Domingo ND, Kamalamma AG (2007) Sustainable groundwater management in Bangkok. Sustainable groundwater management in Asian Cities. Institute for Global Environmental Strategies, pp 26-43

Babel MS (2007) SWMP Study Report on Groundwater Management, Bangkok, Thailand (unpublished). Institute for Global Environmental Strategies

Committee on Comprehensive Countermeasure against Land Subsidence in Osaka (1972) Osaka jiban chinka taisaku shi [History of Countermeasures against Land Subsidence in Osaka] (in Japanese). Isshindo Insatsu, Osaka

IGES (2006) Sustainable groundwater management in Asian Cities - a summary report of research on sustainable water management in Asia. Institute for Global Environmental Strategies

Japan Society of Civil Engineers Kansai Chapter (2002) Chikasui seigyo ga jiban chinka ni oyobosu eikyo hyoka [Assessment of the impact on the geo-environment by groundwater control] (in Japanese). Japan Society of Civil Engineers Kansai Chapter, Osaka

Kataoka Y (2002) Sustainable groundwater management in Asian Cities: lessons from Osaka. International Review for Environmental Strategies 6(2):269-290

Osaka City Waterworks Bureau (2003) Environment Report FY2002

Osaka City Waterworks Bureau (2005) Osaka shi kogyo yousuido 50 knish [History of 50 years of the Osaka city Industrial Water Supply Works] (in Japanese)

Wangsaatmaja S, Sutadian AD, Rachmat A, Prasetiati MAN, Lufiandi (2007) Alternative water resource and recycle programs as effort to strengthen groundwater management in Metropolitan Bandung. Sustainable groundwater management in Asian Cities. Institute for Global Environmental Strategies, pp 44-67

13. Groundwater Contamination in Asian Coastal Cities: Case Study of Ho Chi Minh City, Vietnam

Micha Sigrist

Department of Urban Engineering, the University of Tokyo

13.1 Introduction

There are many large and mega-cities, such as Bangkok, Jakarta and Mumbai located near coastlines. They were built as trading centres and then have evolved to become major cities receiving population from rural areas during the rapid economic growth period in the last couple of decades.

Also Ho Chi Minh City (hereafter HCMC) is one such Asian coastal city that has grown fast over the last few decades. HCMC is now facing similar problems and challenges that cities like Osaka (Japan), Bangkok (Thailand) and Hanoi (Vietnam) have dealt with for many years.

A common feature among all these cities is their proximity to the sea, which has influenced the water bearing formations that we now find in the subsurface. Proximity to the sea can be a reason for natural groundwater contamination, which can be further aggravated by over-exploitation of fresh groundwater occurrences.

To prevent the contamination of valuable freshwater resources, many countries and cities now conduct groundwater vulnerability assessments as part of urban development planning. Providing vulnerability maps enables cities to prevent future groundwater contamination and deterioration.

13.2 Hydrogeological Characteristics of Asian Coastal Cities

13.2.1 Hydrogeological Settings of Asian Coastal Cities

Coastal cities such as Bangkok, Hanoi and Osaka have a tectonically active geological history and have experienced faulting with subsidence, uplift and sea level changes. Rivers have always flowed through these tectonically created valleys and have filled in the basins with clastic material. These fluvial deposits are often intercalated with marine sediments due to the sea level changes. These three cities also receive a good amount of yearly rainfall. Bangkok receives around 1460mm/year, Osaka 1340mm/year (WorldClimate 2005) and Hanoi receives 1800mm/year (Berg et al. 2001).

Looking at the details of the geological settings, each city has a unique setting and boundary conditions. In Bangkok, for example, the elevation of the flood plain is between 1 and 2 m above the mean sea level, and there is a thick clayey layer on top that covers nearly the whole basin. This top

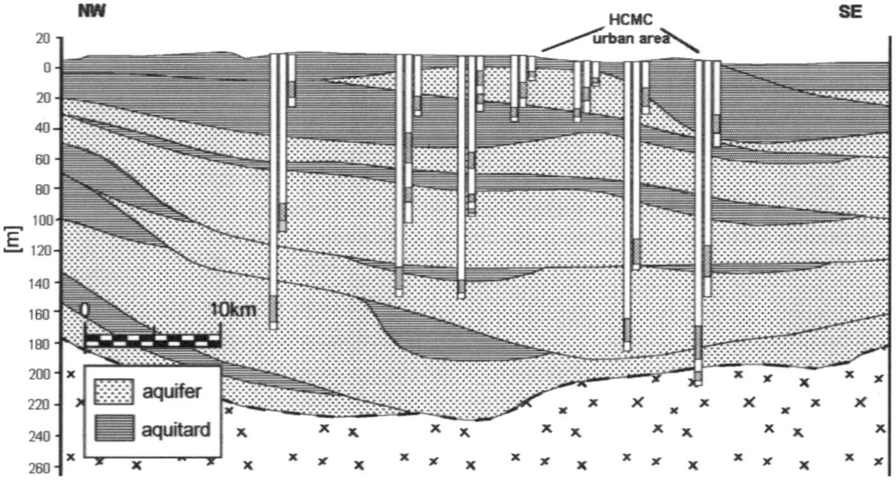

Fig.13.1. Schematic of the aquifer system in HCMC showing the non-continuity of aquitards and connectivity of aquifers.

layer is made of marine deposits and has a thickness between 20 and 30m (Ohgaki et al. 2006). Even though the sedimentary fill reaches more than 2000m in thickness in some places, of primary importance is the uppermost 450m layer, which is divided into eight aquifers. For simplicity's sake, they are often treated as continuous aquifers, separated from each other. The aquifers consist of sandy and gravelly sediments interbedded with stiff to hard clays. The clay layers separate the aquifers over a wide area, but it is expected that there are leakages between them when much groundwater is pumped (Phien-wej et al. 2006). This is common in Asian coastal cities, including HCMC. Fig. 13.1 shows a schematic diagram of the coastal aquifer system in HCMC.

The hydrogeological setting in Hanoi is based on a complicated geological history. Three major faults have been active and created basin structures. Through subsidence, transgressions and river processes the basin was filled in with sediments. The result is a 50 to 90m thick Quaternary formation in the Hanoi area that can be divided roughly into two sequences. The upper sequence is made up of fine sediments like clay, sandy clay and fine sand while the lower sequence consists of gravel with cobbles and coarse sand. The Quaternary sediments are underlain by conglomerated sandstone and siltstone (Berg et al. 2001). The intercalated peat deposits form a naturally anoxic condition that increases the amounts of iron, manganese and arsenic in groundwater. In the Hanoi area, private tube wells mostly tap the groundwater between 12 and 45m in depth, while the groundwater between 30 to 70m in depth is exclusively exploited by urban water treatment plants.

In Osaka, active tectonics also played an important role in the evolvement of a sedimentary basin. The cross-section in Fig. 13.2 shows how the faults split the basement in blocks and created a basin through differential tectonic subsidence. The basement consists of slate and sandstone and is covered by the Osaka group, which consists of alternating layers of freshwater sand deposits that are intercalated with twelve marine clay layers. On top of these alternating layers, an approximately 15m thick deposit of mainly sandy freshwater deposits can be found. Finally, the most recent deposits consist of about 15m marine clay and are exposed to the surface. All these layers are plunging toward the southwest and increase in thickness in the same direction (Yamanaka et al. 2007).

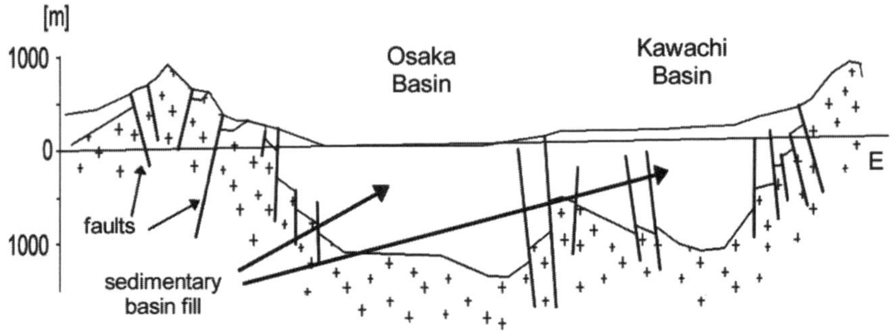

Fig. 13.2. Simplified tectonic setting in the Osaka area. (in Poland 1984)

In the past, the water supply for Osaka came mostly from the Yodo River, especially since groundwater was often too salty to be used for drinking water. However, as industry started to boom, groundwater usage came to play an important role for that sector. Groundwater used for industry and buildings was mainly used for cooling and flashing purposes until abstraction was banned due to severe land subsidence (Ohgaki et al. 2006).

13.2.2 Saltwater Problem in Coastal Cities

Groundwater is a cheap and clean water source that needs little if any treatment. Therefore it is a sought after source of water. In coastal areas, however, overabstraction of groundwater can lead to saltwater intrusion into the aquifers, which make it necessary to remove salts from the groundwater before supplied as drinking water.

Coastal cities have grown tremendously in the last century. Now nearly half of the earth's population is living in coastal cities and eight out of the ten biggest cities are located on the coast (UN Atlas of the Oceans). Due to this huge population growth, the water demand in those places has increased manifold and has lead to the abstraction of valuable fresh groundwater in large quantities. Naturally, the water levels used to be higher than the sea level on land, but due to over-abstraction of groundwater, water levels have fallen below the sea level and have caused seawater to flow inland and contaminate the freshwater sources.

Saltwater intrusion caused by groundwater abstraction is a global phenomenon. In many countries around the world, many wells have had to be abandoned because the water they produce has turned unsuitable for drinking due to a high salt content. Figure 13.3 shows a natural situation of saltwater-freshwater interaction and two different mechanisms of saltwater intrusion caused by groundwater abstraction.

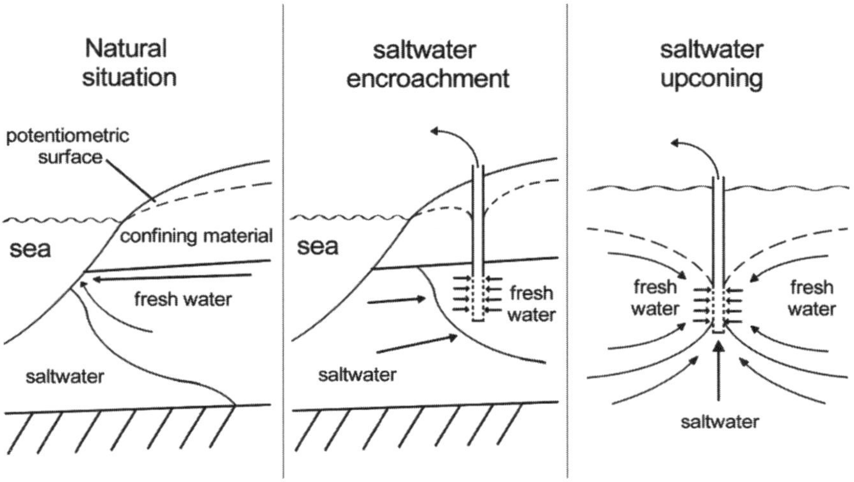

Fig.13.3. Natural coastal situation, saltwater encroachment and upconing

Los Angeles, for example, has a long history of saltwater intrusion (Edwards 2002). Before 1920, groundwater withdrawal was smaller than its recharge but as the population increased, groundwater withdrawal also increased and the groundwater level fell below sea level. The danger of saltwater contamination was recognized in the 1950s. Consequently, barriers were installed in an attempt to halt the intrusion. The barriers consisted of freshwater injection wells that were intended to form a virtual wall to prevent saltwater flow to the production wells. The "barriers" did not work as expected everywhere because underground geologic structures are complex and difficult to predict accurately. The USGS and its co-operators are still working on finding the real flow paths of seawater using techniques like reflection seismology, deep continuous coring, gravity modelling and fluid flow modelling.

Like Los Angeles, Bangkok is facing the problem of saltwater encroachment. The public water supply using groundwater started in 1954 with 8360m^3/d and increased to about 110000 m^3/d by 1960 (Das Gupta 1988). However, the public water supply could hardly meet the demand for industrial and domestic use and therefore private groundwater abstraction grew as well. In 1979, total water abstraction reached 1200000m^3/d. This over-abstraction of groundwater brought about the lowering of water tables in several aquifers. Over a wide area, it dropped to more than 40m below the surface.

Abstraction of these huge amounts of groundwater led to serious problems with land subsidence. Moreover, many wells had to be closed because water levels dropped below the lowest possible pump settings. The overuse of groundwater also led to the salinization of many wells that consequently had to be abandoned because the produced water was no longer suitable for drinking water. Saltwater encroachment is happening only partly because of proximity to the sea. The major reason for the salinization in the Bangkok area can be found in its sedimentary history. The deposition of much of the basin fill took place when Bangkok was below sea level or was subjected to marine conditions after deposition. Hence, saline water bodies were entrapped in the sediment in many places. Due to heavy water abstraction, this ancient water was mobilized and mostly responsible for the salt contamination.

Similar to the Bangkok area, the lower Mekong River basin contains freshwater aquifers at several depths. These are important sources for water supply to the domestic and industrial sectors. However, most of the aquifers are brackish or saline in these areas. The freshwater areas of each aquifer make up 30 to 60% of the total (pers. Comm.). The Mekong delta had been subject to sea level changes. In this period seawater had intruded and was entrapped into the water bearing formations. Some of the saltwater may have been flushed out by recharging freshwater, but due to the small dip of the sediment layers and flat topography saltwater flushing was not complete.

The Mekong River delta area faces another source of saltwater intrusion on the surface. Its terrain is very flat and reaches only 5m of elevation even at a few hundred kilometres from the river mouth on Cambodian terrain. This low elevation makes the area prone to saltwater encroachment from the sea due to the tidal water level changes that have a magnitude of 3 to 3.5m in the East China Sea (White 2002). The saline water intrusion through surface waters becomes more problematic during the dry season (December to May) when the Mekong River has little runoff and is not able to prevent saltwater from intruding the river and canal system. This surface water salt intrusion is at likely to contaminate the local unconfined shallow aquifers that are often used as drinking water sources since they are easily accessible.

13.2.3 Groundwater Contamination Vulnerability

The risk level for contamination or pollution of groundwater is not the same everywhere. The term "groundwater contamination vulnerability" is intended to represent the sensitivity of an aquifer to be negatively affected by a contaminant load that would deteriorate the quality of the groundwater. As most groundwater is recharged by water infiltrating through the soil, it is obvious that the soil, or geologic/hydrogeologic condition, has a significant influence on the vulnerability of the groundwater source.

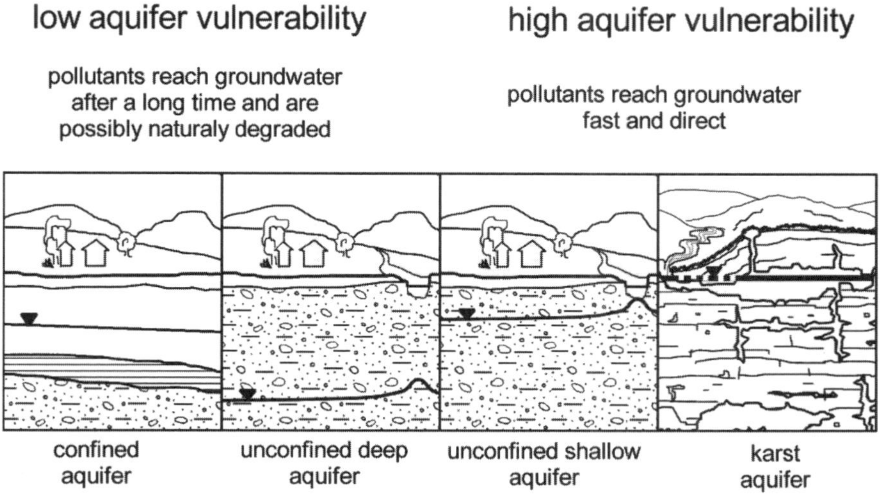

Fig.13.4. Simplified vulnerability classification

A water table close to the surface in an unconfined aquifer is highly vulnerable to contamination, making it possible for any contaminant load to reach the groundwater quickly. On the contrary, a deep water table means there is a long way for any contaminant to go before it reaches the groundwater (Fig. 13.4). This means a lower risk of immediate contamination and also allows for certain contaminants to be eliminated by biochemical degradation and chemical reactions (Foster et al. 2002). These processes, however, depend highly on the site specific characteristics of the soil profiles.

The activities at the land surface have a significant influence on the pollution hazard of the groundwater. Specific to the land use profile, water recharge mechanisms are changed in terms of quantity and quality.

Some typical contaminants connected to specific land uses can be listed as follows (modified after Foster et al. 2002).

Table 13.1: Common groundwater contaminants and associated pollution sources

POLLUTION SOURCE	TYPE OF CONTAMINANT
Agricultural Activities	Nitrates; ammonium; pesticides; faecal organisms
On-site Sanitation	Nitrates; pharmaceuticals; micro organisms; pathogenic micro organisms; organic matter
Petrol Filling Stations & Garages	Aromatic hydrocarbon; benzene; phenols; halogenated hydrocarbons
Solid Waste Disposal	Ammonium; salinity; halogenated hydrocarbons; heavy metals
Metal Industries	Trichloroethylene; tetrachloroethylene; halogenated hydrocarbons; phenols; heavy metals; cyanide
Sewage Sludge Disposal	Nitrates; halogenated hydrocarbons; lead; zinc
Printed Wiring Board (PWB) Manufacture; Semiconductor Manufacture and Component Assembly	Heavy metals; chlorinated volatile organic chemicals (VOC); polybrominated diphenyl ethers (PBDE); nonylphenol; phthalates
Cemeteries	Pathogens; metals; organic matter, increased salinity
Oil & Gas Exploration/Extraction	Salinity (sodium chloride); aromatic hydrocarbons

Since groundwater vulnerability is site specific, depending on the hydrogeologic conditions, it is useful to create aquifer vulnerability maps that can be used for environmental planning and decision making. Several methods for creating vulnerability maps have evolved. Some focus on the hydrogeological setting only, or are based on the delineation of protection zones for groundwater supply systems where the groundwater flow and contaminant processes within the aquifer are considered. Others target the soil and unsaturated zone as well as the aquifer medium (Gogu and Dassargues 2000).

A classic method of creating a vulnerability map is the GOD (Groundwater hydraulic confinement/Overlaying strata/Depth to groundwater table) method. This method considers the soil and unsaturated zone without taking into account the transport processes in the saturated zone.

To create a vulnerability map based on the GOD method, information on the hydraulic confinement, overlying strata (vadose zone or confining beds) in terms of their lithological character and depth to groundwater table (or to groundwater strike in confined aquifers) has to be collected. The spatial information is then indexed as suggested by Foster et al. (2002) for a series of points on a grid where the results can be contoured to show the spatial distribution of the vulnerability indices. This work can be greatly eased by using a GIS (Geographical Information System) for overlaying the different data sources. Fig. 13.5 shows the concept of creating a GOD vulnerability map.

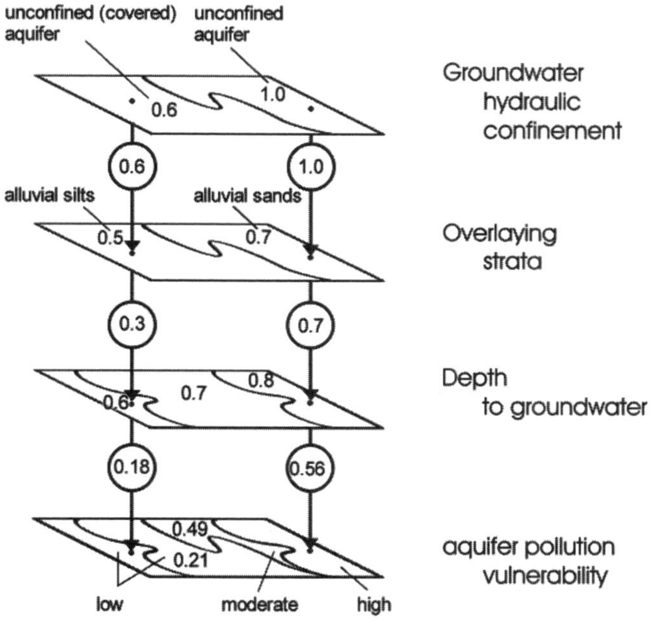

Fig.13.5. Generation of an aquifer vulnerability map using the GOD method (Foster et al. 2002)

The Water Framework Directive of the European Union (EU) requires disclosure of the extent to which groundwater bodies are used and whether they fulfil already defined environmental targets. To follow these directives and laws, the characterization of groundwater vulnerability came to play an important role in the EU member states (Voigt et al. 2004).

In 1987, Portugal became the first EU member state to have its territory mapped for aquifer vulnerability at a scale of 1:500000 (Lobo Ferreira and Oliviera 2003). Since then, different maps have been created at different scales for different areas using the same DRASTIC method. DRASTIC stands for the seven parameters used to create the final model: Depth to water, net Recharge, Aquifer media, Soil media, Topography, Impact of vadose zone media, and hydraulic Conductivity of the aquifer. Later, different methods of vulnerability mapping were applied in some areas and comparisons between the methods became possible.

In the town Murge (Apulia, Italy), intrinsic vulnerability assessment was carried out for a karstic area. Intrinsic is a term used to define the vulnerability of groundwater to contaminants generated by human activity (Gogu and Dassaragues 2000). It is difficult to assess groundwater vulnerability in karst areas because the characteristics of such a groundwater system cannot be easily determined. To account for the special geologic conditions that differ from normal sedimentary settings, the SINTACS vulnerability approach was modified and applied. The results were compared with actual groundwater nitrate concentration, which confirmed the reliability of the modified method (Marsico et al. 2004). The mapping provides a complete picture of pollution risk in the area and is helping environmental protection and the management of natural resources.

The term SINTACS is derived from the Italian terms for depth to groundwater (Soggenza), effective infiltration (Infiltrazione), unsaturated zone attenuation capacity (Non saturo), soil/overburden attenuation capacity (Tipologia della copertura), saturated zone characteristics (Acquifero), hydraulic conductivity (Conducibilita) and topographic surface slope (Superficie topografica). This method is similar to the DRASTIC method, but it includes considerations of the stream network density, connection of aquifer to surface water and land use type. Both DRASTIC and SINTACS are types of the Point Count System Model (PCSM) which are suitable for areas where data is available in sufficient quality and quantity. Enough data are usually available or easy to survey in plane areas. For mountainous and hilly areas, it has proved better to use the Hydrologic Complex and Setting (HCS) method because data density and quality is often low in those areas.

The aforementioned GOD method is a Rating System (RS) which considers only three parameters: depth to water, characteristics of the unsaturated zone and hydrogeological features. A smaller number of parameters does not imply that it is a less convincing method. It is always necessary to compare different approaches and to verify their applicability. In some areas, the less complex method may prove to be more accurate or be equally accurate with less labour.

Cavita and De Maio (2004) explained how an overall approach combining the highly advanced, parametric PCSM method of SINTACS R5 with a basic HCS method. Their method enables making a vulnerability map to cover a large area.

The aquifers underneath Asian coastal cities are vulnerable to different pollutants to different degrees, just like in other cities. The most vulnerable aquifers are the unconfined aquifers near the surface. In some areas they are fairly well protected by a topsoil layer that is less permeable and has enough thickness to prevent quick infiltration of any pollutant. However, these natural barriers are not continuous everywhere and might be cut through by construction work or by natural surface waters so that a direct connection to those aquifers becomes possible. These aquifers are also very vulnerable to contamination by nitrate, coliform and chloride which can originate from on-site sanitation facilities like septic tanks or from industry. Besides these point sources, non-point sources such as excessive application of fertilizers in suburban farms might be a possible cause of higher nitrate concentrations recorded widely (Ohgaki et al. 2006).

Naturally, in coastal cities, there is a source of saltwater at the coast and, depending on the terrain and river conditions, even far inland. This makes aquifers vulnerable to saltwater contamination. The presence of saltwater or brackish water in the aquifers makes it likely that over-abstraction of groundwater will accentuate the problem. Where brackish or salt-water bodies are near, the abstraction of fresh water will often lead to the salinization of the recovered water and abandonment of the wells. This vulnerability exists in all aquifers below coastal cities.

13.3 Case Study in Ho Chi Minh City

13.3.1 Geographic features of Ho Chi Minh City

Climate and Topography

HCMC is located in the south of Vietnam and the city centre lies about 50km from the shoreline (Fig. 13.6). The area is located in the subequatorial and tropical zone and has an average temperature of 27°C. The climate is dominated by the monsoon regime that results in two distinct seasons. The rainy season lasts from May to September and the dry season from October to April. Precipitation is 1800mm/year, of which 80 to 85% falls during the rainy season (Dan et al. 2005).

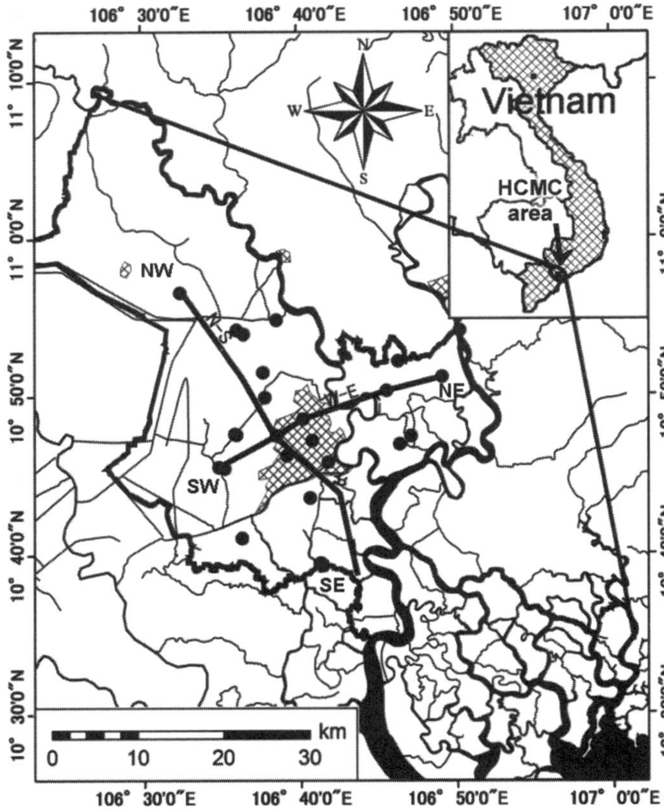

Fig.13.6. Study area with sampling locations and cross-sections

The whole area is at a low elevation and can be divided into three zones. The first one has an elevation below 2 meters above sea level (mabs). The soils to the West and Southwest of the city are often acidic and contain alum. The area comprises the downstream area of the Sai Gon and Dong Nai Rivers and is strongly affected by the tidal movements in those rivers. The second zone is between 2 and 5 mabs, and includes the city centre, the Hoc Mon and Binh Chanh districts and the southern part of Cu Chi district. Dense residential areas or agricultural fields mostly cover these regions. The third zone ranges from 5 to 25 mabs and includes most of Cu Chi district, the northern part of Thu Duc district and the whole Go Vap district (Dan et al. 2005).

Demography and arising problems

HCMC is the largest city in Vietnam and represents a centre for cultural, social and economic activities. The official population was 5.63 million people in 2003 but could be as high as 7.5 million (Dan et al. 2005). The economic and population growth has brought about unprecedented stresses on the freshwater supply and the wastewater disposal system. HCMC is suffering a deficit in urban infrastructure such as road maintenance, drainage networks, sanitary facilities and social facilities (Bolay et al. 1997). In 1997, there were 680 factories and more than 22500 small manufacturing enterprises that had nearly no wastewater treatment applied to their discharge. Another source of pollution in the area is agricultural runoff from the high usage of chemical and organic fertilizers and the wide spread use of pesticides (Nguyen Duc et al. 2003).

The various sources of surface and ground water pollution provided an impetus to analyze these pollution sources and search for fresh groundwater.

13.3.2 Groundwater Resources in HCMC

Water supply in the area originates from three sources: surface water from Dong Nai River, water from the Saigon River and groundwater. Groundwater is abstracted by more than 150000 wells for drinking and industrial purposes and amounts to more than 500000 m^3/d (Dan et al. 2005).

Hydrogeology

Of the five uppermost aquifers in the area, three are the main sources for groundwater abstraction (Fig.13.1 and 13.7). The shallowest aquifer (Pleistocene) ranges between 20 to 50m below ground and has a rather impermeable cover (commonly 5 to 10m thick) consisting of clayey material, which is weathered to laterite in some areas. The aquifer itself consists of fine to coarse-grained sand, and patches of grit and gravel. The second aquifer (upper Pliocene) is not exposed to the surface and is found 40 to 80m below ground. This aquifer is also covered by a low permeable layer, which shows a thickness of 2 to 48m depending on the location. The aquifer itself consists mostly of fine to coarse-grained sand with quartz gravel all over the area. Its thickness ranges from 36 to 82m and it is interbedded with thin clay layers. The third important aquifer (lower Pliocene) is encountered about 125m below surface and consists again of a low permeable zone in the upper area and an aquifer in the lower zone. The low permeable layer is made up of clayey material and ranges from 5 to 10m in thickness. The aquifer consists mainly of fine to middle-grained sand and quartz gravel with coarse sand. This layer is continuous over the whole area and its thickness is everywhere greater than 50m.

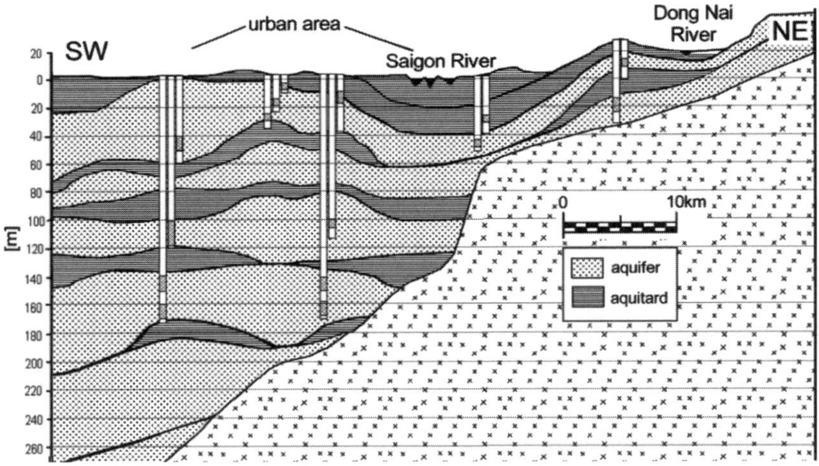

Fig.13.7. Hydrogeological cross-section of HCMC in NE-SW direction

Generally, the aquifers below HCMC are protected from pollution by a rather impermeable topsoil layer that consists of clayey and silty material. However, the cross-section in Fig. 13.7 shows that the cover has a wide range of thickness and is not likely to be continuous over the whole area. Therefore, the uppermost aquifer is especially vulnerable to all kinds of contamination. This includes pollution originating from unlined waste dumpsites and canals that cut through the area and act as an open sewer for all kinds of wastewater. In addition, saltwater intrusion is another hazard that is exacerbated by heavy water abstraction.

In 2006, two sampling campaigns were carried out in order to identify whether the contaminated surface water was a possible source of groundwater contamination and to monitor the variation of groundwater quality in space and time. One sampling campaign was done at the end of the dry season in March and the other at the end of the rainy season in September. A total of 88 surface and groundwater samples were collected. Groundwater samples usually consisted of three samples from different depths at each location.

The parameters that were analyzed were: Fe^{2+}, NH_4^+ Alkalinity, ORP (Oxidation Reduction Potential), DO (Dissolved Oxygen), pH, and EC (Electric Conductivity) at the field site and DOC (Dissolved Organic Carbon), Cl^-, NO_2^-, NO_3^-, PO_4^{2-}, SO_4^{2}, Al, Ca, Fe, K, Mg, Mn, Na and Si in the laboratory. For the September samplings, δ O and δ H were analyzed and temperature profiles along the deepest well at each location were recorded as well.

13.3.3 Brackish Groundwater in HCMC

As in the Mekong Delta, fresh groundwater is not available everywhere in HCMC. Due to this fact, wells have been drilled all over the area with variable success rates. More than a hundred thousand wells are producing fresh groundwater, but over-abstraction has lead to the salinization and abandonment of more than 2300 wells, almost 2.5% of all surveyed wells at the time of reporting (Dan et al. 2005).

The shortage of freshwater and the existence of brackish water in several aquifers led to groundwater investigations over the last two decades in the whole Mekong Delta, including the area of HCMC. At first, a traditional Vertical Electrical Sounding (VES) method was applied, but this rarely reached 200m in depth. For deeper aquifers, costly deep drill-holes were constructed. Pham et al. (2002) presented a combined method using VES and MTS (Magneto Telluric Sounding) together, which gives a good electrical image from the surface down to approximately 800m. These electrical images can then be used for interpretation of water quality.

The analysis for chloride in the HCMC area revealed two locations with brackish water from the surface down to a depth of 200m. One observation well is situated west of HCMC and the other one is south of the city. Between them is a distance of more than 11 km. In the area between the two brackish locations, there are many wells that deliver potable water. Therefore, it can be inferred that there is a freshwater corridor passing under

HCMC. Figure 13.8 shows contours of chloride concentration delineated from the measurements in the aquifers between 150 and 200m in depth.

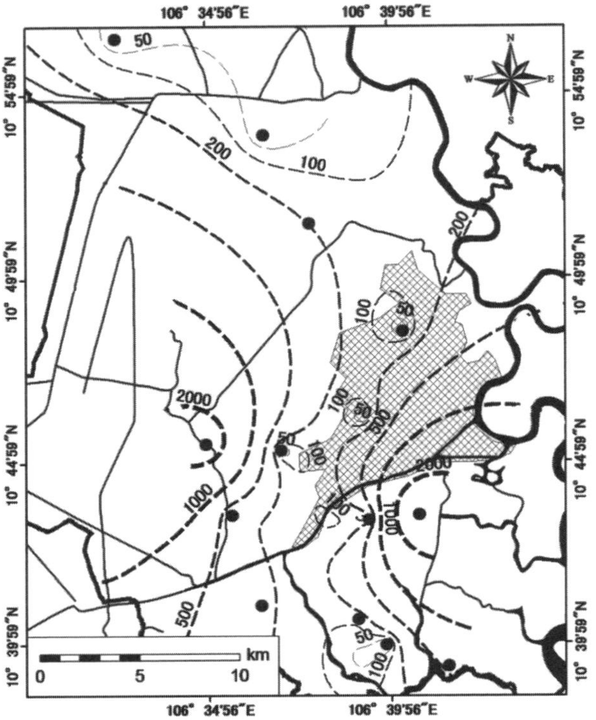

Fig.13.8. Freshwater corridor in the HCMC area in the 150 to 200m depth aquifer (contour lines: mg Cl / L)

Variations of the chloride concentrations between the rainy and dry season were noticed, especially in the two shallow wells. In the western part of HCMC, concentrations varied more than 4000 mg/L and in the south of the city, more than 2000 mg/L. These big variations are most probably due to infiltration and dilution during the rainy season. In Fig. 13.7, it can be seen that the less permeable layer at the surface is rather thin in the western area and might not even be continuous. Therefore, it likely allows direct infiltration of rainwater. In the aquifer between 60 and 200m below ground level, chloride concentration varied only several 100 mg/L, which suggests less interconnection with freshwater bodies.

13.3.4 Ferrous Iron and Manganese

Ferrous iron and manganese are widely detected in the HCMC area. At very low concentrations, 0.3 mg/L for iron and 0.1 mg/L for manganese, there is no problem associated with these ions (White 2005). If, however, they exceed these thresholds, negative effects emerge: laundry will get stained and there will be problems with plumbing fixtures.

Iron and manganese often appear together and are usually at high concentrations in acidic water.

In HCMC, almost all groundwater has a very low pH. During the sampling, pH values were mostly well below 6, which might be associated with the acidic soils present in the area.

Iron

In the case of iron, "iron bacteria" that derive their energy from the oxidation of ferrous iron to ferric iron produce a slimy coating on piping. In addition, water containing oxidized iron has a reddish brown colour and bears an odour. For iron, there is no health-based guideline value established, but due to the negative effects, 0.3 mg/L is set as an acceptability threshold.

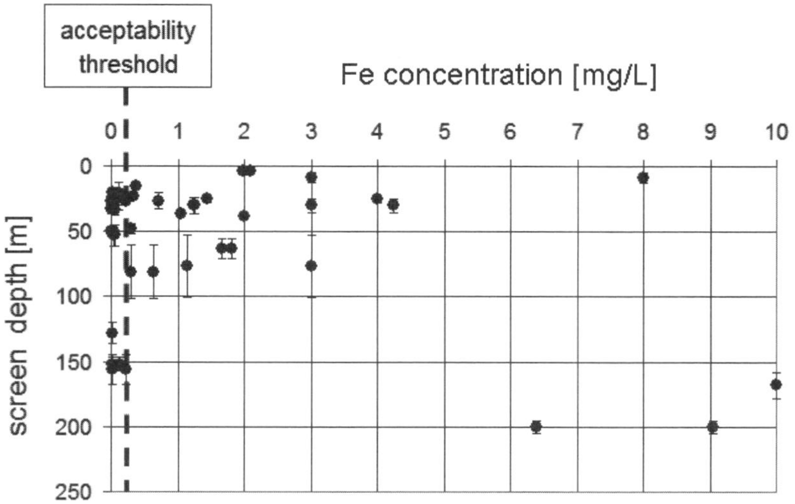

Fig. 13.9. Ferrous iron concentration at each well screen depth (concentration range from 0 to 10 mg/L, error bars indicate the range of screen depth)

Ferrous iron was detected with concentrations up to 100 mg/L. The two brackish locations show extremely high concentrations in particular, but high iron concentration was actually widely spread. As the acceptability threshold is at 0.3 mg/L (Figure 13.9), most of the groundwater needs to be aerated and filtrated before distribution to households.

Manganese

In the case of manganese, a black precipitate forms at concentrations as low as 0.2 mg/L and coats piping delivering the water. It also stains sanitary ware and laundry. Hence, 0.1 mg/L is set as an acceptable concentration. But there is also a health-based concentration limit for manganese, which is set at 0.4 mg/L. Manganese is an essential element for humans but it can have adverse effects. Over-exposure by inhalation or high levels in drinking water is known to have neurological effects.

High manganese concentrations are often associated with acidic water, which is widespread in HCMC. Most of the groundwater shows concentrations above the acceptability threshold of 0.1 mg/L. A considerable number of wells even exceed the health-based guideline value of 0.4 mg/L (Fig. 13.10). The highest values measured were above 20 mg/L in the brackish wells in the south of HCMC.

Manganese in groundwater can be removed by the same process as with iron.

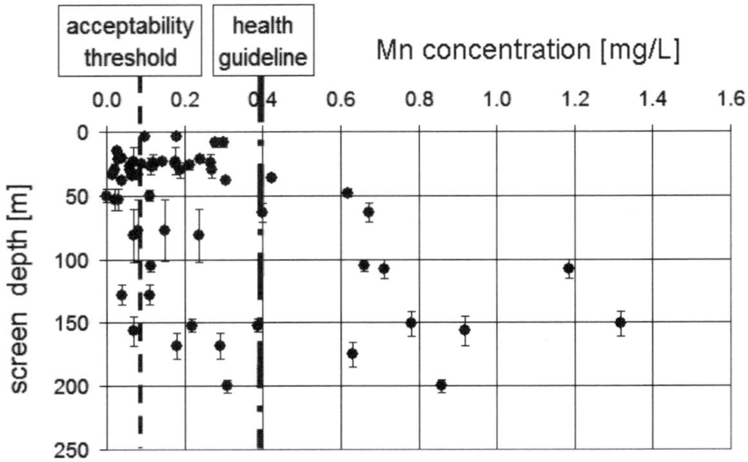

Fig. 13.10. Manganese concentration at screen depth (low concentration range, error bars indicate the range of screen depth).

13.3.5 Nitrate and Ammonia

Groundwater analysis revealed elevated concentrations of nitrate and ammonium. Generally, there was either nitrate or ammonium present depending on the oxic or anoxic condition of the groundwater. NH_4-N ranged from zero to several hundred mg/L and NO_3-N from zero to less than 7 mg/L. Ammonium is not directly a health concern, but it is an indicator of possible bacterial, sewage and animal waste pollution (WHO 2006). In wells adjacent to a closed landfill, NH_4-N concentrations were between 1.58 mg/L and 290 mg/L, clearly indicating that water from the landfill is infiltrating into these shallow aquifers. Similarly, canal water next to another landfill site had 7.5 mg/L NH_4-N when diluted by heavy rains and 115 mg/L NH_4-N in a pond just next to the observation wells. In the latter case, wastewater from a small pig farm was considered to be the source of high ammonia concentration. In contrast, NH_4-N was not detected in the groundwater from 22.5, 29 or 37m depths at this site. This is probably due to the favourable geological situation, including non-permeable materials in the surface layers.

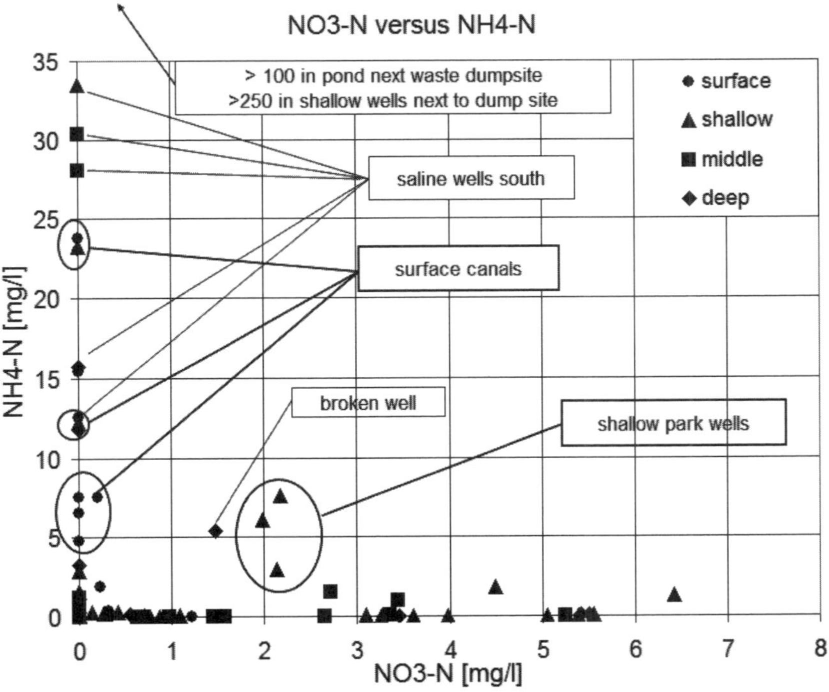

Fig. 13.11. Nitrate-N versus ammonia-N concentration.

At the Dong Thanh landfill site, such problems were investigated in a case study by Dan et al. (2006). They also noticed that the ammonia concentration decreased steadily with distance from the contamination source. This is probably due to the ion exchange capacity of fine-grained, clayey sediments present at the site.

In densely populated areas, the canal water showed NH_4-N concentrations between 5 and 25 mg/L. In a rural area close to a garbage sorting lot, concentrations of 7.5 and 15.4 mg/L NH_4-N were detected. In the main rivers with a high discharge rate, NH_4-N was either not detected or was below 0.5 mg/L.

At one location in the saline area south of the urban area, high NH_4-N concentrations were detected at depths of 36, 129 and 187m (23.2 to 33.4 mg/L in the shallow and middle wells, and 11.7 and 15.7 mg/L in the deep well). Natural background concentrations are usually less than 0.2 mg/L and may contain up to 3 mg/L in anaerobic groundwater (WHO 2006). The source could be anaerobic degradation of organic matter, as ORP is rather low for these samples. However, a high DOC concentration of 17 mg/L

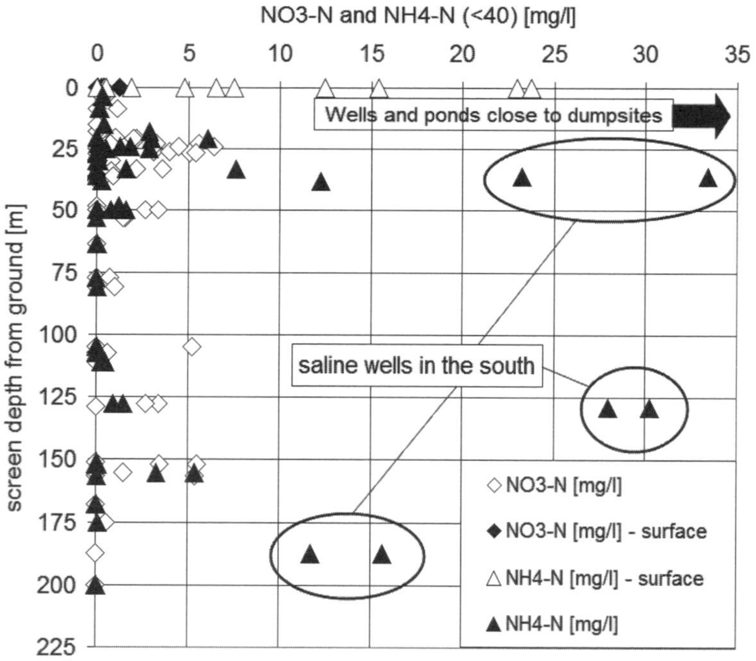

Fig. 13.12. Nitrate-N and ammonia-N concentrations versus well screen depth

was detected only at one site (a shallow aquifer in September). All other DOC measurements for these wells were less than 3 mg/L and, therefore, do not corroborate this theory. Another hypothesis is that the NH_4^+ could be released from clays underground, which could have a possible ion exchange capacity in this concentration range (Roy et al. 2003). Further, ammonium can originate from explosives, and an old fortress can be found just next to the observation station. However, ammonium percolation straight down from the shallow aquifers is unlikely, especially because a recorded temperature anomaly in the middle aquifer zone strongly suggests water movement from an area with a lower water temperature. At the same location, ferrous iron was very high with concentrations between 42 and 103 mg/L and therefore, Fe^{2+} containing minerals should be present. These minerals have the potential to inorganically reduce nitrate to ammonium as shown in some laboratory experiments (Hansen and Koch 1998). Thus, ammonium could possibly result from inorganic nitrate reduction.

NO_3-N was almost not detected in surface water but often detected in groundwater where no ammonium was present. Fig. 13.12 shows that all the detected concentrations, with a maximum of 6.43 mg/L, are below the permissible WHO guideline level of 10 mg/L NO_3-N (WHO 2006). The concentrations tended to be higher in the densely populated area where very little or no agricultural activity was going on. The sources of nitrate must then be something other than agriculture. Possible sources are the use of fertilizers in urban parks and urban agriculture (small-scale farming) and the nitrification of infiltrate from the canals that are used as a general sewer in most of the area. The latter is most likely the main reason for the enhanced NO_3-N concentrations, as the canals cut through the urban area. The elevated concentrations in the deeper groundwater below the urban area may originate with agricultural activities in the rural areas from where they are carried on through the aquifers.

To improve the current situation of nitrate and ammonium contamination of the surface and groundwater, a sewerage system without leakage may be the most important action for HCMC.

13.3.6 Groundwater vulnerability in HCMC

Similar to other coastal cities, HCMC's groundwater system shows a complex structure of aquifers and aquitards, which were mostly deposited under fluvial conditions but have also been influenced by marine sediments due to geologic transgression and regression over time. This has lead to the fact that brackish water bodies are present at different locations in several aquifers. Due to this natural condition, freshwater sources are vulnerable to saltwater encroachment when abstraction of freshwater takes place near the brackish water area. Thousands of wells have already had to be abandoned due to saltwater encroachment. To avoid further abandonment of wells, groundwater flow paths and a more complete understanding of the aquifer system are needed to regulate and control the abstraction of groundwater.

In HCMC, the most vulnerable aquifer to any contamination is the unconfined, uppermost aquifer. Fig. 13.13 shows a very simple vulnerability map based on the two hydrogeological cross-sections and the water analysis presented in this chapter. As mentioned in section 13.3.5, these are

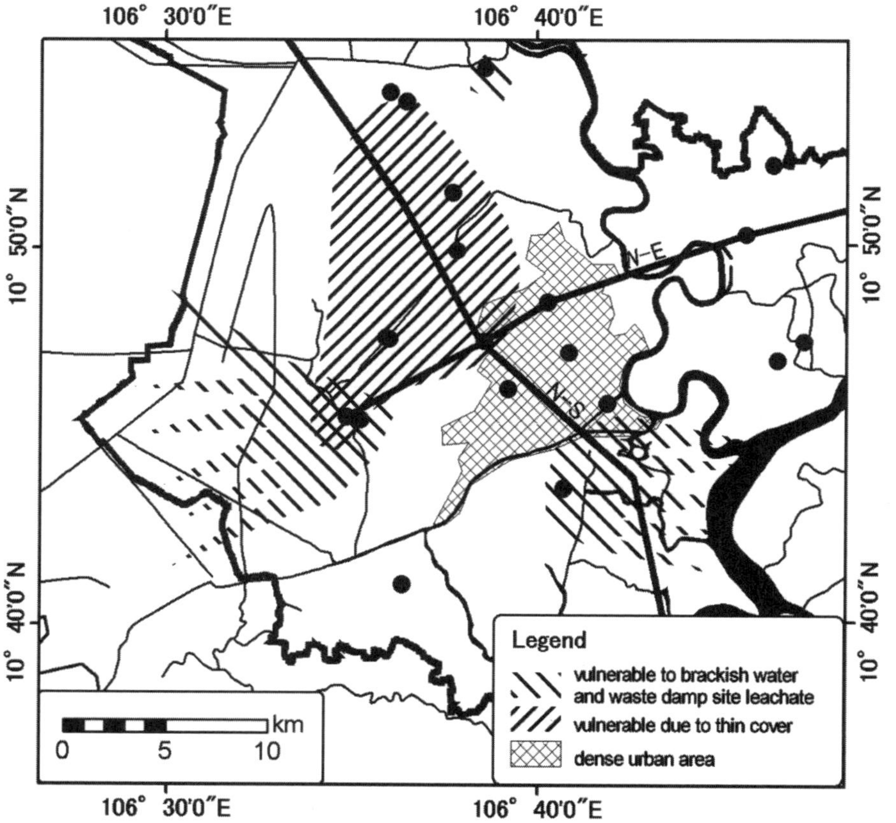

Fig. 13.13. Rough map of areas with an elevated aquifer contamination vulnerability, due to a thin covering layer, brackish water bodies or a waste dumpsite.

in part degraded by nitrate and ammonia. It was also shown that surface water is often contaminated by high ammonia concentrations. The hydrogeological cross-sections (Figures 13.1 and 13.7) suggest that the areas northwest and west of the central urban area are at an increased vulnerability level because the covering, low permeable layer is thin compared to other areas. Hence, the infiltration of surface water into the unconfined aquifer can take place easily. Where the covering low permeable layer is thin or absent, canals may be in direct contact with the aquifer and therefore pose a high risk of contamination. A thin covering layer can also easily be penetrated by construction work, including on-site sanitation facilities like septic tanks that bear the danger of contaminating the aquifer with coliform, chloride and nitrate (Ohgaki et al. 2006).

A thick surface layer with low permeability may prevent quick infiltration of contaminants and protect the aquifer from pollution. If the percolating water has to pass through clayey material, pollutants may be decreased by ion exchange and adsorption. A long travel time may also allow some pollutants to be degraded until they reach the aquifer. The cross-sections clearly show that some areas in the northwest, southeast and underneath the Saigon River are fairly well protected against direct contamination by a thick, less permeable covering layer consisting of silty and clayey material. The vulnerability to pollution from the surface is therefore lower in these areas.

13.4 Conclusion

Many coastal cities around the world share the problem of groundwater quality deterioration. Saltwater intrusion is a common problem in coastal cities, which is brought about by over abstraction of groundwater. It is, however, not easy to understand the flow paths of the saline water due to complex hydrogeological structures in these areas.

Groundwater degradation by nitrate and ammonium is widespread due to wastewater disposal into canals and rivers from which the water infiltrates into the subsurface.

To avoid problems and the abandonment of wells in use, as well as to prevent any kind of groundwater pollution, careful planning and exploitation of the groundwater resources is needed. One of the tools to achieve these goals is aquifer vulnerability assessment where the areas are mapped according to different features and are then combined to create vulnerability maps. Those maps provide a sound tool for city planning and development.

References

Berg M, Tran HC, Nguyen TC, Pham HV, Shertenleib R, Giger W (2001) Arsenic contamination of groundwater and drinking water in Vietnam: a human health threat. Environmental Science & Technology 35(13):2621-2626)

Bolay JC, Cartoux S, Cunha A, Ngoc Du TT, Bassand M (1997) Sustainable development and urban growth: precarious habitat and water management in Ho Chi Minh City, Vietnam. Habitat International 21(2):185-197)

Cavita M, De Maio M (2004) Assessing and mapping groundwater vulnerability to contamination: the Italian "combined" approach. Geofisica International 43:513-532

Dan NP, Ha NTV, Than BX, Nga NV, Khoa LV (2007) Sustainable groundwater management in Asian Cities: a final report of research on sustainable water management policy, water resources management in Ho Chi Minh City. IGES, Hayama, Japan, pp 68-79

Dan NP, Than BX, Khoa LV, Truong BD (2006) Groundwater resource management in Ho Chi Minh City, sustainable groundwater management in Asian Cities, pp 81-87

Dan NP, Thanh BX, Truong BD (2006) Case studies of groundwater pollution in southeast Vietnam. International Review for Environmental Strategies 6(1):361-372

Das Gupta A, Sabanathan S (1988) Saltwater transport in a Heterogeneous formation: a case study. Transport in Porous Media 3:217-256

Duc NH, Truong TP (2003) Water resources and environment in and around Ho Chi Minh City. Vietnam. Electric Green Journal 19

Edwards BD, Evans KR (2002) Saltwater intrusion in Los Angeles area coastal aquifers-the Marine Connection. US Geological Survey, fact sheet 030-02 (http://pubs.usgs.gov/fs/2002/fs030-02/fs030-02.pdf)

Foster S, Hirata R, Gomes D, D'Elia M, Paris M (2002) Groundwater quality protection: a guide for water utilities, municipal authorities and environment agencies. World Bank Group (http://siteresources.worldbank.org/INTWRD/Resources/English_Groundwaterquality_web.pdf)

Gogu RC, Dassargues A (2000) Current trends and future challenges in groundwater vulnerability assessment using overlay and index methods. Environmental Geology 39(6)

Hansen HCB, Koch B (1998) Reduction of nitrate to ammonium by sulphate green rust: activation energy and reaction mechanism. Clay Minerals 33:87-101

Lobo Ferreira JP, Oliviera M (2004) Groundwater vulnerability assessment in Portugal. Geofisica Internacional 43:541-550

Marsico A, Giuliano G, Pennetta L, Vurro M (2004) Intrinsic vulnerability assessment of the south –eastern Murge (Apulia, southern Italy). Natural Hazards and Earth System Sciences 4:769-774

Ohgaki S, Takizawa S, Herath G, Kataoka Y, Hara K (2006) Sustainable groundwater management in Asian Cities. Institute for Global Environmental Strategies
(http://www.iges.or.jp/en/fw/index.html)

Pham VN, Boyer D, Le Mouel JL, Nguyen TKT (2002) Hydrogeological investigation in the Mekong Delta around Ho Chi Minh City (South Vietnam) by electric tomography. Geoscience Journal 334:733–740

Pham VN, Boyer D, Nguyen TKT, Nguyen VG (1994) Deep groundwater investigation by combined VES/MTS methods near Ho Chi Minh City, Vietnam. Vietnam Ground Water 32(4):675–682

Phien-wej N, Giao PH, Nutalaya P (2005) Land subsidence in Bangkok, Thailand. Engineering Geology 82:187-201

Piancharoen C (1976) Ground water and land subsidence in Bangkok, Thailand. In: Proceedings of the Anaheim Symposium. International Association of Hydrological Sciences

Poland JF (1984) Guide book to studies of land subsidence due to groundwater withdrawal, International Hydrological Programme, Working Group 8.4, UNESCO
(http://wwwrcamnl.wr.usgs.gov/rgws/Unesco/)

Roy WR, Glessner JJG, Krapac IG, Larson TH (2003) Possible geological-geochemical sources of ammonium in groundwater: preliminary results. In: Illinois groundwater consortium proceedings
(http://www.siu.edu/orda/igc/proceedings/03/roy.pdf)

UN Atlas of the Oceans
(http://www.oceansatlas.org)

Voigt HS, Heinkele T, Jahnke C, Wolter R (2004) Characterization of groundwater vulnerability to fulfil requirements of the water framework directive of the European Union, Geofisica Internacional, 43:567-574

White I (2002) Water management in the Mekong Delta: changes, conflicts and opportunities. UNESCO Technical Documents in Hydrology No. 61

World Health Organization (2006) Guidelines for drinking-water quality (electronic resource) 1:303-304
(http://www.who.int/water_sanitation_health/dwq/gdwq0506.pdf)

WorldClimate (updated 2005)
(http://www.worldclimate.com)

Yamanaka M, Nakano T, Tase N (2007) Sulfate and sulfide oxidation in anoxic confined aquifers in the north eastern Osaka Basin, Japan. J Hydrology 335:55-67

14. Arsenic Contamination in Hanoi City, Vietnam

Tran Thi Viet Nga

Department of Urban Engineering, the University of Tokyo

14.1 Introduction

In the last decade, arsenic-related health problems have been revealed as a serious problem in terms of the number of people at risk and the degree of the adverse health effects. Since then, a large number of research studies and pilot projects have been implemented to find a solution to the arsenic problem. Some of them have been partially successful, but until now, there seems to be no ultimate solution that can be implemented for low-income households in developing countries. This chapter summarises the current status of the arsenic problem in drinking water, and presents a case study of Hanoi City, Vietnam.

14.2 Toxicity and Occurrence of Arsenic

14.2.1 Arsenic Toxicity and Health Effects

Arsenic is considered one of the most significant hazards in the environment. Acute dosages of 15 mg arsenic/kg of body weight can lead to death. Exposure to the trivalent inorganic form, As(III), and its mono- and dimethylated derivatives, MMA(III) and DMA(III) respectively, are associated with cancers of the skin, lung, bladder, kidney and liver, as well as non-cancer diseases, such as diabetes mellitus, hypertension, and cardiovascular and cerebrovascular diseases (Chen et al. 1988; Tsai et al. 1998).

The latency period in humans of arsenic-related carcinogenesis is considered to be 30–50 years. Therefore, the epidemiological data on arsenic-induced cancer available now are coming from countries where people have consumed contaminated water for more than 50 years, including Taiwan, the U.S., Chile and Argentina (Ferreccio et al. 2000; Bates et al. 1992). The first signs of chronic exposure are skin pigmentation, depigmentation, hyperkeratosis and/or palm, sole and skin lesions. The health effects of arsenic vary among individuals depending on genetic conditions, nutrition levels, and amount and duration of exposure (Brown and Chen 1995; Pontius et al. 1994; Viraraghavan et al. 1994).

The southwest coastal zone of Taiwan was the first area in the world to be identified for the health effects attributed to chronic arsenic exposure. This area is well known for the identification of blackfoot disease, but also a number of other typical health problems, including internal cancers. The affected populations in Taiwan or South America, however, represent a small fraction of the total population. Contrarily, in countries like China, Bangladesh or India, millions of people have been exposed to arsenic contaminated drinking water since the 1980s (Sun et al. 2001; Chowdhury et al. 2001). In these countries, shallow well water or surface water had been used in households. Efforts to improve the quality of drinking water by drilling deeper wells led to unpredicted adverse health effects by increasing the amount of arsenic in their drinking water. Since arsenic poisoning develops over periods of time as long as 30 years, cases of arsenic-related illness may even increase in the future.

14.2.2 Arsenic Chemistry

Arsenic is unique among the toxic trace elements in its ability to mobilise at the pH values typically found in groundwater (pH 6.5–8.5) and in both oxidising and reducing environments. Most toxic trace elements (e.g. Pb, Cu, Ni, Cd, Co, and Zn) generally become more insoluble at higher pH levels and under reducing conditions due to their tendencies to interact with solid phases by precipitation/co-precipitation or adsorption. Arsenic, however, appears to be mobile over a wide range of redox conditions, but especially in reducing conditions.

Arsenic occurs in the environment in several oxidation states (-3, 0, +3, and +5), but in natural water sources it is mostly found in inorganic forms as trivalent arsenite, As(III), and pentavalent arsenate, As(V). The biological availability and physiological and toxicological effects of arsenic depend on its chemical form. As(III) is much more toxic, more soluble and more mobile than As(V). Organic arsenic forms may be produced by biological activities, mostly in surface water, but the concentration is generally low or negligible (Chen et al. 1995). The complex organic arsenic compounds such as tetramethylarsonium salt, arsenocholine, arsenobetaine, dimethyl (ribosyl) arseine oxides and arsenic containing lipids have been identified in marine environments. Organic forms, however, may occur where waters are significantly impacted by industrial pollution.

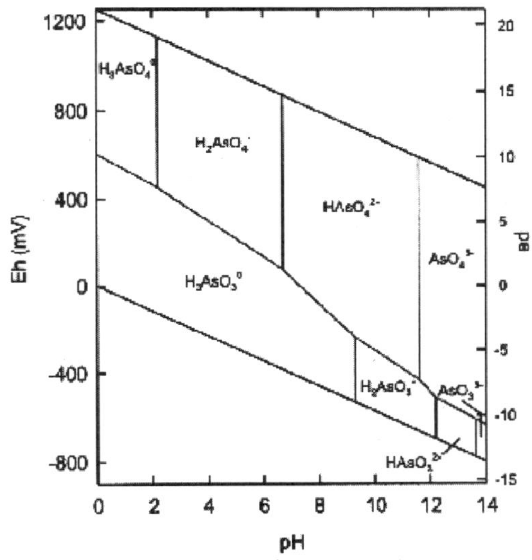

Fig.14.1. Arsenic speciation (arsenite and arsenate) against pH (Smedley and Kinniburgh 2002, copyright Elsevier)

The distribution of the arsenic species as a function of pH and pe is given in Fig.14.1. Redox potential (Eh) and pH are the most important factors controlling arsenic speciation. Under moderate or high redox potential, arsenic can be stabilised as As(V) species (H_3AsO_4, $H_2AsO_4^-$, $HAsO_4^{2-}$, and AsO_4^{3-}). Indeed $H_2AsO_4^-$ is dominant at pH levels less than 6.9, while at higher pH, $HAsO_4^{2-}$ becomes dominant. In strongly reducing aquifers, e.g. sulphate reducing conditions, As(III) typically dominates in the form of H_3AsO_3 (Yan et al. 2000). For example, the As(III)/As(V) ratio is 0.5-0.6 in arsenic-rich groundwater from Bangladesh and 0.7–0.9 in Inner Mongolia (Smedley et al. 1999). In the presence of very high concentrations of sulphide under reducing conditions, dissolved As may be incorporated into sulphide to form precipitations, i.e. orpiment (As_2O_3), realger

(AsS) or other thioarsenite species. Therefore, high arsenic waters are not expected where there is a high concentration of free sulphide.

14.2.3 Sources, Distribution and Mobilisation Pathways

Arsenic is the 20th most abundant element in the Earth's crust (approximately 2–5 mg/kg) (Viraraghavan 1994). It has been used in various fields, i.e., medicine, agriculture, livestock, electronics, industry and metallurgy. It is mobilised in the environment through a combination of natural processes such as weathering reactions, biological activities and volcanic emissions as well though a range of anthropogenic activities.

Arsenic is found at low concentrations in natural water. Background concentrations of arsenic in groundwater are less than 10 μg/L in most countries (Welch et al. 1999). However, values quoted in the literature show a wide difference, from < 0.5μg/L to 5000 μg/L, depending on the sources of arsenic, the amount of eluted arsenic and the local geochemical environment.

Arsenic concentrations are typically very low in surface water sources, including rivers (0.1–2 μg/L), lake water (<1 μg/L in Canada; Azcue and Nriagu 1995) and seawater (15 μg/L). Relatively high concentrations of naturally occurring arsenic in river and lake waters occurs in some areas as a result of inputs from geothermal water or in arid areas where water often has high pH (7–9) and high alkalinity. Significant increases in arsenic concentration in river and lake waters also occur as a result of pollution from industrial or sewage effluents (Andreae 1989) and mining waste and tailings (Azcue and Nriagu 1995; Smedley et al. 1996).

High concentrations of arsenic are found under a variety of environmental conditions, including oxidising conditions at high pH levels (Mexico, Chile, Argentina, and the southwestern region of the U.S.); reducing conditions (in Bangladesh, West Bengal, Taiwan, and Northern China); in areas affected by geothermal waters (the U.S., Japan, New Zealand, Iceland, Kamchatka in Russia, France and Dominica); and as the results of sulphide mineralization and mining-related activities (Austria, Ghana, Greece, India, South Africa, Thailand and the U.S.) (Smedley and Kinniburgh 2002).

Elevated concentrations of arsenic in natural waters are associated with arsenic-rich materials; arsenic is released through various mechanisms depending on the local environment. Among numerous minerals containing arsenic so far reported, iron-oxyhydroxide is considered the most important solid phase for arsenic release. Arsenic is mobilised by the reduction of iron oxyhydroxide or sorbed arsenate by biodegradation of peat deposits. Other mechanisms also involved that explain the arsenic contamination in groundwater are displacement of arsenic by carbonate or by phosphate. In some cases, arsenic is released through sedimentary sulphide oxidation induced by extensive pumping during the dry season.

14.2.4 Arsenic in drinking water

The United States Environmental Protection Agency (US EPA) has estimated that 5% of all U.S. community water systems will have to take corrective action to lower the current levels of arsenic in their drinking water (Ning 2002). Of the various sources of arsenic intake, drinking water poses the greatest threat to human health. High arsenic concentrations are found in groundwater. Generally arsenic concentration in most natural water sources varies between 1 and 2 µg/L, which results in an arsenic intake of 2-4 µg/day, assuming that an adult consumes 2 litres of water per day. This value is very low compared to the exposure through daily food intake, which is in the range of 17-129 µg/day by the World Health Organization's (WHO) estimation (Wu et al 1989). However, it should be noted that arsenic in food, especially in seafood, is mostly present in its organic from such as arsenocholine, arsenobetain or arsenosugars, which are not presently known to be toxic; therefore, drinking water contaminated with inorganic arsenic is the primary route of exposure.

Until recently, arsenic constituents in drinking have not been routinely monitored. In recent years, it has become apparent that arsenic in drinking water quite often exceeds both the WHO guideline value and current national standards. Indeed, arsenic is now recognized as one of the most serous inorganic contaminants in drinking water.

The WHO guideline value for arsenic in drinking water was provisionally reduced in 1993 from 50 µg/L to 10 µg/L. Following the accumulation of evidence for the chronic toxicological effects of arsenic in drinking water, the WHO recommended that many authorities lower their regulatory limits. For example, Japan, Europe, and the U.S lowered the arsenic standard from 50 to 10 µg total As/L. However, many developing countries still implement at the present standard of 50 µg/L (Smedley and Kinniburgh 2002).

14.2.5 Arsenic Contamination in Asian Countries

Arsenic occurring in groundwater at concentrations above 50 µg/L and its health impact on humans has already been reported in more than twenty regions in different parts of the world (Rahman et al. 2006). In addition, many new cases have been discovered recently. The magnitude of this problem is severe in Bangladesh (Chowdhury et al. 2001; Raychowdhury et al. 2002; Van Geen et al. 2003) and in West Bengal, India (Rahman et al. 2001). Arsenic contamination and associated health effects have also been reported from many parts of Asia such as Taiwan, China (Sun 2001; Xia and Liu 2004), Nepal (Shrestha et al. 2003), and Vietnam (Berg et al. 2001; Nga et al. 2001, 2003). In recent years, evidence of arsenic contamination has also emerged in other Asian countries including Pakistan, Myanmar, Cambodia and Laos (ESCAPE-UNICEF-WHO 2001).

Bangladesh and India (West Bengal)

Arsenic problems in groundwater from the alluvial and deltaic aquifer of Bangladesh and West Bengal were first documented in 1991 (PHED 1991), and are now known to be the most serious cases in the world. Concentrations in groundwater from affected areas are widely ranged from <0.5 μg/L to 3200 μg/L (GBS/PDHE 2000). In Bangladesh and West Bengal, at least 28 million and 6 million people, respectively, drink water containing more than 50μg/L of arsenic, which is the current Bangladeshi Drinking Water Standard . Arsenic-related health problems were first identified in West Bengal in the late 1980s, but the first diagnosis in Bangladesh was not made until 1993. Skin disorders including hyper/hypopigmentation changes and keratosis are the most common external manifestation, although skin cancer has also been identified.

The aquifers containing high arsenic concentrations are generally less than 100–150m in depth and of the Holocene age. In most of the areas with high arsenic groundwater, the sediment is capped by a layer of clay or silt of various thicknesses. This, together with abundance of organic matter deposited within the sediments, has resulted in the development of highly reducing environments which favour the mobilisation of arsenic. Most of the deep wells, tapped deeper than 150–200 m, invariably have low arsenic concentrations, usually less than 0.5 μg/L. The chemical features of the high-arsenic groundwater are high concentrations of Fe (>0.2 mg/L), Mn (>0.5mg/L), bicarbonate (>500 mg/L) and often phosphorus (>0.5 mg/L), and low concentrations of chloride (<60 mg/L), sulphate (<1 mg/L), nitrate and fluoride (<1 mg/L), with pH values close to or greater than 7. However, there is usually no correlation between these dissolved elements. One commonly observed relationship in the groundwater is a generally negative correlation between arsenic and SO_4^{2-} concentrations (DPHE 2001). This inverse association suggests that arsenic mobilisation takes place under the most strongly reducing conditions, wherein sulphate reduction also takes place.

Three mechanisms to explain arsenic contamination of groundwater in Bangladesh were proposed as follows:

- Pyrite oxidation. As is released by oxidation of As-bearing pyrite in alluvial sediments (Mallick and Rajagopal 1996; Mandal et al. 1998).
- Ion exchange. Arsenic anion sorbed to aquifer minerals are displaced into solution by competitive exchange with phosphate occurring due to over-application of fertilizer on the surface soils.
- Iron reduction. As that is sorbed to Fe (hydro)oxides (FeOOH) is released when the iron oxyhydroxide is reduced under anoxic conditions caused by organic sediment (Bhattacharya et al. 1997).

The first and second hypotheses imply human influence due to over-abstraction of groundwater, whereas the third one finds the cause of arsenic leaching in natural processes. British Geological Survey (BGS), after groundwater investigation, rejected the first and second hypotheses and supported the third one. This process is driven by microbial fermentation of buried peat deposits and possibly organic waste from pit latrines, which also contain ammonia (< 23 mg/L) and phosphorus (5 mg/L).

Fazal et al. (2001) carried out similar investigations as BGS and got similar results. However, they contend that their study, like that of BGS, is not definitive enough to make any conclusions on the causes of arsenic contamination because the data varied in time and space. They concluded that the pyrite oxidation hypothesis should still not be totally rejected.

Taiwan

The southwest coastal zone of Taiwan was perhaps the first area in the world where health effects attributed to the chronic arsenic exposure could be identified. This place is well known for the identification of blackfoot disease, and a number of other typical health problems related to arsenic, including internal cancers, have been identified there. Blackfoot disease, first observed in the 1930s and which peaked in the 1950s, was reported to correlate with the consumption of groundwater by the local people. Awareness of the arsenic problem began in the 1960s (Tseng et al. 1968) and arsenic related health problems have been well documented since (Chen et al. 1985).

High arsenic concentrations are found in samples from deep artesian wells (mostly 100-280m) whose sediments include fine sands, mud and black shale (Tseng et al. 1968). Arsenic concentration in groundwater samples from affected areas ranges between 10 µg/L and 1800 µg/L, and half the samples analysed had concentrations between 400 µg/L and 700 µg/L (Kuo 1968). The predominant arsenic species in the well waters of the affected area is As(III), with an average As(III)/As(V) ratio of 26 (Chen et al. 1994). In this area, the groundwater sources are likely to be in strongly reducing environments, and, hence, analogous to groundwater in the affected areas of Bangladesh and West Bengal. However, both the geochemistry of the groundwater and the mineral sources in Taiwan are poorly defined at present.

Northern China

High arsenic concentrations in groundwater and associated health problems have been identified in several parts of northern China, especially, Inner Mongolia (e.g. Ba Mang, Tumet Plain, Huhhot Basin, Baiyingao), Xinjiang and Shanxi Province. The affected areas lie on an alluvial basin of the Holocene age. The alluvial sediments are composed of a mixture of sands, silts and clays derived from a formed lake and marginal alluvial fan deposits eroded from the surrounding mountains.

The observed range of arsenic concentrations in groundwater varies from <1 µg/L to 1400 µg/L. The arsenic rich groundwater is highly reducing with low NO_3^- and high NH_4^+ concentrations. Similar to Bangladesh, arsenic is present in both oxidised As(V) and reduced As(III) forms, though As(III) is dominant. The highest As concentrations were found in the low-lying region of the basin where the conditions are most reducing and groundwater flow is most sluggish.

14.2.6 Arsenic Removal Processes

As many countries have lowered the acceptable limits of arsenic in drinking water, processes to selectively remove the excess arsenic from drinking water, industrial wastewater and mining discharge must be urgently established. Various treatment technologies have been developed for arsenic removal from drinking water. The conventional physical-chemical processes used for arsenic removal can be classified on the basis of the arsenic removal mechanisms involved: (a) precipitation, (b) adsorption, (c) ion exchange and (d) membrane technology (see also Chapter 2).

The commonly used technologies include coagulation and precipitation with iron and aluminium salts (Edwards 1994; Hering et al. 1996), adsorption onto activated alumina, activated carbon and activated bauxite (Gupta and Chen 1978), ion exchange and reverse osmosis (Clifford 1999). Some recent treatment technologies based on oxidation and adsorption are green sand filtration (Christen 2000), iron oxide coated sand (Joshi and Chaudhuri 1996), ferruginous manganese ore (Chakravarty et al. 2002), ferrihydrite (Jain et al. 1999), clay minerals (Manning and Goldberg 1997) and zero-valent iron (Farrel et al. 2001). Coagulation, precipitation and other adsorption techniques such as activated alumina, and activated carbon, have been found to be not as efficient for As(III) removal as for As(V) removal. Therefore, for efficient arsenic removal, oxidation of As(III) to As(V) is necessary. Despite the fact that a variety of treatment methods are available, the efficiency of these processes is not completely known (Krishna et al. 2001). Many of them have been reported to be capable of removing arsenic to levels lower than 50 μg/L. With the revision of the permissible levels to 10 μg/L or lower, however, it is necessary to investigate treatment processes that would consistently provide drinking water with arsenic levels less than 10 μg/L.

Kumar et al. (2003) have shown that the electro-coagulation processes is able to remove more than 99% of arsenic from arsenic contaminated water and can meet the drinking water standard of 10 μg/L with iron electrodes. Furthermore, As(III) was more efficiently removed in electrocoagulation than in chemical coagulation, whereas As(V) removal rates of both electro- and chemical coagulations were nearly the same.

The technology of adsorption is a process using materials that have a strong affinity for dissolved arsenic. Arsenic is attracted to the sorption site on the adsorbent's surface and is removed from water. This technology has been shown to be an effective way to remove arsenic from a solution (Lorenzen at al. 1995). Activated carbon is by far the most widely used adsorbent, though it is rather expensive.

Ion exchange is the exchange of arsenic anions for chloride or other anions at an active site bound to a resin. Some ion exchange processes are very effective for removing arsenic from water, but this process has the disadvantage of releasing noxious chemical reagents into the environment (Shih 2005). Membrane technology removes arsenic by allowing water to pass through the membranes, which physically retain the impurities present in water. Membrane technologies can effectively remove portions of all of the dissolved minerals. They also prevent microorganisms from passing through the membrane. The membrane itself does not accumulate arsenic, so that disposal of used membranes would be simple. Maintenance and operation requirements are minimal, and no chemicals need be added.

Pressure-driven membrane filtration is classified into two categories: high-pressure membranes, such as reverse osmosis (RO) and nanofiltration (NF), and low-pressure membranes, such as microfiltration (MF) and ultrationfiltration (UF). High-pressure membranes normally have a relatively small pore size compared to low-pressure membranes since smaller pore size requires more force to drive the fluids across the membrane.

RO and NF membrane processes can remove arsenic more effectively compared to UF and MF membrane processes (US EPA 2000). The operating conditions, such as the membrane material, water source, and pH of the solution, also impact the efficiency of arsenic removal. Overall, RO and NF membrane processes have excellent arsenic removal efficiency. RO membranes can remove more than 90% of arsenate, though the removal efficiency is much lower for arsenite, ca. 70% (Ning 2002). Furthermore, oxidation is not an easy way to improve efficiency since oxidants could damage the RO membrane (Koch Membrane Systems 2000). Membrane technologies could be the best choice for household filters to help drinking water meet the arsenic standard.

Recently, many research projects have been conducted to identify technologies for arsenic removal that can be applied in rural areas. For low-income people, RO or NF technology seems difficult to apply due to its high-energy consumption. Oh et al. (2000) have proposed an RO system (HR3155 membrane, Toyobo Co, Ltd) coupled with a bicycle pump system, as well as an NF system (ES-10 membrane) coupled with a bicycle pump system. It was found the arsenate removal efficiency was over 95% and arsenite removal efficiency was about 55% by the RO system, while the arsenic removal efficiency was about 90% for the NF system. Therefore the RO/NF process coupled with a bicycle pump device could be used for the removal of arsenic and satisfy the low operation costs, especially in areas where the power supply is not stable or is too expensive (Oh et al. 2000).

14.3 Arsenic Contamination in Hanoi City

14.3.1 Public Water Supply in Hanoi City

Location

Hanoi Province is located in the Red River Delta (or Bac Bo Plain) of northern Vietnam and has 11 million inhabitants. The study area is Hanoi City, which is in the centre of Hanoi Province. The area of the city is ca. 583 km^2 including all eight urban districts (Ba Dinh, Hoan Kiem, Hai Ba

Trung, Dong Da, Cau Giay, Thanh Xuan and Tay Ho) and the population is 1.5 million (JICA 1996).

Hydrology and Meteorology

The Red River Basin stretches from 20°00' to 25°30'N and 100°00' to 107°10'E and has a gross catchment area of 169,000 km^2. The total length of the river is 1,150 km. The river runs through 20 km of Hanoi City from the north to the southeast. Hanoi has a tropical monsoon climate and is subject to rainy (May to September) and dry seasons (October to April). The average temperature of Hanoi is 23.4°C, and the average rainfall is 1,800 mm/year. The flow rate of the river is highly variable throughout the year, with an average annual rate of approximately 4,100 m^3/sec. In the rainy season, the river may reach a water discharge rate of 9,500 m^3/sec.

The altitude of the southern part of Hanoi urban area is a few meters lower (4–5 m above the mean sea level) than the central part. At present, there is no central sewage treatment plant in the city. All domestic wastewater is discharged into open canals and rivers and flows to the south of Hanoi City. There are also numerous fishponds in the south and nutrient-rich fish bait and human and animal excreta are put into these ponds daily.

Hydrogeology

The Red River delta has a complicated geological history with transgressions, erosion, and river processes that have formed a thick Quaternary formation (50–90m in Hanoi). The sediment falls in the categories of alluvial river and lake deposits from the Plio-Pleistocene to the Late Holocene. It consists of sand and gravel infilled with fine material such as clay and silt, and often contains organic matter. The Quaternary sediments are underlain by Tertiary deposits of the Neogene age that are composed of sandstone, slay and silt stone. In total, the deposits exceed 400m in depth.

Based on the lithological characteristics and the permeability of sediments, hydrogeologists have identified two water-bearing aquifers in the Hanoi area (JICA 1999). The two aquifers are: (1) a water-bearing porous aquifer in Holocene sediment and (2) a water-bearing porous aquifer in the middle to upper Pleistocene.

- Water-bearing porous aquifer in Holocene sediment: This is the first aquifer beneath the surface. This aquifer usually consists of clay-sand, sand-clay and clay with thickness ranging from several meters to 34 meters. The water table changes from 2 m to 5 m, but in the rainy season the water table is near the surface. Groundwater in this porous aquifer has free surface in some cases. The chemical type of this aquifer's water is HCO_3^--Ca-Mg. The mineral content of groundwater in this aquifer is usually lower than 0.5 g/L.

- Pleistocene water-bearing aquifer: The lithological content of this water-bearing aquifer is characterized by sand and gravel. The thickness varies from 25–30 m to 70–80 m. In some places, such as along the bottom of the Red River, the layer separating these two aquifers is absent and the two aquifers are directly connected. The water level changes depending on the season. In the dry season, the groundwater level is about 2–4 m from surface. In the rainy season, the groundwater rises up to the surface in the region nearest the river, and the groundwater level becomes 1–2m above the surface. Mineral contents in this aquifer are always less than 0.3 g/L. The chemical type of the water is either HCO_3^--Ca-Na or HCO_3^--Ca-Mg.

The direction of groundwater flow in different regions is very different and changes depending on time. It is also influenced by the amount of groundwater pumped from the region. In the southern part of the Red River (including Hanoi City), the groundwater flow is from the northwest to the southeast. In the central districts of the city, the intensive exploitation of groundwater has formed a big water level depression.

Hanoi Public Water Supply

Groundwater has been easily exploitable all around Hanoi City and has been abstracted as a municipal water source since the beginning of the 20th century. Presently, the only water source used for Hanoi municipal water supply system is groundwater exploited from a confined aquifer (so-called Pleistocene or Lower aquifer) in a Quaternary deposit 40–70 m below ground level.

There are 105 groundwater wells (50–70m deep) that provide water to Hanoi City. These wells are located around eight major water treatment plants (WTPs) managed by the Hanoi Water Business Company (HWBC). These WTPs were either renovated or newly built through a grant-in-aid program provided by the Finish International Development Agency in the late 1980s. All WTPs apply the same conventional water treatment processes, i.e. tray-aeration (sedimentation), contact basin rapid sand filtration and chlorination (see Chapter 2). The treatment process is designed for the removal of iron and manganese, which are predominantly present in the groundwater in Hanoi, and for the disinfection of filtrated water.

14.3.2 Arsenic Contamination in Groundwater in Hanoi City

Occurrence of Arsenic in Groundwater

The groundwater in Hanoi City has been historically believed to be of good quality, according to a few hydrogeochemical surveys before the first construction of the WTPs, such as Yen Phu (in 1909), Ngoc Ha (in 1939), Ngo Si Lien (in 1944), and Gia Lam (in 1953). All these WTPs are located in the north of the city. Groundwater in this region generally satisfies the requirements for the water supply, except for high manganese content.

Since 1954, Hanoi has been rapidly expanding to the south. Some WTPs such as the ones in Ha Dinh, Tuong Mai, Phap Van and a number of other small water stations have since been constructed in the southern part of the city. Since then, high concentrations of arsenic, iron, ammonia and organic matter have been observed in these three WTPs (HWBC 2000; Le Giao 1995; Anderson and Norrman 1998). Berg et al. (2001), in a study of arsenic contamination in Hanoi City and its surrounding rural districts, reported arsenic concentrations ranging 240–320 μg/L in three WTPs in the south of the city. These reports raised people's concerns regarding the safety of their drinking water and groundwater contamination has become a priority issue in Hanoi City. This research was undertaken to identify and analyse sources and mechanisms of groundwater contamination by arsenic.

Materials and methods

Groundwater samples were collected from 45 deep wells in eight major water treatment plants (WTPs) over three sampling periods (August 2000, December 2000, and September 2001). Water samples were collected into three 100 ml-plastic bottles and brought back to the Environmental Engineering Laboratory at the University of Tokyo for analysis. All samples were stored at 4°C. Acidified samples by HNO_3 (1mL of HNO_3 1:1) were used to determine major cations, trace metals and arsenic species by ICP-AES (Optima 3000, Perkin-Elmer) and ICP-MS (HP4500, Yokogawa). Non-acidified samples were used to analyse major anions by ion-chromatography. Acidified samples by HCl (1ml of concentrate HCl) were used to determine NH_4^+, dissolved organic carbon (DOC), and total nitrogen (TN). Arsenic speciation was performed by liquid chromatography (Shimidzu LC-10A) coupled with ICP-MS to determine four arsenic species: As(III) (H_3AsO_3, $H_2AsO_3^-$), As(V) ($H_2AsO_4^-$, $HAsO_4^{2-}$), methylarsonic acid (MMAA; $(CH_3)AsO_2H_2$, $(CH_3)AsO_2H^-$) and dimethylarsenic acid (DMAA; $(CH_3)_2AsO_2H$, $(CH_3)_2AsO_2^-$).

Soil sampling was carried out in September 2001. Four bore cores were drilled within the Phap Van, Ha Dinh, Mai Dich and Luong Yen well fields. The locations were chosen because groundwater quality in these well fields was very different from each other. The bore cores were drilled next to the existing well, from which we took groundwater samples. Sediment cores were taken at different depths between 2 m and 40 m. Immediately after the sediment cores were collected, they were placed in plastic bags and then put in a plastic box under anaerobic condition using AnaeroPak (Mitsubishi Gas Chemical), and were kept in the dark at 4°C. A total of 68 sediment cores from four boreholes were collected. After being brought back to Japan, a part of each core was crushed and dried for 4–5 days at room temperature (22°C). The air-dried cores were crushed again to less than 2 mm in size, and homogenized by thorough mixing. The remaining fresh cores were still kept in an anaerobic box at 4°C until further analysis.

The solid-phases of arsenic were determined by applying the sequential extraction procedure. The procedure was modified from several reported references (Ahmann et al. 1997 and Zobrist et al. 2000) and was used to target different solid phases of arsenic as exchangeable, metal-oxide-bound, organic and sulphide-bound, well-crystallized (hydro)oxides of metal (Fe, Al, Mn)-bound and residuals. An aliquot of sediment (1 gram) was placed in a 50mL plastic bottle and 20mL of the extraction solution were added sequentially. After every extraction step, the bottle containing the soil and the extractant were centrifuged for 5 minutes at 3,500rpm. The solution was filtered through a 0.45 μm cellulose acetate filter paper and diluted with water to a volume of 30mL, then kept in a plastic bottle at 4°C prior to analysis by ICP-MS. The residual sediment was used for the subsequent extraction steps.

Arsenic in Groundwater

Arsenic levels in groundwater ranged from less than the detection limit to 112 μg/L over the three sampling periods (Fig. 14.2). In the northern area, groundwater collected from Mai Dich, Ngoc Ha, and Ngo Si Lien well fields had low levels of arsenic (less than 2 μg/L). Groundwater collected along the Red River (Luong Yen and Yen Phu well fields) had moderately high arsenic levels (< 30 μg/L). The rest, including groundwater collected from the Tuong Mai, Phap Van and Ha Dinh well fields in the south, had high arsenic concentrations of 43-112 μg/L, which were above the WHO guidelines for drinking water (10 µg/L). The concentration ranges of arsenic detected in this study are about one-third the results of another report (Berg et al. 2001), though the spatial distribution and seasonal variations of arsenic in the lower aquifer was the same as the previous report.

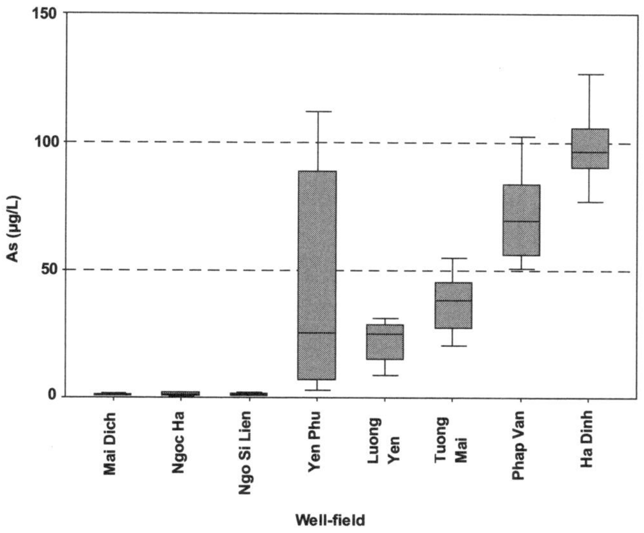

Fig. 14.2. Box plots of arsenic concentration in eight well fields in Hanoi City. The line across the box represents the median. The bottom and top of the box show the location of the first and third quartile values. Data obtained in August 2000, December 2000, and September 2001.

One-way analysis of variance (ANOVA) applied to the data set of three sampling periods revealed that the arsenic concentration was higher in the rainy season (August 2000 and September 2001) than in the dry season (December 2000) ($p<0.05$).

Samples having arsenic concentration above 30 µg/L were selected for analysis of four arsenic species, namely arsenite As(III), arsenate As(V), MMA and DMA. The total dissolved arsenic and the concentration of its species are given in Table 14.1. The main arsenic species detected in the groundwater were As(III) and As(V), with the ratio of As(III) and total arsenic varying greatly from 0.49 to 1.00.

Table 14.1. Arsenic species in groundwater[1]

Well label	Fe^{2+} mg/l	NH_4^+ mg/l	ORP	As(III) µg/L	As(V) µg/L	T-As µg/L	As(III) / ΣAs
Y20	7.60	3.5	-98	67.4	57.1	99.6	0.54
Y22	3.16	2.2	-65	61.1	62.7	85.7	0.49
P2	7.55	13.0	-60	39.7	15.1	56.3	0.73
P5	7.30	12.0	-100	36.0	6.7	51.1	0.84
P6	8.15	21.0	-65	37.7	7.7	50.6	0.83
P8	8.15	29.0	-92	81.3	ND	79.5	1.00
P9	8.10	19.0	-94	48.1	18.2	69.5	0.73
T11	6.85	4.5	-77	36.0	ND	55.5	1.00
T15	6.25	5.1	-77	23.8	ND	33.1	1.00
T16	7.80	11.0	-71	32.2	22.3	43.9	0.59
H6	20.00	9.4	-83	61.5	33.0	92.0	0.65
H10	13.65	7.5	-80	50.3	27.1	77.1	0.65
H17	8.20	5.0	-90	49.4	31.5	77.3	0.61
H18	10.25	7.2	-80	68.9	24.6	83.1	0.74
H14	8.45	6.6	-80	71.2	17.1	90.2	0.81

Note: ΣAs: As(III) + As(V), analysis by LC-ICP-MS (MMAA and DMAA are under detection limits)
T-As : mean analysis by ICP-MS
1. Data from December 2000 sampling

Arsenic in sediment cores

Arsenic concentration varied from less than 1 mg/kg to 20 mg/kg in different sediment layers in all four boreholes. The highest arsenic concentrations of 4–20 mg/kg were associated with the organic-rich peat, followed by greenish-grey clay (3–12 mg/kg), clayey silt and silt (1–9 mg/kg) and red brown clay and sand (up to 7 mg/kg).

The core samples were subject to sequential extraction, and each extract was analyzed for arsenic to allow detection of a single solid-phase arsenic associated with sediments. The $(NH_4)_2SO_4^-$ extractable arsenic is exchangeable arsenic, which is ionically bound to sediment particles, and was found at a low level of 0.08 mg/kg (max.: 0.26 mg/kg; min.: 0.02 mg/kg). Extraction with $NH_4H_2PO_4$ targeted arsenic bound to amorphous iron oxyhydroxides yielded about 0.03–5.34 mg/kg (mean: 1.43 mg/kg). The amount of arsenic released by digestion with heated H_2O_2, intended to oxidize organic matter and sulphides and thus release associated species, was slightly less, between 0.02–5.61 mg/kg (mean: 0.99 mg/kg). However, we found that the samples of peat have contents of arsenic bound to organic matter comparable to arsenic bound to iron oxyhydroxides. Oxalate-extractable arsenic accounted for the largest portions of extractable arsenic in the soils, and ranged from 0.01–9.33 mg/kg.

Relationship between Arsenic, Iron, Ammonia, and Organic Matter

Nga et al. (2003) have characterized the hydrogeochemistry of groundwater in Hanoi using multivariate data analysis and successfully identified three groundwater zones in the lower aquifers beneath the city, i.e. the northwest zone including Mai Dich, Ngoc Ha and Ngo Si Lien well fields, the riverain zone including Yen Phu and Luong Yen well fields, and the south zone covering the area of Tuong Mai, Phap Van and Ha Dinh well fields. The northwest groundwater is very good quality and the riverain groundwater quality is similar to the Red River water. The south zone, however, has high concentrations of iron, arsenic, ammonia and organic matter (Nga et al. 2003). Following this study, groundwater quality data were classified into three groups corresponding to the three distinguishable groundwater zones.

Reducing environments, indicated by negative ORP values, were observed in the aquifers along the Red River (Luong Yen and Yen Phu well fields) and in the south (Phap Van, Ha Dinh and Tuong Mai well fields). Hydrogen sulphide (H_2S), a pronounced indicator of reductive conditions, was only detected in groundwater collected along the Red River. No detection of either SO_4^{2-} or H_2S in the groundwater from the South indicates that sulphate reduction is taking place under strongly anaerobic conditions and sulphide produced may combine with ferrous iron Fe(II) or dissolved As to form insoluble FeS or As sulphosalt. These processes might be the cause of the absence of detectable sulphide in groundwater rich in arsenic and ammonia.

Elevated concentrations of NH_4-N (4.4–29 mg/L) and DOC (2.8–13.1 mg/L) were found in the groundwater of the Tuong Mai, Phap Van and Ha Dinh well fields in the south of Hanoi City. Fig.s 14.3a and 14.3b show that HCO_3^- increases with the increase of DOC and NH_4-N. These correlations may be explained by the anaerobic degradation of soil organic carbon that release DOC and NH_4-N. The DOC can be subsequently degraded to methane and bicarbonate in fermentation processes. As the anaerobic degradation processes of organic matter are slow, and since some DOC is not biodegradable, it is common to detect both DOC and HCO_3^-. Further, in most of the samples, about 80% of the TN was NH_4-N and NO_3-N concentration in the groundwater was very low, between the undetectable level and 0.22 mg/L. It was found that under the strongly reducing conditions, NH_4-N resulting from the degradation of organic matter remains in a stable form.

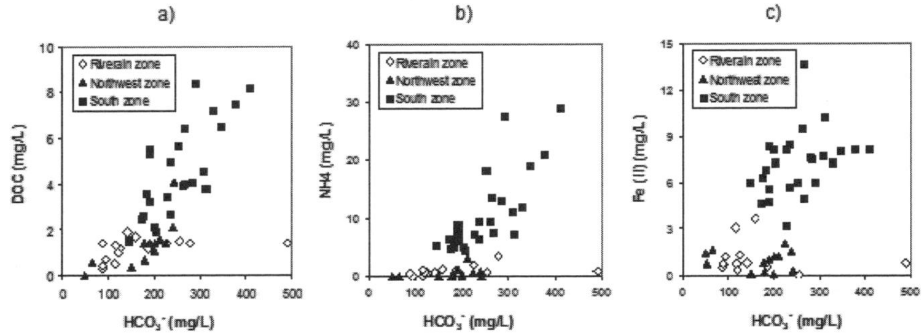

Fig. 14.3. Correlation between groundwater quality parameters: a) Dissolved organic carbon (DOC) and bicarbonate (HCO_3^-); b) ammonia nitrogen NH_4-N and HCO_3^-; and c) ferrous iron Fe(II) and HCO_3^- Data obtained in December 2000 and September 2001.

The groundwater in the south contained very high concentrations of iron (7–38mg/L), whereas a very low iron content was detected in other places. Fe(II) accounted for about 40-50% (3.8–13.6 mg/L) of the total iron and increased with the total iron concentration. These findings suggest that reduction of Fe(III) took place under a reducing condition. A high concentration of Fe and degradation products of organic matter (DOC, NH_4-N and HCO_3^-), as well as good correlations among these parameters (Fig 14.3a, 14.3b and 14.3c) strongly support the model proposed by Nickson et al. (2000) regarding the iron reduction process. In this model, reduction and dissolution of iron are driven by the degradation of OM under a reducing condition:

$$4FeOOH + CH_2O + 7H_2CO_3 \rightarrow 4Fe^{2+} + 8HCO_3^- + 6H_2O \quad (1)$$

Arsenic reduction and mobilisation

Fig. 14.4a shows that high arsenic concentration is associated with high Fe(II) content in the groundwater from the south of Hanoi City, indicating that iron reduction may be linked with arsenic mobilisation. These results are consistent with many previous studies (Nickson et al. 2000). In addition, results from arsenic speciation showed that As(III) was predominant compared to As(V) in these groundwater samples. Under such reducing environments as Hanoi groundwater, reduction of As(V) to As(III) could take place. As(III) is known to be more mobile than As(V) (Zobrist et al. 2000; Langner et al. 2000; Ahmann et al. 1997), thus making the arsenic concentration in groundwater even higher.

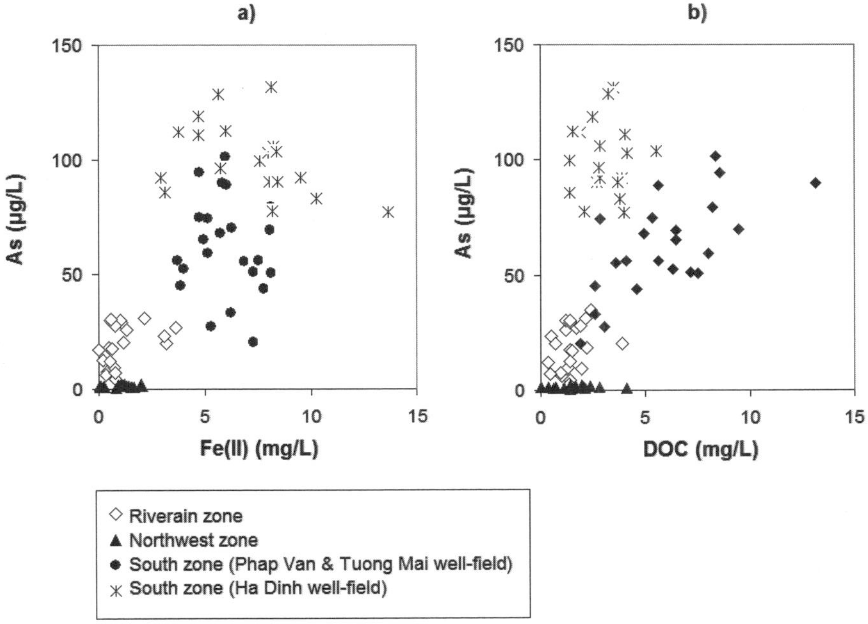

Fig. 14.4. Correlation between groundwater quality parameters: a) arsenic and ferrous ion Fe(II) and b) arsenic and dissolved organic carbon (DOC). Data obtained in August 2000, December 2000 and September 2001.

We have also found a strong correlation between arsenic and some organic degradation products, such as HCO_3^-, P, and NH_4^+ (Fig.14.4b). The strong correlation between dissolved arsenic and HCO_3^- concentration in the groundwater has also been identified by other researchers (Nickson et al. 2000; Kim et al. 2000). From Eq. 1, the ratio of released Fe(II) and HCO_3^- is 2, but the ratio found in our study was about 24, which is 12 times higher. It implies that only a small fraction of DOC in groundwater is used to reduce Fe(III) to Fe(II). Nickson et al. (2000), in their study of groundwater pollution in Bangladesh, also found that released HCO_3^- was by far greater than the concentration predicted by Eq.1, and they estimated that most of HCO_3^- came from natural mineral dissolution. Our results, on the contrary, suggest that most of HCO_3^- came from the degradation of soil organic matter. These findings also suggest that degradation of organic matter plays an important role in arsenic mobilisation.

Peat as an arsenic source

Solid-phase arsenic and peat layers are associated as shown in Fig. 14.5a. Organic-matter-bound arsenic increases with the carbon content and NH_4-N content in the cores taken from the Ha Dinh borehole. These results were consistently observed in all cores in all four boreholes. Particularly, high contents of organic-matter-bound arsenic were associated only with peat samples. These results suggest that organic matter in these soils is an important solid-phase source for arsenic.

In a similar study on arsenic contamination in the Red River Delta, Berg et al. (2001) reported a correlation between extracted arsenic and extracted iron from sediments, suggesting that arsenic could be associated with iron phases. Our results, however, did not show any significant correlation between these two parameters in all cores (Fig. 14.5a and 14.5b). The iron contents ranged from 7.2–102.8 mg/g and were associated with arsenic only in the shallow aquifer (<10m). There is, however, no apparent correlation between iron contents and the elevated arsenic level found in some sediment layers such as peat or green clay. These findings suggest that Fe minerals do not have significant influence on the distribution of soil arsenic.

Furthermore, results from sequential extraction revealed that, in the peat cores, 70–100% of the total sedimentary arsenic is comprised of iron-bound and organic-matter-bound arsenic. It indicates that peat is the sink and source of sedimentary arsenic. Namely, peat sequesters arsenic through two major solid phases: arsenic bound to organic matter and arsenic associated with iron oxyhydroxides. Results obtained from water monitoring were consistent with our model, and were indicated by the coincidence of geographical distribution of arsenic rich-wells with the distribution of peat layer.

As mentioned earlier, both degradation of organic matter and reduction of iron oxyhydroxides are responsible for the mobilisation of arsenic. However, the correlation between different water parameters does not show a similar trend when we look at each well field independently (Fig. 14.4a and 14.4b). Fig. 14.4b shows a good correlation between total arsenic and DOC for the groundwater from Tuong Mai and Phap Van well fields. Even though these groundwater sources have relatively high levels of iron, the release of arsenic has a close link with the degradation of organic matter and not with the reduction of iron. The groundwater from the Ha Dinh well field, nevertheless, does not show such correlation. DOC and arsenic concentrations were within a narrow range. These results imply that the predominant process of arsenic release is determined depending on different hydrogeological conditions. While degradation of organic matter is the prerequisite condition for arsenic mobilisation, it also controls the leaching of arsenic in the Phap Van and Tuong Mai well fields, whereas the reduction of iron plays a predominant role in the Ha Dinh well field.

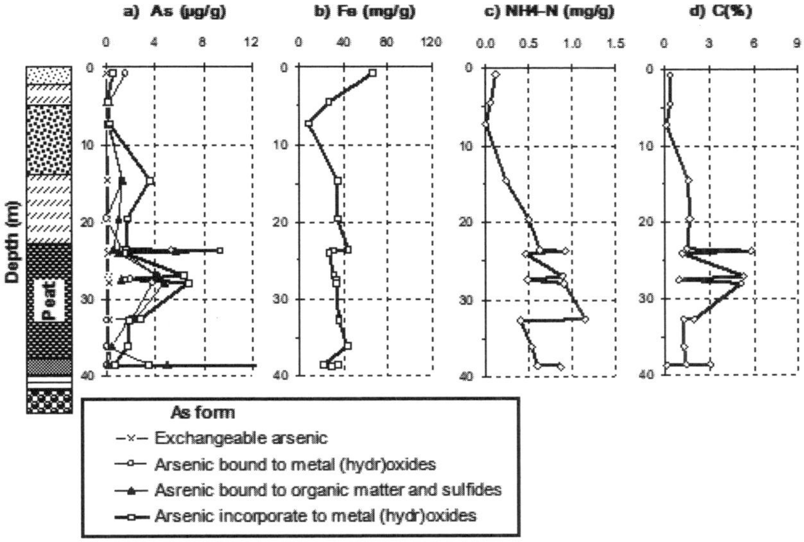

Fig. 14.5. Vertical profiles of geochemical characteristics in the Ha Dinh borehole: a) arsenic dissolved by the sequence of extractions that target different solid phases; b) soil iron; c) soil ammonia nitrogen; and d) carbon contents. The peat layer (C: 3-8%, N: 01-045%) was found at depths of 24-36m.

14.4 Conclusion

The study suggested that the reduction and dissolution of arsenic is taking place in the lower aquifer in the south of Hanoi City under a strongly reducing condition. The people living in the southern part of the city are therefore exposed to a risk of chronic arsenic poisoning. In a study to investigate the health effect of arsenic in Hanoi City and its outskirts, Berg et al. (2007) showed that the arsenic content in hair samples collected from people drinking groundwater with arsenic levels > 50 μg/L are significantly higher than control groups (arsenic less than 50 μg/L). A recent study on health risk assessment of arsenic contamination in groundwater in Hanoi City (Asuga 2006) indicated that the cumulative arsenic ingestion in the residents was estimated to be 0.04-2.41mg, much lower than the levels that can cause internal cancer (Bates 2004). So far no case of arsenic-related illness has been reported in the arsenic-affected area in Hanoi Province. This could possibly be attributed to the fact that, in Vietnam, arsenic contaminated groundwater has been used only for the past 7-10 years (Berg et al. 2007). Previous experiences show that it can take many years for the first arsenic poisoning symptoms to become apparent. Considering the magnitude of arsenic pollution, early mitigation measures are urgently needed to protect the people from arsenic-related health problems.

References

Ahmann D, Krumholz LR, Hemond, HF, Lovley DR, and Morel FMM (1997) Microbial mobilization of arsenic from sediments of the aberjona watershed. Environmental Science and Technology 31:2923-2930

Anderson L, Norrman J (1998) Ammonium contamination of groundwater in Hanoi area, Vietnam. Master Thesis, Geologiska Institutionen

Andreae MO, Andreae TW (1989) Dissolved arsenic species in the Schelde estuary and watershed, Belgium. Estuarine Coastal and Shelf Science 29:421-433

Astolfi E, Maccagno A, Fernandez JCG, Vaccara R, Stimola R (1981) Relation between arsenic in drinking water and skin cancer. Biological Trace Element Research 3:133-143

Asuga T, Kunito T, Fujihara J, Kubota R, Tu BM, Pham TKT, Iwata H, Subramanian A, Pham HV, Tanabe S (2006) Contamination by arsenic and other trace elements in tube-well water and its risk assessment to humans in Hanoi. Vietnam Environmental Pollution 139:95-106

Azcue JM, Nriagu JO (1995) Impact of abandoned mine tailings on the arsenic concentrations in Moira Lake, Ontario. Journal of Geochemical Exploration 52:81-89

Bates MN, Smith AH, Hopenhayn-Rich C (1992) Arsenic ingestion and internal cancer: a review. American Journal of Epidemiology 135:462-476

Berg M, Stengel C, Trang PTK, Pham HV, Sampson ML, Leng M, Samreth S, Fredericks D (2007) Magnitude of arsenic pollution in the Mekong and Red River Deltas-Cambodia and Vietnam. Science of the Total Environment 372:413-425

Berg M, Tran HC, Nguyen TC, Pham HV, Schertenleib R, Giger W (2001) Arsenic contamination of groundwater and drinking water in Vietnam: A human health threat. Environmental Science and Technology 35:2621-2626

BGS/DPHE (2000) Groundwater studies of arsenic contamination in Bangladesh. (Technical report to Department for International Development, United Kingdom)

Bhattacharya P, Chatterjee D, Jacks G (1997) Occurrence of arsenic-contaminated groundwater in alluvial aquifers from the Delta Plain, Eastern India: options for a safe drinking water supply. Water Res Dev 13:79-92

Biagni RE (1972) Epidemiologic study of arsenic poisoning in the city of Monte Quemado, Argentina. La Semana Medica 141:812-816

Borgono JM, Greiber R (1972) Epidemiological study of arsenicism in the city of Antofagasta. In: Hemphill DC (ed) Trace Substances in Environmental Health. Proceedings of the 5th Annual Conference Columbia. University of Missouri, pp 13-24

Brown KG, Chen CJ (1995) Significance of exposure assessment to analysis of cancer risk from inorganic arsenic in drinking water in Taiwan. Risk Analysis 15(4):475-484

Cebrian ME, Albores A, Aguilar M, Blakely E (1983) Chronic arsenic poisoning in the north of Mexico. Human Toxicology 2:121-33

Chakraverty S, Dureja V, Bhattacharyya G, Maity S, Bhattacharjee S, (2002) Removal of arsenic from groundwater using low cost ferruginous manganese ore. Water Research 36:625-632

Chen CJ, Kuo TL, Wu MM (1988) Arsenic and cancers. (Letter). Lancet 1:414–15

Chen SL, Dzeng SR., Yang MH, Chlu KH, Shieh GM, Wal CM (1994) Arsenic species in groundwaters of the Blackfoot disease areas, Taiwan. Environmental Science & Technology 28:877-881

Chen SL, Yeh SJ, Yang MH, Lin TH (1995) Trace element concentration and arsenic speciation in the well water of a Taiwan area with endemic Blackfoot disease. Biological Trace Element Research 48:263-274

Chowdhury UK, Rahman MM, Mandal BK, Paul K, Lodh D, Basu GK, Chanda CR, Saha KC, Mukherjee SC, Roy S, Das R, Kaies I, Barua AK, Palit SK, Quamruzzaman Q, Chakraborti D (2001) Groundwater arsenic contamination and human suffering in West Bengal - India and Bangladesh. Environmental Sciences 8:393-415

Christen K (2000) Cleaning technologies can remove arsenic, but at cost. Environmental Science and Technology 34:75-79

Clifford DA (1999) Ion-exchange and inorganic adsorption. In: Water Quality and Treatment: A Handbook of Community Water Supplies, 5th edn. American Water Works Association, McGraw-Hill, New York

DPHE (2001) Groundwater studies for arsenic contamination in Bangladesh. (Final report by Mott MacDonald and British Geological Survey for Department of Public Health Engineering, Government of Bangladesh)

Edwards M (1994) Chemistry of arsenic removal during coagulation and Fe-Mn oxidation. Journal of American Water Works Association 86: 64-78

ESCAP-UNICEF-WHO (2001) Economic and Social Commission for Asia and the Pacific; Geology and Health: solving the arsenic crisis in the Asia Pacific Region. Report, ESCAP-UNICEF-WHO Expert Group Meeting, Bangkok

Farrell WJ, Peggy OD, Colklin M (2001) Electrochemical and spectroscopic study of arsenate removal from water using zero-valent ion media. Environmental Science and Technology 31:315-320

Fazal MA, Kawachi T, Ichion E (2001) Extent and severity of groundwater arsenic contamination in Bangladesh. Water International 26:370-79

Ferreccio C, Gonzalez C, Milosavjlevic V, Marshall G (2000) Lung cancer and arsenic concentrations in drinking water in Chile. Epidemiology 1:673-679

Gupta SK, Chen KY (1978) Arsenic removal by adsorption. Journal Water Pollution Control Federation 50:493-506

Hering JG, Chen PY, Wilkie JA, Elimelech M (1996) Arsenic removal by ferric chloride. Journal of the American Water Works Association 88:155-167

Hopenhayn-Rich C, Biggs ML, Smith AH (1998) Lung and kidney cancer mortality associated with arsenic in drinking water in Cordoba, Argentina. International Journal of Epidemiology 27:561-569

HWBC (2000) Monitoring data for 1989-1999 (Unpublished data). Hanoi Water Business Company

Jain A, Ravene KP, Loeppert RH (1999) Arsenite and arsenate adsorption on ferrichydrite: surface charge reduction and net OH^- release stoichiometry. Environmental Science and Technology 33:1179-1184

JICA (1996) The study on water supply systems in the Socialist Republic of Vietnam. Main Report

JICA (1997) The study on Hanoi water supply systems in the Socialist Republic of Vietnam. Final Report

JICA (1999) The study on groundwater development in the rural provinces of northern part in the Socialist Republic of Vietnam. Progress Report

Joshi A, Chaudhuri M (1996) Removal of arsenic from ground water by iron-oxide coated sand. Journal of Environmental Engineering 122:769-776

Kim MJ, Nriagu J, Haack S (2000) Carbonate ions and arsenic dissolution by groundwater. Environmental Science and Technology 34:3094-3100

Koch Membrane Systems (2001) Reverse osmosis membrane remove arsenic and salt from water supplies. Membrane Technology 140:10-11

Korte N (1991) Naturally occurring arsenic in groundwaters of the Midwestern. United States Environmental Geology and Water Sciences 18:137–141

Krishna B, Chadrasekaran K, Karunasagar D, Arunachalam J (2001) A combined treatment approach using Fenton's reagent and zero-valent iron for the removal of arsenic from drinking water. Journal of Hazardous Materials 84:229-240

Kumar PR, Chaudhari S, Khilar KC, Mahajan SP (2003) Removal of arsenic from water by electrocoagulation. Chemosphere 55:1245-1252

Kuo TL (1968) Arsenic content of artesian well water in endemic area of chronic arsenic poisoning. Institute of Pathology of the National Taiwan University, pp 7-13

Langner HW, Inskeep WP (2000) Microbial reduction of arsenate in the presence of ferrihydrite. Environmental Science and Technology 34:3131-3136

Le G (1995) Ammonium pollution in groundwater of Hanoi City and a case study for removal of ammonia by the zeolite adsorption process. Master Thesis, Asian Institute of Technology

Lorenzen L, van Deventer JSJ, Landi WM (1995) Factors affecting the mechanism of the adsorption of arsenic species on activated carbon. Minerals Engineering 8: 557-569

Mallick S, Rajagopal NR (1996) Groundwater development in the arsenic-effected alluvial belt of West Belgan-Some questions. Current Science 70:956-958

Mandal BK, Chowdhyry TRG, Samanta G, Muhkerjee D, Chanda CR, Saha KC, Chakraborti D (1998) Impact of safe water for drinking on five families for 2 years in West Bengal, India. Science Total Environment 22:432-437

Manning BA, Goldberg S (1997) Arsenic(III) and arsenic(V) adsorption on three California soils. Soil Science 162:886–895

McArthur JM, Ravenscroft P, Safiulla S, Thirlwall MF (2001) Arsenic in groundwater: Testing pollution mechanisms for sedimentary aquifers in Bangladesh. Water Resources Research 37:109-117

Nga TTV, Inoue M, Khatiwada NR, Takizawa S (2001) Groundwater contamination due to arsenic, iron, ammonia, and organic matter in Hanoi City, Vietnam. In: Proceedings of the IWA 2nd World Water Congress, Berlin

Nga TTV, Inoue M, Khatiwada NR, Takizawa S (2003) Heavy metal tracers for the analysis of groundwater contamination: case study in Hanoi City, Vietnam. Water Science & Technology: Water Supply 3:343-350

Nickson RT, McArthur JM, Ravenscroft P, Burgess WG, Ahmed KM (2000) Mechanism of arsenic release to groundwater, Bangladesh and West Bengal. Applied Geochemistry 15:403-413

Ning RY (2002) Arsenic removal by reverse osmosis. Desalination 143:237-241

Oh JI, Yamamoto K, Kitawaki H, Nakao S, Sugawara T, Rahaman MM, Rahaman MH (2000) Application of low-pressure nanofiltration coupled with a bicycle pump for the treatment of arsenic-contaminated groundwater. Desalination 132:307-314

Pataky J and Lusztig G (1958) Arsenic liver cirrhosis in several members of a family (in Germany). Z Gesamte Inn Med 17:668–671

Peng FF, Di P (1994) Removal of arsenic from aqueous solution by adsorbing colloid flotation. Industrial & Engineering Chemistry Research 33:922-928

PHED (1991) National Drinking Water Mission: Arsenic pollution in groundwater in West Bengal. Final report, Steering Committee, Arsenic Investigation Project, Gov. West Bengal

Pontius FW, Brown KG, Chen CJ (1994) Health implications of arsenic in drinking water. Journal of the American Water Works Association 86:52-63

Rahman M, Axelson O (2001) Arsenic ingestion and health effects in Bangladesh: epidemiological observations. In: Chappell WR, Abernathy CO, Calderon RL (eds.) Arsenic exposure and health effects. Elsevier, Amsterdam, pp 193-199

Rahman M (2006) Arsenic contamination accidents around the world. In: Naidu R, Smith E, Owens G, Bhattacharya P, Nadebaum P (eds) Managing Arsenic in the environment - From soil to human health. Science Publishers

Ramana A, Sengupta AK (1992) Removing Selenium (IV) and Arsenic (V) oxyanions with tailored chelating polymers. Journal of Environmental Engineering 118:755-775

Raychowdhury T, Uchino T, Tokunaga H, Ando M (2002) Survey of arsenic in food composites from an arsenic-affected area of West Bengal, India. Food and Chemical Toxicology 40: 611-1621

Robins RG, Nishimura T, Singh P (2001) Removal of arsenic from drinking water by precipitation, adsorption or cementation. In: Ahmed MA, Ali MA, Adeel Z (eds) BUET-UNU International workshop on technologies for Arsenic removal from drinking water, Dhaka, Bangladesh, pp 59-68

Shih MC (2005) An overview of arsenic removal by pressure-driven membrane process. Desalination 172:85–97

Shrestha RR, Shrestha MP, Upadhyay NP (2003) Groundwater arsenic contamination, its health impact and mitigation program in Nepal. Journal of Environmental Science and Health. Part A-Toxic/Hazardous Substances & Environmental Engineering 38:185-200

Smedley PL, Edmunds WM, Pelig-Ba KB (1996) Mobility of arsenic in groundwater in the Obuasi area of Ghana. In: Appleton JD, Fuge R, McCall GJH (eds) Environmental geochemistry and health, Geological Society Special Publication No 113. Geological Society, London, pp 163-181

Smedley PL, Kinniburgh DG (2002) A review of the source behaviour and distribution of arsenic in natural waters. Applied Geochemistry 17:517-568

Smedley PL, Nicolli HB, Luo ZD (1999) Arsenic in groundwaters from major aquifers: sources, effects and potential mitigation. BGS Technical Report WC/99/38. British Geological Survey, Keyworth

Smith AH, Arroyo AP, Guha Mazumder DN, Kosnett MJ, Hernandez AL, Beeris M, Smith T, Moore LE (2000) Arsenic-induced skin lesions among Atacameno people in northern Chile despite good nutrition and centuries of exposure. Environmental Health Perspectives 108:617-20

Smith AH, Goycolea M , Haque R, Biggs ML (1998) Marked increase in bladder and lung cancer mortality in a region of northern Chile due to arsenic in drinking water. American Journal of Epidemiology 147:660-669

Sun GF (2001) Current situation of endemic arsenicosis in China. Environmental Science 8:425-434

Tondel M, Rahman M, Magnuson A, Chowdhury IA, Faruquee MH, Ahmad SA (1999) The relationship of arsenic levels in drinking water and the prevalence rate of skin lesions in Bangladesh. Environmental Health Perspectives 107:727-729

Tsai SM, Wang TN, Ko YC (1998) Cancer mortality trends in a blackfoot disease endemic community of Taiwan following water source replacement. Journal of Toxicology and Environmental Health, Part A 55:389-404

Tseng WP, Chu HM, How SW, Fong JM, Lin CS, Yeh S (1968) Prevalence of skin cancer in an endemic area of chronic arsenicism in Taiwan. Journal of the National Cancer Institute 40: 453-463

US EPA (2001) National primary drinking water regulations: Arsenic and clarifications to compliance and new source contamination monitoring

US EPA (2000), Technologies and cost for removal of arsenic from drinking water. Office of water, EPA 815-R-00-028 van Geen A, Zheng Y, Veersteg R, Stute M, Honerman A, Ahmed AM (2003) Local analysis helps avoid arsenic-contamination well water. Water Resources Research 10

Viraraghavan T, Subramanian S, Swaminathan V (1994) Drinking water without arsenic: a review of treatment technologies. Environmental Systems Reviews 37:1-35

Welch AH, Helsel DR, Focazio MJ, Watkins SA (1999) Arsenic in ground water supplies of the United States. In: Chappell WR, Abernathy CO, Calderon RL (eds) Arsenic Exposure and Health Effects. Elsevier, Amsterdam

Wu MM, Kuo TL, Hwant YH, Chen CJ (1989) Dose-response relation between arsenic concentration in well water and mortality from cancers and vascular diseases. American Journal of Epidemiology 130:1123-1132

Xia Y, Liu J (2004) An overview on chronic arsenism via drinking water in PR China. Toxicology 198:25-9

Yan XP, Kerrich R, Hendry MJ (2000) Distribution of arsenic (III), arsenic (V) and total inorganic arsenic in porewaters from a thick till and clay-rich aquitard sequence, Saskatchewan, Canada. Geochimica et Cosmochimica Acta, 64:2637-2648

Zobrist J, Dowdle PR, David JA, and Oremland RS (2000) Mobilization of arsenite by dissimilatory reduction of adsorbed arsenate. Environmental Science and Technology 34:4747-4753

15. Health Risks of Fluoride in the Chiang Mai Basin, Thailand

Tomoko Takeda and Satoshi Takizawa

Department of Urban Engineering, the University of Tokyo

15.1 Introduction

Fluorine is the most abundant halogen in the earth's crust and the thirteenth most abundant among all elements. The fluorine atom has a molar weight of 18.998, a van der Waals radius of 1.47Å and is isosterically similar to a hydroxyl group. It has an electronegativity value of 4.0, which causes the molecule to be very reactive. Because of its high reactivity, fluorine is rarely found in pure form, but exits in various compounds or as the monovalent anion fluoride.

Fluoride occurs naturally in groundwater, and is one of the emerging problems in groundwater use. Due to limited availability and contamination of surface waters, more regions in the world are turning to groundwater usage. It follows that an increasing number of people may suffer from fluoride poisoning in the future.

The Chiang Mai Basin in one such region where groundwater contains very high levels of fluoride. Many residents of the basin suffer from chronic fluorosis. An international, collaborative study has been conducted in the Chiang Mai Basin to elucidate the distribution of fluoride in groundwater and to find out the health risks of fluoride intake among residents in this region. This chapter describes the results of the 3-year study, in addition to discussions of fluoride toxicity and global fluoride occurrence.

15.2 Fluoride Toxicity and Occurrence in Groundwater

15.2.1 Fluoride in Groundwater Environment

Fluoride is contained in natural water. Seawater contains fluoride at a level of about 1.4mg/L (Gupta, 1978) while fresh surface water contains less than 0.5mg/L. High salt water concentration reduces ionic activities; thus salty water including seawater can dissolve higher amounts of fluoride. Gao et al. (2006) found that in the Yuncheng Basin of China, salty lake water intrusion causes high fluoride contents in neighbouring shallow aquifers. Groundwater in some regions of the world contains well over 10mg/L of fluoride. Fluoride concentration in natural water is determined by the interaction between water and the minerals contained in the local geology. Therefore, groundwater in contact with fluoride-containing minerals contains higher amounts of fluoride than seawater or surface water.

Fluorides are found in a wide variety of minerals, including fluorspar, rock phosphate, cryolite, apatite, mica, hornblende, and fluorite (Fawell et al. 2006). Among these minerals, fluorite (calcium fluoride, CaF_2) is abundant in igneous and sedimentary rocks, and is commercially exploited as a fluoride-containing mineral. Rock phosphate is used to produce phosphorus fertiliser and consists of a few percent fluorite, which is either discharged or used to add fluoride into public water supply in some countries such as the United States. Fluoride is also associated with geothermal waters; for example, some Japanese hot spring waters contain more than 15mg/L of fluoride.

Fluoride concentration in natural water sources is limited by the solubility of fluoride-containing minerals. Among them, fluorite (CaF_2) often limits the solubility of fluoride due to the wide availability of calcium ions. The solubility product of fluorite is:

$$K = [Ca^{2+}][F^-]^2 = 10^{-10.57} \quad \text{(at 25°C)}$$

Hence, if pure calcite is dissolved in pure water, then the saturation concentration of fluoride is 5.7mg/L. By adding more calcium such as gypsum ($CaSO_4 \cdot 2H_2O$), we can reduce the concentration of dissolved fluoride (Appelo and Postma 2005). On the contrary, if calcium in groundwater is removed by ion exchange with other cations, then the concentration of fluoride can increase. In fact, in many fluoride-laden groundwater sources, calcium concentration is very low (Matsui et al. 2006a, b).

Elevated concentrations of fluoride in groundwater are associated with some specific geological and climate conditions including:

- volcanic activities and geothermal waters such as hot springs
- a dry climate where the evaporation rate is extremely high
- presence of fluoride mines, such as fluorite
- presence of igneous and metamorphic rocks containing fluoride-bearing minerals
- presence of an aquifer with high cation-exchange capacity, which sequesters Ca^{2+} from groundwater

15.2.2 Toxicity of Fluoride

Fluoride Intake Sources

Although drinking water is considered to be the main source of fluoride intake, food and beverages prepared with high-fluoride water often have high fluoride contents (Zohouri and Rugg-Gunn, 1999). Other than water-related intake sources, sources such as toothpaste (500-2000ppm), tea (up to 371mg/g, Lung et al. 2003; Fawell, 2006), chili pepper, and beverages may contribute significantly to the total intake. Warnakulasuriya et al. (2002) measured the fluoride contents of several European beers and found that the fluoride contents vary between 0.23mg/L and 1.12mg/L.

In Guizhou Province, China, the major intake source of fluoride is coal combustion. Because coal in China contains high amounts of fluoride and arsenic, indoor combustion of coal is a serious health threat in rural communities. An estimated 9.9 million people, about 52% of the provincial population in Guizhou, are affected by fluoride intake via inhalation of airborne fluoride that arises from indoor coal combustion (Zheng et al. 2007).

Some of the ingested fluoride in accumulated in bones and teeth, while the rest is excreted in urine. It is reported that in cases of acute ingestion of fluoride, the administration of calcium and magnesium can relieve the toxicity of fluoride (Kao et al. 2004). In the presence of high concentration of calcium, fluoride can be excreted as calcium fluoride in faeces. Hence, the health effects of chronic fluoride intake may be also related with nutrition, especially calcium intake.

Dental Fluorosis

The most well documented health effects of fluoride ingestion are dental and skeletal fluorosis. Dental fluorosis proceeds with white spots on the teeth surface, and then develops into brown mottled teeth (Fig. 15.1). Although dental fluorosis is often dismissed as just an aesthetic problem, it is actually an indicator of health effects on other human organs. Because dental fluorosis is easier to diagnose than other symptoms of fluoride ingestion, inspection by dentists is commonly carried out to diagnose fluorosis among the people living in fluoride-laden areas. The damage in tooth development due to over-exposure to fluoride is most likely to occur between 6 months and 5 years of age. Fluoride causes damages on the enamel-forming cells of the teeth, which results in a mineralisation disorder, whereby the porosity of the sub-surface enamel is increased. Teeth are composed of hydroxyapatite and carbonated hydroxyapatite, but when fluoride is present, fluorapatite is also created.

There are several indices to diagnose the severity of dental fluorosis, including Dean's Index, which was the first index developed in 1942. Dean's Index is a six-point ordinal measurement scale from level 0 to level 5: namely, normal, questionable, very mild, mild, moderate, and severe. This index is the most widely used, but there are limitations of this index due to vagueness at the questionable end and the absence of discrimination at the severe end. In addition, it is dependent on the expertise of the dentists who make diagnoses, and it does not give the numbers of teeth affected by fluorosis (Whelton et al. 2004). DI level 1 stands for "questionable," which in itself is a dubious term, as pointed out by earlier researchers (Fejerskov et al. 1996). Although there are some shortcomings of this index in comparison with other indices proposed later, it is still commonly used for fluorosis diagnosis because it is the most conventional index and makes it possible to compare the results of fluorosis diagnosis with other regions or with past results. (Committee on Fluoride in Drinking Water; National Research Council, 2006).

For example, in order to determine the optimum dose of fluoride in drinking water, Khan et al. (2004) performed a study on dental fluorosis (using Dean's Index) and dental caries, measured by decayed-missing-filled-teeth (DMFT) among 1,020 twelve-year-old school children in nineteen cities in Pakistan. Based on the annual mean maximum temperature of 29°C in Pakistan, it was calculated by the commonly used equation of Galagan and Vermillion (1957) that the optimum fluoride concentration in drinking water was 0.7mg/L. However, at fluoride concentrations over 0.4mg/L, it was found that the percentage of fluorosis among the school children abruptly increased and severe cases of dental fluorosis appeared.

Hence they determined the optimum fluoride concentration in Pakistan should be 0.39mg/L instead.

Colquhoun (1985) conducted a survey on dental heath among children living in fluoridated and non-fluoridated water supply areas in Auckland, New Zealand. He found that social class factors rather than the presence or absence of water fluoridation had more significant effects on the dental health of the children.

Inaba et al. (2002) reported that the enamel fluoride contents of Japanese school children using 0.025% of NaF mouth rinse for two years was 1810ppm, whereas those children without using such mouth rinse was 1113ppm.

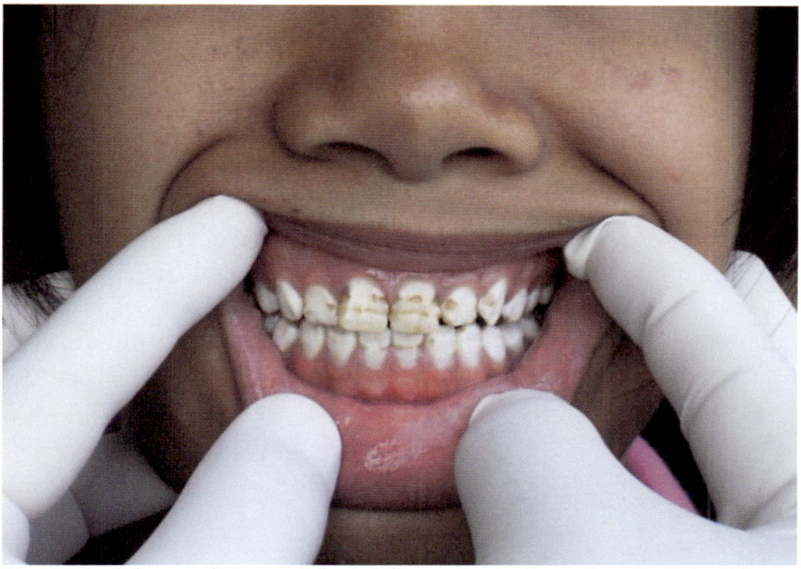

Fig. 15.1. Advanced stage of dental fluorosis; teeth surface is stained brown

Skeletal Fluorosis

Skeletal fluorosis is also associated with the high intake of fluoride. The basic component of human bones is hydroxylapatite, $Ca_{10}(PO_4)_6(OH)_2$, of which hydroxyl groups can be replaced by fluoride to produce fluoroapatite, $Ca_{10}(PO_4)_6F_2$, increasing the brittleness and decreasing the solubility of the bone structure. Human bones are constantly resorbed and redeposited during one's lifetime and high fluoride disturbs the homeostasis of bone mineral metabolism (Fordyce et al. 2007). Fluoride induces the production of an altered organic matrix and reduces mineralisation; thus fluoride reduces bone strength. Patients of skeletal fluorosis suffer from bone pain, stiffness, rigidity and possible deformities of the spine and stiffness (Chavassieux et al. 2007).

Typical symptoms of skeletal fluorosis are: osteosclerosis, osteomalacia, osteoporosis, and exostosis formation. Skeletal fluorosis is classified into the following four stages (Committee on Fluoride in Drinking Water 2006; Groves 2001):

- Preclinical: asymptomatic; slight radiographically detectable increases in bone mass
- Clinical stage I: sporadic pain; stiffness of joints; osteosclerosis of pelvis and spine
- Clinical stage II: mobility is slightly affected; chronic joint pain, arthritic symptoms, slight calcification of ligaments, and osteosclerosis of the cancellous bones
- Clinical stage III: crippling fluorosis. Limitation of joint movement; calcification of ligaments/neck, spinal column; crippling deformities of spine and major joints; muscle wasting; neurological defects/compression of spinal cord

In the United States, the Maximum Contaminant Level Goal (MCLG) of fluoride in drinking water was determined by the United States Environmental Protection Agency in 1985 at 4mgF/L so as not to cause crippling skeletal fluorosis, or stage III skeletal fluorosis. Drinking high-fluoride water causes sporadic pain and stiffness of joints similar to arthritis. Fluoride also weakens the bone structure, causing more frequent bone fractures. It is likely that people drinking high fluoride water of 4mgF/L are more likely to suffer from bone fractures than those drinking 1mgF/L water.

Other Health Effects

Other health effects of fluoride include kidney, carcinogenic, neurotoxic, and reproductive health effects. Borke and Whiteford (1999) reported that chronic ingestion of an elevated concentration of fluoride lowered calcium transport in the kidneys of rats. Bassin (2001) investigated the effects of fluoride ingestion from drinking water in the U.S. on osteosarcoma, a rare but fatal bone cancer. There were a total of 103 cases in people less than twenty years of age and 215 control cases that matched the age and residential history of the case subjects. The results of age-specific epidemiological study suggested that for males less than twenty years old, fluoride levels in drinking water during growth are associated with an increased risk of osteosarcoma, with a peak in the odds ratios from ages six to eight (adjusted odds ratio: 5.46, 95% CI 1.50-19.90 at age 7 years).

Mullenix (1995) carried out a study on the effects of fluoride intake on rat behaviour using a computer pattern recognition system. Fluoride exposures caused sex- and dose-specific behavioural deficits among rats. The severity of the behaviour disorder was correlated with plasma F levels and F concentrations in specific brain regions of the rats. These findings imply similar effects on humans because those humans exposed to high levels of fluoride have a similar rage of plasma F levels as the rats tested in this study.

Verner et al. (1998) investigated the effects of aluminium fluoride and sodium fluoride on neural and cerebrovascular integrity. Male rats were separated into three groups and were administered one of the three types of drinking water: water containing 0.5ppm AlF_3, water containing 2.1ppm of NaF, which contains the equivalent amount of fluoride as in the AlF_3, and distilled deionised water. Although the average body weights of the rats were not different among the three groups, more rats in AlF_3 group died than in the control, i.e. distilled water, group. The Al levels of both AlF_3 and NaF groups were higher than the control group, and cerebrovascular and neuronal integrity were altered by AlF_3, and to a lesser degree by NaF.

Xiang et al. (2003) carried out a survey on the linkage between children's intelligence and fluoride in their drinking water. 512 children drinking either high (2.47mgF/L) or low (0.36mgF/L) fluoride water were selected from Jiangsu Province in China, and their Intelligence Quotient (IQ) was measured. Those drinking high fluoride water had a mean IQ of 92.0±13.0, which was significantly lower than the mean IQ of the children drinking low fluoride water (100.4±13.2). Higher concentrations of fluoride in drinking water was associated with mental retardation (IQ<70); 4% of the children drinking high fluoride water scored IQ<70, whereas there was no student who scored IQ<70 among the low fluoride group. The Benchmark Concentration (BMC) for the concentration-response relationship between IQ<80 and drinking water fluoride concentration was 2.32mgF/L. A similar result was reported by Wang et al. (2007).

15.3 Fluoride Occurrence in Asian Countries

Asian countries reported to have high fluoride contents in groundwater include China, India, Indonesia, Iran, Laos, Pakistan, Japan, Saudi Arabia, Sri Lanka, Thailand, Turkey, and Vietnam. It is likely that there are more countries that may have the problem of elevated fluoride concentrations in their groundwater. Among these countries, China and India are the most seriously affected by the health effects of fluoride intake.

China

The estimated population exposed to high fluoride varies in the literature. As of the year 2000, it was estimated that there were about 34 million people in China living in areas with high fluoride concentration in the groundwater (Zheng et al. 2007). The Chinese Ministry of Health reported that the affected population numbered 45 million (Wang et al. 2002). According to the United Nations Environment Programme, more than 80 million people in China are exposed to high fluoride in their drinking water. In endemic areas of China, 26% and 1.6% of the people suffer from dental and skeletal fluorosis, respectively (UNEP 2007). Endemic fluorosis is widespread in 30 provinces, with an estimated population of 26 million suffering from dental fluorosis (Wang et al. 2002; Fawell et al. 2007). The problem of endemic fluorosis is especially grave in arid and semi-arid regions of northern, eastern and northwestern China.

Yong and Hua (1991) summarised the environmental factors that give rise to the fluoride concentrations in the groundwater of northern China. Three environmental factors that increase the fluoride concentration are:

- presence of granite and alkaline volcanic rocks
- soda salinised soil
- low precipitation and high evaporation rates

In the Kuiteng Basin, fluoride concentrations can go as high as 28mg/L due to high evapotranspiration.

Wang et al. (2002) investigated the fluoride adsorption rates in different types of Chinese soils. They reported that fluoride adsorption capacity decreased from soils in the humid region to soils in the arid region, and they estimated that low fluoride adsorption by arid soils may be one of the causes of high fluoride concentration in shallow groundwater in arid regions in China.

Shallow groundwater in an area of over 3,156 km^2 in the Yuncheng Basin contains more than the Chinese standard of 1.0mgF/L. Linyi County has the most serious cases of endemic fluorosis, and 22% of villagers in 13 villages have lost labour capability due to fluorosis (Gao et al. 2007).

Zhang et al. (2003) investigated groundwater use and fluoride contents in groundwater in Jilin Province. They found that the local people generally used unconfined groundwater, which had higher levels of fluoride than groundwater from confined aquifers. The occurrence of dental and skeletal fluorosis is correlated with the concentration of fluoride in their drinking water.

India

Endemic skeletal fluorosis has been reported in India since the 1930s. The estimated population at risk for high fluoride intake is 62–66 million people, living in 19 out of 32 states (Jacks et al. 2005). Among those states, the northwestern and southern states are the most seriously affected by fluorosis. In Rajasthan, Gujarat and Andhra Pradesh, more than 70% of the districts are affected by fluoride in groundwater. At present, the Indian Standard for Fluoride in drinking water is 1.5mg/L, but the maximum concentration of fluoride in groundwater in India was reported to be 68mg/L in Rajasthan (Fawell et al. 2006).

Ravandra and Garg (2007) analysed fluoride contents in groundwater in Hisar City, Haryana State. A total of 127 groundwater samples were obtained from residential and commercial areas, and 42% of the samples exceeded 1.5mgF/L, with a maximum concentration of 16.7mgF/L.

Saxena and Ahmed (2003) analysed 58 groundwater samples collected from sites known for high fluoride, and found the range of fluoride concentration to be between 1.7 and 6.1mg/L. They found groundwater with high fluoride contents were characterised by alkaline pH (7.4 – 8.8) and high electric conductivity (up to 2,680μS/cm), and thus estimated that a high concentration of sodium bi-carbonate can increase the fluoride concentration by the following reaction:

$$CaF_2 + 2NaHCO_3 \rightarrow CaCO_3 + 2Na^+ + 2F^- + H_2O + CO_2$$

Kundu et al. (2001) investigated the effect of hot spring water containing more than 10 mg/L of fluoride on fluoride concentration of groundwater in Orissa State. They found that fluoride concentration was positively correlated with sodium, bicarbonate and pH, but negatively correlated with calcium and magnesium. Statistical analysis of groundwater quality data suggested that the in-mixing of hot spring water was the cause of elevated concentrations of fluoride in the local groundwater.

Other Asian Countries

Sri Lanka is a tropical country with a wet climate in the southwest of the country, and an arid climate elsewhere. A majority of the rural population lives in areas with no central waterworks system, and relies on groundwater or surface water, which are affected by the local geological conditions (Dissanayake and Chandrajith, 2007). Van der Hoek (2003) reported up to 5.9mg/L of fluoride among 416 shallow wells in Uda Walawe province. High fluoride in groundwater in Sri Lanka is associated with the climate;

in the dry region the fluoride levels are higher than 1.5mg/L, whereas in the wet region near Colombo fluoride levels are lower. Among 518 14-year-old students, 43.2% suffered from dental fluorosis.

Kim and Jeong (2005) conducted a study on factors affecting fluoride occurrence in groundwater in the southeastern part of Korea. 10% of 422 public water supply wells inspected in their study contained fluoride at concentrations higher than 1.5mg/L. Fluoride rich groundwater sources were distributed along with the major faults, and therefore, it was estimated that groundwater is coming up along the fault plane. The groundwater is enriched with fluoride while at deep layers through prolonged contact with granitic rocks, followed by calcium removal by precipitation of calcite.

15.4 Case Study: Fluoride Intake and Health Effects in Chiang Mai Basin, Thailand

15.4.1 Fluoride in the Groundwater of Chiang Mai Basin

Due to high fluoride concentrations in the groundwater, the people living in Lamphun Province, which is located in the Chiang Mai Basin, Thailand, have long suffered from dental and skeletal fluorosis. The origin of fluoride in the groundwater of the Chiang Mai Basin has not yet been clarified, but there are 2 hypotheses. One is that it comes from geothermal water containing high fluoride, while the other is that fluorite and/or other fluoride-bearing mineral deposits near the high-fluoride groundwater are the causes. Na-Ca exchange causes fluoride to escape from the Ca-F bondage, and in turn leads to high fluoride concentrations. The presence of faults in the Chiang Mai Basin might provide a channel for the deep groundwater, which contains high fluoride, to flow upward and mix with shallow groundwater.

Premgamone (1980) proposed a two-step process in which gaseous fluoride forms fluorite deposits near the surface, and then the hot springs dissolve, transport, and redeposit the minerals containing fluoride. Noppakun et al. (2000) found 18.9mgF/L in a geothermal source, but less than 0.5mg/L in other sources. Their results seem to support the hypothesis that hot spring water is the source of fluoride. However, definitive evidence has not yet been presented because the number of their samples was small and the sampling location and well depth was not accurately reported.

A three-year research project was conducted from 2005 to 2007 to accomplish the following objectives:

- to monitor the fluoride concentration in groundwater in the region, and illustrate the distribution of fluoride concentrations in the groundwater
- to investigate the current status of the health effects of fluoride intake and to identify the sources of fluoride intake among the people living in Lamphun Province
- to investigate the effectiveness of fluoride removal by the reverse osmosis membrane plants installed in Lamphun Province.

Fig. 15.2 shows the concentration of fluoride in groundwater in Lamphun Province. The groundwater samples were obtained from the wells of village waterworks and private wells. The groundwater in the northeastern area had higher fluoride concentrations than in the southwestern area. In addition, the high fluoride groundwater sources are located in a straight line, which implied that there might be a fault underneath this line. However, because the well logs were not available, the depth of the wells is not known. Therefore, the fluoride concentration changes along the depth of wells cannot be estimated.

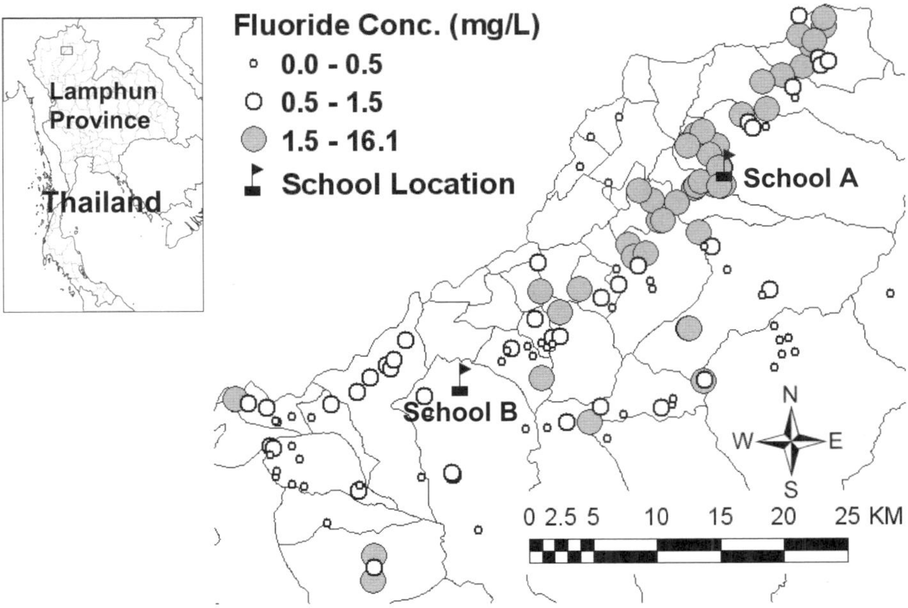

Fig. 15.2. Location of the study area and fluoride concentrations in groundwater

Most of the village waterworks built in this area use unconfined groundwater, many of which contain high concentrations of fluoride, as illustrated in Fig. 15.2. Hence, until the residents were warned of the health risks of drinking fluoride-laden groundwater, they had been drinking water from those village waterworks regularly. Recently, many of the residents stopped drinking groundwater from their wells or village waterworks, and opted for bottled water sold by private companies or for water from the community waterworks equipped with reverse osmosis membranes.

15.4.2 Fluoride Intake and Health Effects in the Chiang Mai Basin

Occurrence of Dental Fluorosis

The relationship between fluoride in drinking water and urine of 130 students from Schools A and B (Fig. 15.2) was investigated in November 2005. The samples of morning urine, household drinking water, and cooking water were collected by each student in plastic containers. Fluoride contents in the urine samples were tested by the ion-selective electrode (F125, TOA-DKK) immediately after collection, and then 5mL of each sample was frozen for the analysis of creatinine by direct colorimetry using a creatinine analysis kit (Cayman Chemical Company). The validity of the urine samples was tested by measuring the urinary creatinine excretion of the subjects and comparing them with standard values of 500–2000mg/24 hours (Cayman Chemical Company, 2006). Dental fluorosis among the students was diagnosed according to the Dean's Index (DI) by a local dentist experienced in fluorosis research.

The number of subjects in each DI level is shown for the high (School A) and low (School B) fluoride areas, respectively, in Table 15.1. For the subjects living in the high F area, 42.2% (27 subjects) fell into DI categories 4 and 5, 26.6% (17 subjects) in DI levels 2 and 3, and 26.6% (17 subjects) in DI level 1. Only 3 subjects (4.7%) had a level 0. Those subjects living in the high F area but with low DI levels may have low intake values of fluoride by choosing low fluoride-containing waters, or may have been living in the high-fluoride area for just a short time.

Table 15.1. The number of subjects under each DI level

Dean's Index level	0	1	2	3	4	5	Total
School A	3	17	9	8	14	13	64
School B	36	11	15	2	2	0	66
Total	39	28	24	10	16	13	130

In the low fluoride area near School B, no subjects fell into DI level 5, but 4 subjects (6.1%) had DI levels 3 and 4. Although this area was named as the low fluoride area, there are sporadically high levels of fluoride in the groundwater. These subjects may be living in those areas, or could be receiving other sources of fluoride, such as cooked food, which can be transported from the high fluoride area. 15 subjects (22.7% from School B) had a DI level of 2, very mild.

In Ireland, where tap water is fluoridated, the prevalence of fluorosis had increased in fluoridated areas; nonetheless, only 4% of them had fluorosis higher than DI level 2 (Whelton et al. 2004). Even in the low fluoride area in Lamphun Province, i.e. School B, the percentage of the people who had DI levels above 2 was significantly higher than that for the people living in the fluoridated areas in Ireland. This finding indicated the severity of fluorosis in the Lamphun Province.

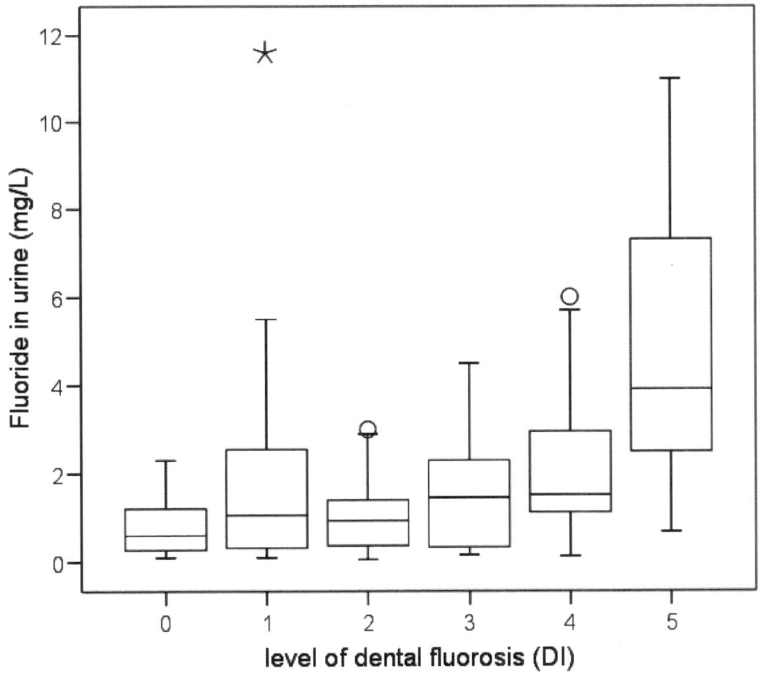

Fig. 15.3. Level of dental fluorosis and fluoride concentrations in urine

The horizontal line in the middle of each box marks the median of the sample. The edges of each box represent the 25th and 75th percentiles. The whiskers show the range of values within 1.5 box-lengths of each edge of the box. Cases with values over 3 box-lengths from 75th percentile are marked by an asterisk (*), while cases with values between 1.5 and 3 box-lengths are marked by an open circle (o).

The relationship between the levels of dental fluorosis and fluoride concentrations in urine is shown in Fig. 15.3. Although there is variation among the subjects in each DI level, especially in DI level 1, the average fluoride concentration in urine samples were shown to be higher for the subjects with DI levels 3–5 than in those subjects with DI levels 0–2. The relationship is most obvious for the subjects with DI level 5. The reason why some subjects with DI level 1 had high fluoride concentrations in their urine samples has not yet been clarified, but they could have moved in this area recently so that they have not yet developed any symptoms of dental fluorosis. However, it can be said that there are more students who are at risk of fluoride health effects than those who have already been identified to be affected by fluoride in water.

Fluoride Intake and Excretion

Al-Saleh and Al-Doush (2000) have researched the correlation between fluoride concentration in drinking water and urine samples of residents in Riyadh, Saudi Arabia. Fluoride concentration in the urine samples of 130 subjects aged 16–70 years old ranged between 0.16–4.68mg/L with an average of 0.93 mg/L. 22.3% of the subjects had urinary fluoride concentration higher than 1.2mg/L, which poses a high possibility of causing fluorosis. They found a statistically significant correlation between fluoride concentration in the drinking water and fluoride concentration in the urine samples. However, there was no clear correlation between concentrations in the drinking water and in the urine of junior high school students in Lamphun. This is due to the facts that a large number of residents were drinking groundwater which was defluoridated by reverse osmosis membranes installed by the government and that drinking water was not the only source of fluoride intake (Takeda et al. 2006a).

To determine the main intake sources and the intake-excretion balance of fluoride among the population in the study area, 21 junior high school students (ages 13–15) and their 21 parents/grandparents (one for each student; ages 39–83) were selected from School A (Fig. 15.2), which is located in the high-fluoride area. From School B, located in the low-fluoride area, seven students (ages 12–15) and their seven parents/grandparents (ages 34–50) were selected for comparison.

The fluoride intake was calculated from individual food and beverage samples. Foods and beverages consumed by the subjects were collected for 3 days. A double portion of each meal was collected by taking the same amount as the eaten portion with the same utensils used for eating. The latest day with three complete meals was selected for the estimation of fluoride intake. On the other hand, subjects with incomplete meal samples were excluded from the discussion.

The fluoride contents in foods were measured according to the AOAC official method 975.04 with minor modifications, using the fluoride-sensitive electrode. A portion of the food was dried for 48 hours in an oven at 80˚C, and then homogenized by a food processor. 0.5g of the food powder was measured into a 50mL wide-mouth plastic bottle; firstly, 20mL of 0.05mol/L HNO_3 was added and magnetically stirred for 20 minutes; secondly, 20mL of 0.1mol/L KOH was added and magnetically stirred for twenty minutes. Thirdly, 5mL of 0.2mol/L HNO_3 was added. The TISAB-II buffer was added to adjust pH and ion strength before fluoride analysis by the electrode.

All other dishes except rice were termed side dishes, and separated from rice. One-day samples of all side dishes were homogenized together, and then measured by the above-mentioned AOAC method. Rice was homogenized separately, and also measured by the AOAC method. Two kinds of rice (sticky and jasmine rice, with sticky rice being the most popular staple food) are eaten in this area; consequently, both kinds of rice were measured separately. Cooked rice samples taken from households were dried in an 80°C oven for 48 hours, then ground into fine powder by a food processor before the analysis. Samples were kept in a refrigerator until analysis. The fluoride content in food was calculated on the wet-weight basis, and the daily intake was estimated by multiplying the fluoride contents in food (mg/kg-wet) by the weight of food (kg-wet). The total daily fluoride intake was estimated by adding the total fluoride of all side dishes, rice, and beverages (Takeda et al. 2007a,b).

Fig. 15.4 shows the difference in fluoride concentrations between cooking and drinking water. Although seventeen households drank bottled water containing 0-0.85mg/L of fluoride, only four households used bottled water for cooking. The majority of households (seventeen out of 28) used village waterworks as the main source for cooking water. In the low-fluoride area (School B), none of the households had high fluoride concentrations in their cooking or drinking water (Takeda et al. 2007a). This result revealed that the local residents use water containing fluoride for their cooking either knowingly or unknowingly of the risk of fluoride in their cooking water. Too much emphasis put on the word "drinking water" may sometimes mislead the people to think that drinking water is the only source of fluoride intake. On the other hand, using bottled water for all the purposes may be too expensive for them. Therefore, development of an educational program on fluoride risks and sources for the local people is advisable.

In Riyadh, Saudi Arabia, bottled water contained significantly higher fluoride levels (0.52mg/L) than tap water (0.10 mg/L) (Al-Saleh and Al-Doush, 2000). This may be because the tap water comes from desalinated seawater, whereas most of the bottled waters are produced from groundwater containing fluoride. It should be noted that the maximum fluoride concentration of fluoride in bottled water in Lamphun, i.e. 0.8 mg/L, was even higher than that in Riyadh.

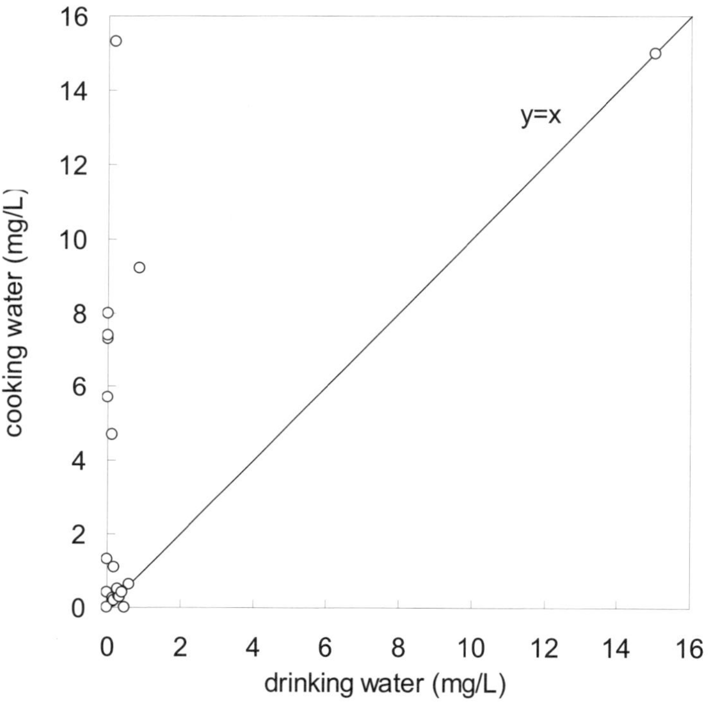

Fig. 15.4. Fluoride in drinking and cooking water

Figure 15.5 shows the relationship between fluoride intake and excretion. In School B, which is located in the low fluoride area, the students had very low fluoride intake, whereas their parents had a little higher intake. Most of the plots lie near the y=x line or lie below it, indicating fluoride intake to be equal to, or more than, the fluoride excretion. This is reasonable, since not all of the fluoride ingested is excreted through the urine, and some of the fluoride is inevitably carried to body parts with a slower metabolism (Fejerskov et al. 1996). There is one outlier at {5.1, 0.1}, which is a parent living in the low fluoride area. As shown by the high fluoride intake of this subject, it should be noted that even in the low fluoride area, there are fluoride-laden groundwater sources that pose a high risk of health effects. One of the reasons why the fluoride excretion of this subject was so low is that this subject handed in only 80mL as the 24-hour urine sample, which is too low in quantity. Hence, it seems that the fluoride excretion of this subject was significantly underestimated.

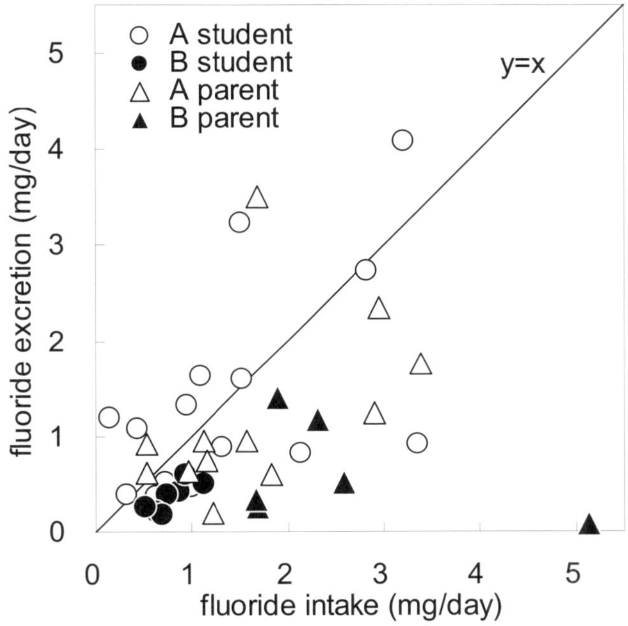

Fig. 15.5. Intake and excretion of fluoride

The estimated intake ranges for each subject group are shown in Table 15.2. The estimated intake range for adolescents (7–15 years) living in an area of China with high levels of fluoride in water supplies were 1.16–4.57mg/day. In adults (16+ years), this was 1.61–7.51mg/day. In another study, the estimated intake range was 0.3–1.49mg/day in adolescents (mean age of 14 years) living in Hungary with drinking-water containing levels of fluoride ranging from 0.06 to 3.1mg/L (Fawell et al. 2006; op. cit.). The results shown in Table 15.2 are comparable to the intake amounts in the previous reports.

Table 15.2. The range of fluoride intake in subjects

	Intake (mg/day)	
Subjects	School A	School B
Student	0.15–3.35	0.51–1.12
Parent	0.19–3.51	1.67–5.14

The urinary excretion ranges are shown in Table 15.3. In Spain, Torra et al. (1998) found the excretion rates to be 0.16-1.90mg/day in healthy adults (15-90 years). Ruiz-Payan et al. (2005) found the maximum urinary fluoride levels of 2.22±1.16 mg/g-creatinine in adolescents (15-20 years) living in Mexico. The results shown in Table 15.3 indicated that some of the people living in the high fluoride area (School A) are excreting higher amounts of fluoride than the people in the Mexico study. This illustrates that there are people at very high risk of fluoride intake and fluoride-related illness among the population living in the high fluoride area. The results of a two-way ANOVA showed that the excretion rate was significantly related to the residential area ($p<0.05$), but not to the age group.

Table 15.3. The range of urinary F excretion in subjects

Subject	School A	School B
Student	0.37-4.09 (0.15-4.90)	0.17-0.59 (0.40-0.94)
Parent	0.19-3.51 (0.32-5.23)	0.09-1.41 (0.60-2.57)

Above: urinary excretion (mg/day)
Below, in parentheses: urinary F adjusted to creatinine (mg/g creatinine)

Fluoride Contents in Rice

Sticky rice is the main staple of households in Lamphun Province and was a major intake source of fluoride for some subjects. Fig. 15.6 shows a mother in a typical household soaking rice before steaming. In each household, 3–5 L of rice is cooked every day. Rice is washed several times with cooking water, and then put in a bowl to soak in water overnight. As illustrated in Fig. 15.6, water is filled up to the first joint of the index finger above the rice.

Sticky rice is steam-cooked, and as fluoride does not evaporate from water, it follows that not much fluoride is taken into rice in the steaming process and that most of the fluoride in rice comes from the soaking process before steaming. To measure the fluoride uptake by rice in the soaking process, rice-soaking experiments were conducted in the laboratory. 50g of sticky rice were soaked into 65mL and 100mL of water with F concentrations of 0, 1, 5, 10, and 15mg/L for 24 hours.

Fig 15.6. A mother soaking rice.

Fig. 15.7 shows the relationship between fluoride in cooking water and in rice. The lab results (shown in white circles) show a very strong correlation between fluoride in cooking water and fluoride in cooked (soaked) rice. The fluoride concentrations in rice taken from households in Lamphun (black triangles in Fig. 15.7) are more scattered. Furthermore, on the whole, field data tend to lie beneath the regression line obtained from the lab data. Hence, it was estimated that, in some households, the soaking time in water was shorter than the laboratory experiment or they used water that contain low fluoride such as rainwater. In addition, the streaming process increases the water content of rice; it thus reduces the fluoride content on a wet-weight basis.

These experimental results revealed that rice absorbs a significant amount of fluoride if it is cooked with water containing fluoride. As rice is the main staple food in many Asian countries including Thailand, fluoride in cooking water is a serious health hazard in these countries. The interview survey conducted in Lamphun Province revealed that the local people's knowledge of fluoride sources and its effects is very limited, and sometimes they have misunderstandings regarding fluoride. In order to counteract the misconceptions, an educational program on fluoride sources and selection of water for each purpose was developed by the author's research team, and was implemented in both elementary and junior high schools in Lamphun Province (Takeda and Takizawa 2007).

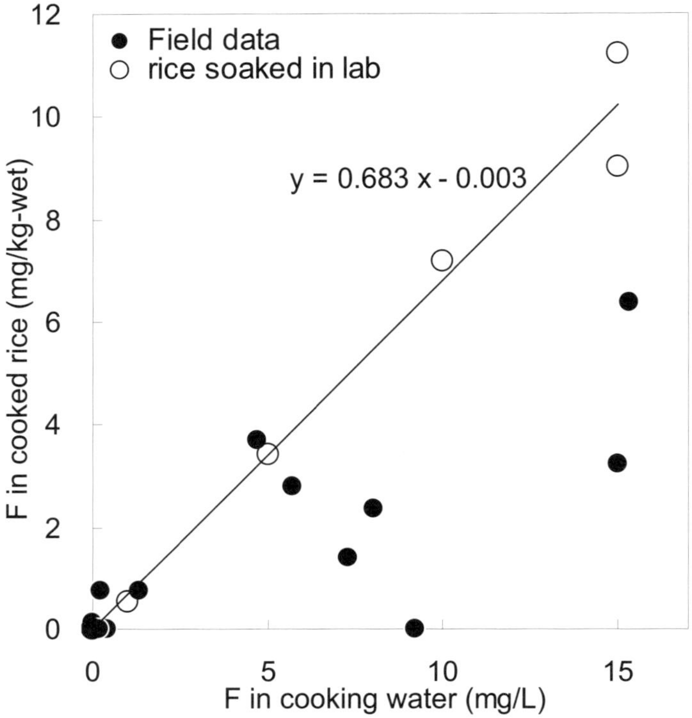

Fig. 15.7. Relationship between Fluoride in cooking water and in rice
The correlation for the lab data are shown by line y=0.683x-0.003

15.4.3 Fluoride Removal by Reverse Osmosis Membrane Plants

Among various advanced defluoridation processes, the reverse osmosis membrane process is gaining popularity because of its comparatively easy operation and its lower cost compared to other advanced processes. The most important advantage of membrane defluoridation processes is high fluoride removal efficiency. It can be applied even in such regions where fluoride concentration in raw water is higher than 10mg/L.

In response to the requests of the local residents, the Thai Government has built low-pressure reverse osmosis (RO) membrane plants in this area to provide fluoride-free drinking water. So far, about 30 plants have been built. The locations of ten plants where water samples were obtained to analyse the water quality are illustrated in Fig. 15.8.

A typical flow sheet demonstrating the operation of the RO membrane plants is shown in Fig. 15.9. The plants are operated on an intermittent filtration system, where the membrane-feeding pump is turned on and off according to water levels in the permeate tank. The operational conditions and specification of the membrane are shown in Table 15.4. The membrane plants are operated either by private companies or by the community. The water production capacity of the privately operated plants was larger than the community-operated ones, but the privately operated plants were operated only seasonally to meet the high demand of bottled water. The operational pressure was 0.8–1.5MPa, and the overall water recovery rate was 30–50%. The production rate was 4–10m^3/day.

Temperature, pH, electric conductivity (EC), NH_4^+, and alkalinity were analysed immediately at the sampling sites. DOC, F^-, Cl^-, NO_2^-, NO_3^-, PO_4^{3-}, and SO_4^{2-} were analysed in the laboratory by TOC-5000 and ion chromatography.

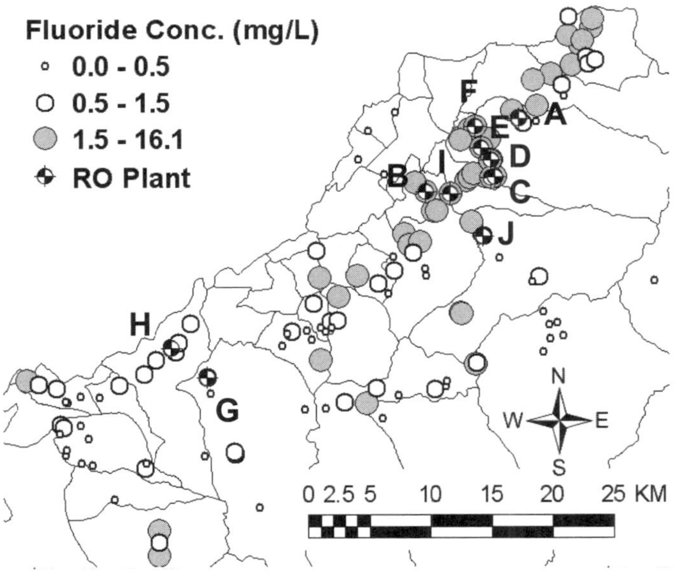

Fig. 15.8. Location of membrane plants

Table 15.4. Operation and specifications of the RO membrane plants.

Operation	Production	Service Population	Delivery	Membrane*	Operation Pressure
	m^3/d	households	bottle size	NaCl rejection	M Pa
Public	1-5	150-200	1L or 20L	99.4%	0.8-1.5
Private	50-100	seasonally varied	1L or 20L	99.4%	0.8-1.5

*: polyamide membrane (Matsui et al. 2006a, b)

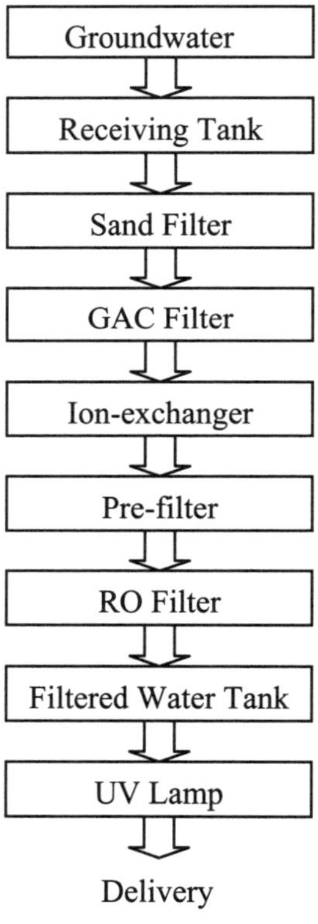

Fig. 15.9. Low-pressure reverse osmosis membrane process for fluoride removal, Lamphun, Thailand

Anions, Dissolved Organic Carbon (DOC), and electric conductivity (EC) of the raw water taken from each site are shown in Fig. 15.10. The numbers shown in this figure are concentrations of fluoride, which ranged from 0.7 to 13.4mg/L. As shown in Fig. 15.10, RO membrane Plants G and I were installed in such locations where the fluoride in groundwater was low. In addition, Plant B was not in operation, due to the poor quality of the treated water. The groundwater of Plant B had low oxygen reduction potential (ORP), high pH and temperature; it also released a sulphur smell. This water was likely affected by geothermal water. DOC, NO_3^-, and NH_4^+ levels were rather low in all groundwater sources, which indicated that there is no groundwater contamination due to infiltration of surface water. For the four parameters shown in Fig. 15.10, the removal rates were higher than 85%. It should be pointed out that fluoride removal rates were high even in Plants A, G and I, where fluoride concentration in groundwater were less than 1.3mg/L. DOC removal rates ranged between 32% and 81%. The low DOC removal rates in G and H are due to very low DOC in the groundwater.

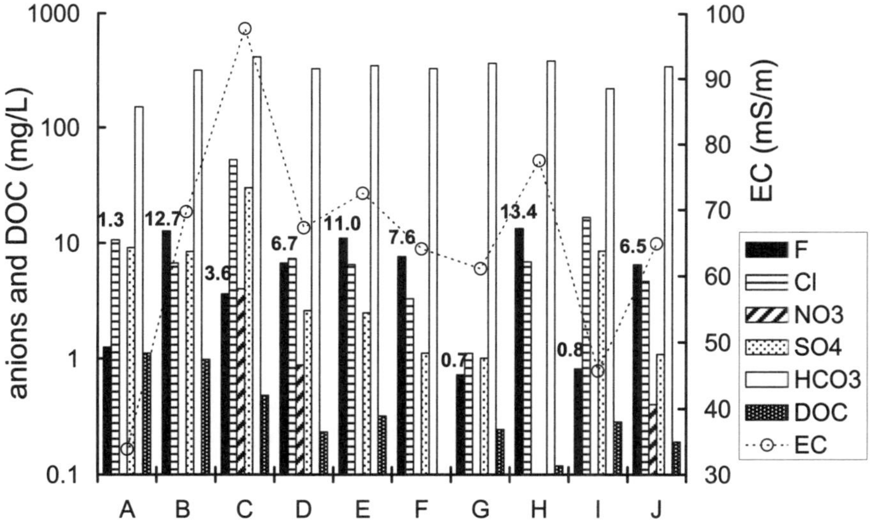

Fig. 15.10. Anions, DOC, and EC of raw water. (Takeda et al. 2006b)

At these sites, the prices for RO-treated water were 0.3–0.75 Baht/L for bottles containing up to 20L. (As of September 2007, 1 Baht was ca. 0.03 US Dollars.) Assuming that the consumption is 2L/day/capita, one year's worth of drinking water would amount to 219.0–547.5 Baht/capita. The average income is 32,900 Baht/year/capita (Lamphun Province, 2006), which means that 0.7–1.7% of the income would be spent for the RO treated water in an average household. If it is recommended to use bottled water for cooking as well as drinking, each household would use 10–20 L/capita of bottled water each day. Hence, they would have to spend 3.5–17% of their income on bottled water. At present, this spending may be too much for most of the households in Lamphun. However, if the economy in Lamphun Province and other fluoride-affected areas of the world gets better, it is highly probable that more areas will turn to RO as a cost effective treatment of fluoride-rich water.

15.5 Conclusion

Fluoride naturally occurs in groundwater and causes health problems in people who drink fluoride-laden groundwater in many Asian countries. In this case study of Lamphun Province, Thailand, it was revealed that a large number of people are suffering from dental fluorosis. Even the residents in the low fluoride area had a high percentage of fluorosis. Despite the low fluoride concentration in their drinking water, the fluoride concentrations in their urine were high. This was because their cooking water contained high concentrations of fluoride. The rice soaking study demonstrated that rice uptakes a high amount of fluoride while being soaked in fluoride-containing water. Since rice is the main staple here, as well as in other parts of Asia, fluoride intake via cooked rice could be the cause of fluorosis in Asia. Hence, people living in the areas with high fluoride levels in groundwater should be given the right knowledge on fluoride sources. An educational program on fluoride sources and selection of water sources for each water use would also help to reduce the risks arising from ingestion of fluoride.

Sampling and analyses of groundwater and permeate from low-pressure RO processes in the Chiang Mai Basin demonstrated that RO membranes are quite efficient for fluoride removal from groundwater. There is, however, one plant which was not operated because the permeate quality from the RO plant was unsatisfactory due to the fact that the groundwater was affected by geothermal water. It is recommended that a hydrogeological and groundwater quality survey be carried out before installation of the membrane plants.

References

Al-Saleh, Al-Doush (2000) Urinary fluoride levels in Saudi individuals drinking tap and bottled water. Int'l J Environ Health Res 10:21-26

Appelo CAJ, Postma D (2005) Geochemistry, groundwater and pollution, 2nd edn. Balkema

Bassin EB, Wypij D, Davis RB, Mittleman MA (2006) Age-specific fluoride exposure in drinking water and osteosarcoma (United States). Cancer Causes Control 17:421-428

Borke JL, Whitford GM (1999) Chronic fluoride ingestion decreases 45Ca uptake by rat kidney membrane. Nutrient Interactions and Toxicity 1209-1213

Carton RJ (2006) Review of the 2006 United States National Research Council Report: Fluoride in Drinking Water. Fluoride 39(3):163-172

Cayman Chemical Company (2006) Creatinine Assay Kit, Catalog No. 500701

Chavassieux E, Seeman E, Delmas PD (2007) Insights into material and structural basis of bone fragility from diseases associated with fractures: How determinants of the biomechanical properties of bone are compromised by disease. Endocrine Reviews 28(2):151-164

Colquhoun J (1985) Influence of social class and fluoridation on child dental health. Community Dent Oral Epidemiology 13:37-41

Committee on Fluoride in Drinking Water, National Research Council (2006) Fluoride in drinking water: a scientific review of EPA's standards. The National Academies Press, Washington DC

Dissanayake CB, Chandrajith R (2007) Medical geology in tropical countries with special reference to Sri Lanka. Environ Geochem Health 29:155-162

Fawell J, Bailey J, Chilton J, Dahi E, Fewtrell L, Magara Y (2006) Fluoride in drinking-water. WHO Drinking-water Quality Series, IWA Publishing.

Fordyce FM, Vrana K, Zhovinsky E, Povoroznuk V, Toth G, Hope BC, Iljinsky U, Baker J (2007) A health risk assessment for fluoride in Central Europe. Environ Geochem Health 29:83-102

Galagan DJ, Vermillion JR. (1957) Determining optimum fluoride concentrations. Public Health Rep 72:491–3.

Gao X, Wang Y, Li Y, Guo Q (2007) Enrichment of fluoride in groundwater under the impact of saline water intrusion at the salt lake area of Yuncheng basin, northern China. Environ Geol 53(4):795-803

Groves B (2001) Fluoride–Drinking ourselves to death?. Newleaf, Dublin.

Gupta RS, Naik S, Singbal SYS (1978) A study of fluoride, calcium and magnesium in the Northern Indian Ocean. Marine Chemistry 6: 25-141

Inaba D, Kawasaki K, Iijima Y, Taguchi N, Hayashida H, Yoshikawa T, Furugen R, Fukumoto E, Nishiyama T, Tanaka K, Takagi O (2002) Enamel fluoride uptake from mouthrinse solutions with different NaF concentration. Community Dent Oral Epidemiol 30:248-53

Jacks G, Bhattacharya P, Chuadhary V, Singh KP (2005) Controls on genesis of some high-fluoride groundwaters in India. Appl Geochem 20(2):221-228

Kao WF, Deng, JF, Ching SC, Heard K, Yen DHT, Lu MC, Kuo BIT, Kuo CC, Liu TY, Lee CH (2004) A simple, safe, and efficient way to treat severe fluoride poisoning-Oral calcium or magnesium. Clinical Toxicology 42(1):33-40

Khan AA, Whelton H, O'Mullane D (2004) Determining the optimal concentration of fluoride in drinking water in Pakistan. Community Dent Oral Epidemiol 32:162-72

Kim K, Jeong GY (2005) Factors influencing natural occurrence of fluoride-rich groundwaters: a case study in the southern part of the Korean Peninsula. Chemosphere 58(10):1399-1408

Kundu N, Panigrahi MK, Tripathy S, Munshi S, Powell MA, Hart BR (2001) Geochemical appraisal of fluoride contamination of groundwater in the Nayagarh District of Orissa, India. Indian Environ Geol 41: 451-460.

Lamphun Province (2006) As of 3 September 2007 (http://www.lamphun.go.th)

Lung SCC, Hsiao PK, Chiang KM (2003) Fluoride concentrations in three types of commercially packed tea drinks in Taiwan. J Exposure Analysis and Environ Epidemiology 13(1):66-73

Matsui Y, Takeda T, Takizawa S, Wongrueng A, Wattanachira S (2006a) Application of nanofiltration to fluoride removal from groundwaters in Chiang Mai Basin, Thailand. In: Proceedings of 32nd WEDC International Conference, Colombo, Sri Lanka, pp 585-592

Matsui Y, Takeda T, Takizawa S, Wongrueng A, Wattanachira S (2006b) Evaluation of nanofiltration process for fluoride removal from groundwaters in the Chiangmai Basin (in Japanese). Jour JSCE 62(4):403-414

Mullenix PJ, Denbesten PK, Schunior A, Keren WJ (1995) Neurotoxicity of sodium fluoride in rats. Neurotoxicology and Teratology 17(2):169-177

Premgamone C (1980) Fluid inclusion studies on Fluoride deposits, northern Thailand. Master thesis; Chiang Mai University, Thailand

Ravindra K, Garg VK (2007) Distribution of fluoride in groundwater and its suitability assessment for drinking purpose. Int'l J Environ Health Res 16(2):163-166

Saxena VK and Ahmed S (2003) Inferring the chemical parameters for the dissolution of fluoride in groundwater. Environ Geol 43:731-736

Takeda T, Matsui Y, Takizawa S (2006a) Study on Fluoride distribution of groundwater and the effect on health impact in the Chiang Mai Basin (in Japanese). In: Proceedings of the 40th Annual Conference of Japan Society on Water Environment, p 257

Takeda T, Matsui Y, Takizawa S, Wongureng A, Wattanachira S (2006b) Fluoride removal from groundwater by low pressure reverse osmosis membrane: Case study in the Chiang Mai Basin, Thailand. In: Proceedings of the 57th Annual Conference of Japan Waterworks Association, pp 740-741

Takeda T, Takizawa S, Supak C, Wattanachira S (2007a) Fluoride intake pathways among the population living with fluoride-laden groundwater in the Chiang Mai Basin, Thailand. In: Proceedings of the 16th Symposium on Environmental Chemistry, p 296

Takeda T, Wongrueng A, Takizawa S, Choompolkul W, Chaimongkol S, Wattanachira S (2007b) Estimation of fluoride intake sources and fluorosis among the population living in the Chiang Mai Basin. In: Proceedings of the XXVIIth Conference of the International Society for Fluoride Research, Beijing, China, pp 12-13

Takeda T, Takizawa S (2007c) Reducing the impact of fluoride intake in Thailand. Asian Water, pp 22-26

UNEP (2007) Workplan - Thematic Working Group (TWG) on water supply, sanitation and hygiene. UNEP and WHO (http://www.rrcap.unep.org/envhealth/twg/Workplan_water.pdf)

Van Der Hoek W, Ekanayake L, Rajasooriyar L, Karunaratne R (2003) Source of drinking water and other risk factors for dental fluorosis in Sri Lanka. Int J Env Health Res 13(3):285-293

Vandeputte M, Cock JD, Dryon L, Vercruysse A, Alexander F, Massart DL (1977) A contribution to the study of fluoride excretion. Clinica Chimica Acta 75:205-212

Varner AV, Jensen KF, Horvath W, Isaacson RL (1998) Chronic administration of aluminum- fluoride or sodium-fluoride to rats in drinking water: alterations in neuronal and cerebrovascular integrity. Brain Research 784:284-298

Wang SX, Wang ZH, Cheng XT, Li J, Sang ZP, Zhang XD, Han LL, Qiao SY, Wu ZM, Wang ZQ (2007) Arsenic and fluoride exposure in drinking water: children's IQ and growth in Shanyin County, Shanxi province, China. Environmental Health Perspectives 115(4):643-647

Wang W, Li R, Tan J, Luo K, Yang L, Li H, Li Y (2002) Adsorption and leaching of fluoride in soils of China. Fluoride 35(2):122-129

Warnakulsuriya S, Harris C, Gelbier S, Keating J, Peters T (2002) Fluoride content of alcoholic beverages. Clinica Chimica Acta 320:1-4

Whelton HP, Ketley CE, McSweeney F, O7Mullane DM (2004) A review of fluorosis in European Union: prevalence, risk factors and aesthetic issues. Community Dent Oral Epidemiol 32(1): 9-18

Xiang Q, Liang Y, Chen, L, Wang, C, Chen B, Chen X, Zhou M (2003) Effect of fluoride in drinking water on children's intelligence. Fluoride 36(2):84-94

Yong L, Hua ZW (1991) Environmental characteristics of regional groundwater in relation to fluoride poisoning in North China. Environ Geol Water Sci 18(1):3-10

Zhang B, Hong M, Zhao YS, Lin XY, Zhang XL, Dong J (2003) Distribution and risk assessment of fluoride in drinking water in the west plain region of Jilin province, China. Environ Geochem and Health 25(4):421-431

Zheng B, Wu D, Wang B, Liu X, Wang M, Wang A, Xiao G, Liu P, and Flinkelman RB (2007) Fluorosis caused by indoor coal combustion in China: discovery and progress. Environ Geochem Health 29:103-108

Zohouri FV, Rugg-Gunn AJ (1999) Fluoride concentration in foods from Iran. Int J Food Sci Nutr 50:265-274

Index

X

Y

Z